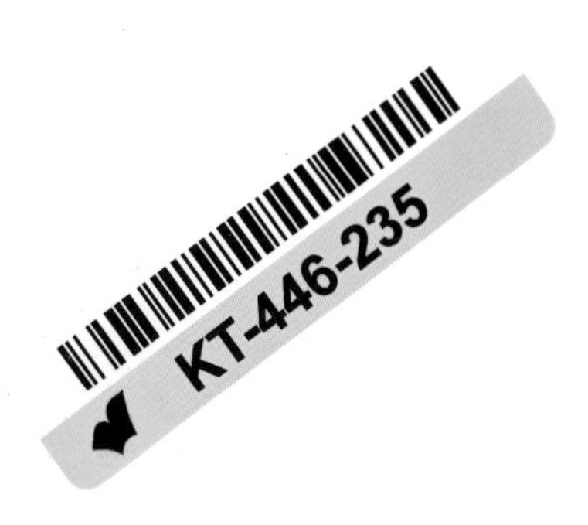
KT-446-235

B.S.U.C. - LIBRARY
00294611

Rhythm and Transforms

William A. Sethares

Rhythm and Transforms

William A. Sethares, Ph.D., Professor in Electrical Engineering
Department of Electrical and Computer Engineering
College of Engineering
University of Wisconsin-Madison
2556 Engineering Hall
1415 Engineering Drive
Madison, WI 53706
USA

British Library Cataloguing in Publication Data
Sethares, William A., 1955-
Rhythm and transforms
1. Musical meter and rhythm 2. Signal processing - Digital techniques 3. Music - Acoustics and physics
I. Title
781.2'24
ISBN-13: 9781846286391

Library of Congress Control Number: 2007926810

ISBN 978-1-84628-639-1 e-ISBN 978-1-84628-640-7 Printed on acid-free paper

9 8 7 6 5 4 3 2 1

Springer Science+Business Media
springer.com

Prelude

Rhythm and Transforms contrasts two ways of understanding temporal regularities in the world around us: directly via perception and indirectly via analysis. "Rhythm" alludes to the perceptual apparatus that allows people to effortlessly observe and understand rhythmic phenomena while "transforms" evokes the mathematical tools used to detect regularities and to study patterns.

Music has always been part of my life. Just as I don't remember learning to speak, I don't remember learning to distinguish the sound of one instrument from another or learning to tap my foot in time with the music.

I do recall being perplexed one day when my elementary school teacher demonstrated how to identify the sounds of individual instruments. To me, the sounds of a clarinet, a trumpet, and a guitar were as distinct as the colors red, green, and blue. Many years later, I was similarly mystified by Ann's experiences in *Ballet I.* The class was taught by a professional dancer named Vivian and, as is traditional in ballet, had live piano accompaniment. From Ann's perspective a typical drill began with Vivian cuing the pianist. After a few twiddly notes, Vivian would call out instructions.

The class had been practicing basic jumps for several weeks when Vivian announced that from now on the class should go "down on four and up on one." Ann was mystified. It was Vivian who called out the numbers

"One—and up!—two—and up!"

"One" was where Vivian starting counting. The idea that the piano was somehow involved was a foreign concept. How do you know when the piano gets to four? Or one? Ann asked me later.

Until this time I had been only vaguely aware that there were people in this world who could not find "one." As clear as it seemed to me, I found it difficult to describe in words exactly what "one" is and how to find it. I became aware that much of my perception of music, and rhythm in particular, was colored by training and practice. The ability to tap the foot in time to the music and to find "one" is a cognitive event, a learned behavior. What other aspects of rhythmic perception are learned?

At the intersection of music, signal processing, and psychology lies an area I call "perception-based audio processing." My first book, *Tuning, Timbre, Spectrum, Scale* explored the relationships between the timbre (or spectrum) of a sound and the kinds of musical intervals that musicians throughout the world use. *Rhythm and Transforms* explores the temporal and rhythmic relationships between sounds and the structure of music, between biological/perceptual aspects of the human auditory system and the sound machines (musical synthesizers, effects units, drum machines, musical computer software, etc.) we design.

People commonly respond to music by keeping time, tapping to the beat or swaying to the pulse. Underlying such ordinary motions is an act of cognition that is not easily reproduced in a computer program or automated by machine. The first few chapters of *Rhythm and Transforms* ask – and answer – the question: How can we build a device that can "tap its foot" along with the music? The result is a tool for detecting and measuring the temporal aspects of a musical performance: the periodicities, the regularities, the beat.

The second half of *Rhythm and Transforms* describes the impact of such a "beat finder" on music theory and on the design of sound processing electronics such as musical synthesizers, drum machines, and special effects devices. The "beat finder" provides a concrete basis for a discussion of the relationship between the cognitive processing of temporal information and the mathematical techniques used to describe and understand regularities in data. The book also introduces related compositional techniques and new methods of musicological analysis. At each stage, numerous sound examples (over 400 minutes in total) provide concrete evidence that the discussion remains grounded in perceptual reality. Jump ahead to Sect. 1.7 on p. 21 for an overview of the audio contents of the CD.

Think about it this way. Humans are very good at identifying complex patterns. The auditory system easily senses intricate periodicities such as the rhythms that normally occur in music and speech. The visual system readily grasps the symmetries and repetitions inherent in textures and tilings. Computers are comparatively poor at locating such patterns, though some kinds of transforms, certain statistical procedures, and particular dynamical systems can be used in the attempt to automatically identify underlying patterns. A variety of computer programs in MATLAB® are also provided on the accompanying CD for those who wish to explore further.

Rhythm and Transforms will be useful to engineers working on signal processing problems that involve repetitive behavior, to mathematicians seeking clear statements of problems associated with temporal regularities, and to musicians and composers who use computer-based tools in the creation and the recording process. It will be useful to those interested in the design of audio devices such as musical synthesizers, drum machines, and electronic keyboards and there are clear applications to the synchronization of audio with video. Finally, there are tantalizing tidbits for those interested in the way the ear works and how this influences the types of sound patterns we like to listen to.

Madison, WI, Dec. 2006 *William A. Sethares*

Acknowledgments

This book owes a lot to many people.

The author would like to thank *Tom Staley* for extensive discussions about rhythmic phenomenon. *Ian Dobson* has always been encouraging, and his enthusiasm for rhythm, rhythm, rhythm, is contagious, contagious, contagious. I owe great thanks to *Robin Morris* for initiating me into the likelihooded elite of the Bayesian Brotherhood and for holding my hand through the dark and mysterious corridors of priories and posteriories. *Diego Bañuelos'* remarkable hard work and insightful analyses are matched only by his generosity in allowing use of his thesis *Beyond the Spectrum*. You can find it on the CD! *Phil Schniter* and *Ruby Beil* were especially courageous in allowing me to play with their *Soul*. Bwahaahaa. It, too, appears on the CD. *Mark Schatz* probably doesn't realize what a dedicated following he has, but he deserves it... he even let Julie Waltz home with me. You can read about her in Chap. 11 and hear her on the CD.

I have been blessed with a group of dedicated and insightful reviewers. Jim and Janet read so many drafts of *Rhythm and Transforms* that they would probably have it memorized if I hadn't kept changing it! Jacky Ligon provided the much needed viewpoint of a drummer in the wild. Phil and Diego (yes, the same Phil and Diego as above) read and provided detailed comments in a vain attempt to keep me from saying ridiculous things. Bob Williamson's careful and thoughtful comments were more helpful than he probably imagines, and he does have a good imagination. *Marc Leman's* insights into psychoacoustics and *Gerard Pape's* thoughts on composition helped me refine many of the ideas. The students at CCMIX in the fall of 2005 were both receptive and skeptical; this tension helped make the project successful. Thanks to everyone on the "Alternate Tuning Mailing List" and the "MakeMicroMusic list" at [W: 2] and [W: 26] who helped keep me from feeling isolated and provided me with challenge and controversy at every step. Thanks to my editors *Anthony Doyle*, *Kate Brown*, *Simon Rees*, and *Sorina Moosdorf* for saying "yes" to a book that otherwise might have fallen into the cracks between disciplines. *Ingo Lepper* came to the rescue with a fix for my LaTex referencing woes.

I am also grateful to the Dept. of Electrical and Computer Engineering at the University of Wisconsin for allowing me the freedom to pursue this book during my sabbatical leave in the fall 2005 and spring 2006 semesters. Many thanks also to *Christos* who hosted me in Pythagorion, *Σαμοσ*, one of the birthplaces of musical thought.

The very greatest thanks go to *Ann Bell* and the *Bunnisattva.*

Contents

1

What is Rhythm?

How can rhythm be described mathematically? How can rhythm be detected automatically? People spontaneously clap in time with a piece of music – but it is tricky to create a computer program that can "tap its foot" to the beat. Some peculiar features of the mind help us to internalize rhythms. Teaching the computer to synchronize to the music may require some peculiar mathematics and some idiosyncratic kinds of signal processing.

Rhythm is one of the most basic ways that we understand and interact with time. Our first sensory impression as we float in our mother's womb is the rhythmic sound of her heart. The time between the opening and closing of the heart's valves is a clock that measures the passing of our lives. Breathing is our most lasting experience of rhythm.

Rhythms occur at all time scales. The motion of waves against a beach, the daily flow of the sun and moon, the waxing and waning of the year; these occur at rates much slower than the heartbeat. Pitched phenomenon such as the rotation of an engine, the oscillations of a string, and the vibrations of our vocal chords occur at rates much faster than the heartbeat. At the slowest rates, we conceive of the rhythm via long term memory. At rates near that of the heartbeat, we perceive the repetition directly as rhythm. At the fastest rates, the repetitions blur together into the perception called "pitch." Thus we perceive rhythmic patterns differently at different time scales.

Figure 1.1 shows several seconds of the rhythmic beating of a healthy heart. Large spikes recur at approximately regular intervals and at approximately the same height. After each peak is a dip followed by two smaller bumps which vary slightly in position and size from beat to beat. This kind of regularity with variation, of repetition with change, is a characteristic feature of rhythmic patterns in time. The sound of a heartbeat can be heard in example [S: 1].[1]

It is not easy to measure irregularities in the heartbeat by eye. The scraggly circular object in Fig. 1.1(b) contains the same heartbeat signal plotted so that

[1] References marked [S:] point to the sound examples contained on the CD. They are described in detail starting on p. 295. References to printed matter [B:] are in the bibliography on p. 309 and references to recordings [D:] are in the discography on p. 321. Web references [W:] are detailed on p. 323 and are also available on the CD as direct links.

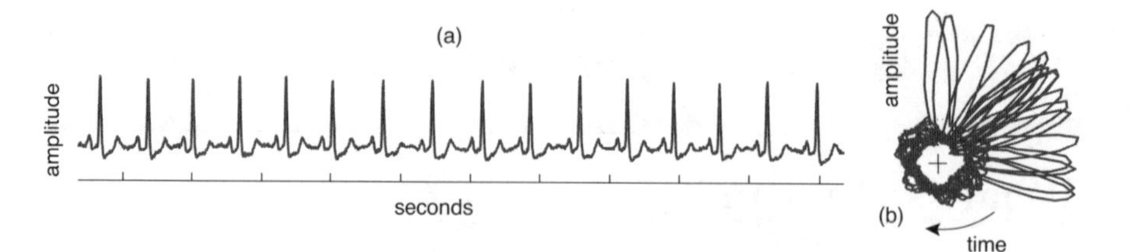

Fig. 1.1. The steady rhythm of a heartbeat is among our earliest sensations; the "lub-dub" sound is caused by the rhythmic closing of the heart's valves. (a) is a plot of the waveform over time while (b) is a polar plot in which time moves around the circle at a rate of approximately one beat per revolution. Listen to the heartbeat in sound example [S: 1].

time moves around the circle at a rate of approximately one heartbeat each cycle. This overlays the various beats, and the differences in the location of the peaks emphasizes the variability of the timing.

All animals engage in rhythmic motions: their hearts beat, they breathe, their brains exhibit periodic electrical activity, and they walk, run, or swim. Indeed, many such physical activities are easier to do rhythmically. It is almost impossible to walk without rhythm. The rhythms inherent in these behaviors are not direct responses to the environment, they are internally generated.

Though rhythms surround us in both the physical world and the animate world, humans are among the only creatures who create rhythmic patterns consciously.[2] From the age of about 3–4 years, a child can tap along with a metronome or a song. This is significant not because of the regularity of the tapping but because of the synchrony: the child learns to recognize the regularity of the pulse and then anticipates successive pulses in order to tap at the same time. Most animals never develop this capability, suggesting that much of our experience of rhythm "comes from the mind, not from the body" [B: 19]. It is not easy to emulate this kind of synchronization in a computer, and Chaps. 5 – 7 discuss the problem in depth. Perhaps our highest level of rhythmic achievement occurs in music, where rhythmic phenomena play in complex ways with our perceptions of time, sequence, and pattern.

A computational approach to the study of rhythm builds a model or a computer program that attempts to mimic people's behavior in locating rhythms and periodicities. Such a "foot-tapping machine" is diagrammed in Fig. 1.2. To the extent that the model can duplicate or imitate human behavior it can be judged successful. When it fails (as any model will eventually fail to capture the full range of human abilities), it helps identify the limits of our understanding. Understanding rhythm well enough to create a foot-tapping machine involves:

[2] Recent work [B: 91] suggests that bottlenose dolphins may be capable of spontaneously employing rhythmic patterns in communication.

Fig. 1.2. A foot-tapping machine designed to mimic people's ability to synchronize to complex rhythmic sound must "listen" to the sound, locate the underlying rhythmic pulse, anticipate when the next beat timepoint will occur, and then provide an output

Cognition: High level mental abilities allow us to recognize, remember, and recall complex patterns over long periods of time.

Perception: The information that enters the senses must be organized into events, chunks, and categories. See Chap. 4.

Mathematics: Scientists searching for patterns and repetitions in data have developed sophisticated mathematical tools (for example: *transforms*, *adaptive oscillators*, and *statistical models*) that help to uncover regularities that may exist. See Chaps. 5–7.

Signal processing: Using various kinds of "perceptual preprocessing" (which attempt to extract significant features of a sound from its waveform) leads to new classes of algorithms for the manipulation of signals and songs at the beat level. See Chaps. 8 and 9.

Music: The human spirit engages in an amazing variety of rhythmic behaviors. An overview of the world's music in Chap. 3 reveals several common threads that broaden our view of what rhythm is and how it works, and Chap. 10 shows how beat-synchronized methods can be used in musical (re)composition.

1.1 Rhythm, Periodicity, Regularity, Recurrence

> Mathematics helps explain the patterns and rhythms of the universe. Music helps us synchronize to those patterns. (paraphrased from [B: 70])

Many words have meanings that are intimately related to rhythm. Repetition embodies the idea of repeating, and exact repetition implies periodicity. For example, the sequence

$$\ldots\ c\ 3\ \sharp\ \aleph\ \mu\ \heartsuit\ c\ 3\ \sharp\ \aleph\ \mu\ \heartsuit\ c\ 3\ \sharp\ \aleph\ \mu\ \heartsuit\ c\ 3\ \sharp\ \aleph\ \mu\ \heartsuit\ c\ 3\ \sharp\ \aleph\ \mu\ \heartsuit\ c\ 3\ \sharp\ \aleph\ \ldots$$

recurs at predictable intervals, in this case, every six elements. Thus one component of periodicity is the rate of recurrence, called the *period*. But what is it that is repeating? Is it the six element pattern $c\,3\,\sharp\,\aleph\,\mu\,\heartsuit$ or is it the pattern $\sharp\,\aleph\,\mu\,\heartsuit\,c\,3$? Evidently, the two descriptions are on equal logical footing since they are identical except for the starting element.

While any of the possible six element patterns may be equal on a logical footing, they may not be equal on a perceptual footing. Look again at the complete sequence. For many observers, the $\heartsuit$ stands out from the others, creating a visually compelling reason to place the $\heartsuit$ symbol at the start of the sequence. Thus, perceptually, the sequence is perhaps best described by a repetition of the pattern

$$\heartsuit\,c\,3\,\sharp\,\aleph\,\mu,$$

though some might wish to place the $\heartsuit$ at the end. Thus the starting ambiguity may be resolved by choosing the position of one prominent element. This defines the *phase* of the sequence (by analogy with the phase of a sinusoid).

Analogous effects occur in the auditory realm. The rhythmic pattern shown in Fig. 1.3 is presented in sound examples [S: 2](0–7). In the first case, all notes are struck identically. Which note appears to start the pattern? The succeeding examples with $N = 1$ to $N = 7$ each single out one note (the notes are labeled N in the figure) to emphasize. In many cases (but not all) this changes the apparent starting note of the pattern.

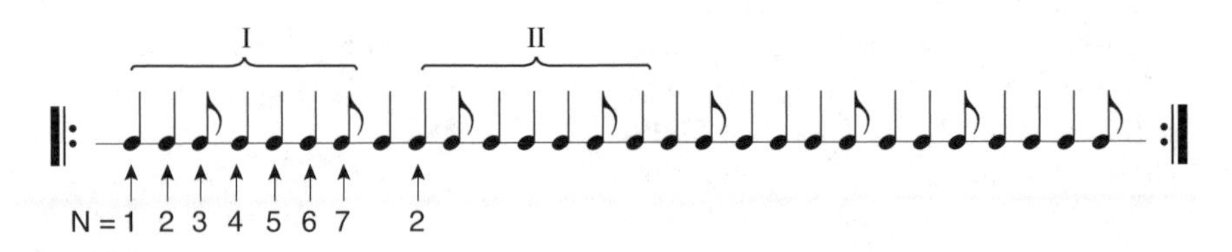

Fig. 1.3. The rhythmic pattern repeats every seven notes. Where does the pattern start? Listen to the sound examples in [S: 2] and find out.

An important feature of periodic phenomena is that they readily support hierarchical structures. For example, by emphasizing every other c, the visual focus changes from the $\heartsuit$ to the $\boxed{\text{c}}$.

$$\ldots\ \boxed{\text{c}}\,3\,\sharp\,\aleph\,\mu\,\heartsuit\,c\,3\,\sharp\,\aleph\,\mu\,\heartsuit\,\boxed{\text{c}}\,3\,\sharp\,\aleph\,\mu\,\heartsuit\,c\,3\,\sharp\,\aleph\,\mu\,\heartsuit\,\boxed{\text{c}}\,3\,\sharp\,\aleph\,\mu\,\heartsuit\,c\,3\,\sharp\,\aleph\ \ldots$$

Pairs of the six-term sequences become perceptually linked, creating a simple two-level hierarchy where each repetition (of twelve elements) consists of a pair of lower-level six-element terms. This is a visual analogy of metrical structure, as discussed in Sect. 3.2.

Exact repetition and perfect periodicity are absent in an imperfect world. For example, in the visual pattern $\ldots\heartsuit\,c\,3\,\sharp\,\aleph\,\mu\ldots$ there may be occasional mistakes. Maybe one out of every hundred symbols is randomly changed. Clearly, the sequence is still "mostly" periodic, and this kind of corruption is often described by saying that the periodic sequence has been contaminated with *noise*. (Observe that this use of noise is quite different from the use of noise in the sense of an annoying or undesirable sound, as in a hiss-filled recording or a noisy environment.) Another kind of deviation from periodicity occurs when the number of elements changes; perhaps one out of every hundred symbols is randomly deleted, or extra ones may be added.

In the auditory realm, this corresponds to a kind of jitter in the exact timing of events and of the underlying period. For example, in the final sound example in the series [S: 2](8), the notes are all struck identically, but each note is displaced in time by a small random amount. Such small timing deviations may change the rhythm that is perceived.

Yet another kind of change in a periodic structure occurs when the period changes. Perhaps the six-element sequence is augmented to include a seventh element, increasing the underlying period to seven. Analogous changes are quite common in the auditory realm where they may be perceived as changes in tempo: faster tempos occur when there is a decrease in the period, slower tempos correspond to an increase in the underlying period. Alternatively, changes in period may be perceived as changes in pitch: lower pitches correspond to longer periods and higher pitches correspond to shorter periods.

In summary, periodic phenomenon are characterized by

(i) period
(ii) phase or starting point
(iii) the elements that are ordered within a single repetition

Deviations from periodicity may occur in several ways:

(i) elements may (occasionally) change
(ii) elements may jitter in time (or the number of elements in a period may temporarily increase or decrease)
(iii) the period may increase or decrease

Moreover, periodic sequences may not be perceived straightforwardly: they may be perceived in different ways depending on the rate of presentation and they may be interpreted in hierarchical terms. Deviations from periodicity may influence (and be influenced by) the perceived hierarchies. As will become clear in later chapters, these basic notions of periodic sequences underlie many of the analytical tools that can be used to locate rhythmic phenomena. Deviations from periodicity will cause the bulk of the difficulties with these methods.

Subjective Rhythm

The constant interaction between the physically measurable properties of a sequence and the human perception of those properties is fascinating. Think of a periodic sequence of identical clicks separated by a constant time interval, as is shown schematically in Fig. 1.4. Now listen to the *Regular Interval 750* sound example [S: 3] where the time between successive clicks is exactly 750 ms.

Does it *sound* perfectly regular? Almost invariably, listeners begin "counting" the clicks. The inner ear may count 1-2-1-2 or perhaps 1-2-3-1-2-3. Many people report hearing 1-2-3-4-1-2-3-4. Almost no one hears this as it truly is: a repetition of exactly identical clicks.[3] Consider the implication; the ear creates an ordered pattern (the apparent repetition of the clicks in units of 2, 3, or 4) where none objectively exists.

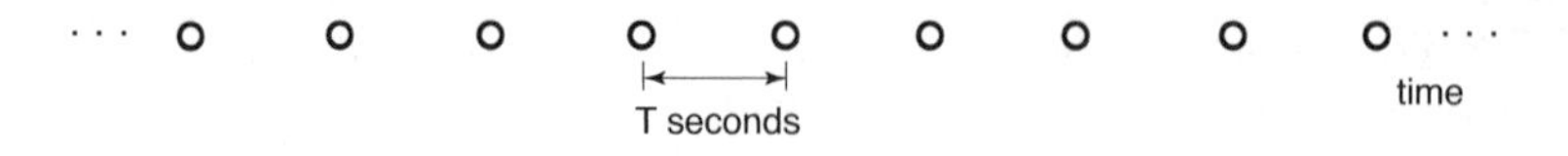

Fig. 1.4. A completely regular sequence of identical clicks is perceived as clustered into groups with 2, 3, or 4 elements even though there is no objective (physical) basis for such grouping. Listen to [S: 3].

Psychologists have been fascinated by this subjective rhythm for more than a century [B: 17]. It presents a clear cautionary message to researchers attempting to simulate human responses to music; physical measurements alone cannot fully reveal how a sound is perceived.

1.2 Perception and Time Scale

Cognitive scientists describe memory as operating on three time scales: echoic, short term, and long term [B: 219]. Echoic memory operates on a very short time scale (up to about a second) where features are extracted from sensory impressions and events are fused together to form individual objects of perception. For example, consider the sound of a stick hitting a drum. It may seem as if the sound is a single object; in reality it is a complex pattern of pressure waves that impinge on the ear. Binding all the necessary impressions into a single entity requires considerable cognitive activity.

Similarly, the puff of air at the start of a flute note becomes bound to the (relatively) steady oscillations that follow. This new entity must interact with long term memory for the listener to "recognize" and name the sound, creating the illusion that it represents a familiar object: *the flute sound.* Moreover, because the flute generates more-or-less periodic sound waves of a certain kind that we recognize as pitch, the auditory system integrates this information and we "hear" *the flute playing C* as a single object of perception. Similarly, complex sense impressions such as those that represent phonemes in speech, simultaneous musical intervals, timbres (the sound of the guitar in contrast to that of the flute playing the same note), and the boundaries between such events are aggregated together into coherent auditory events [B: 18]. Figure

[3] Logically, the counting should be 1-2-3-4-5-6-7... or perhaps 1-1-1-1-1-1-1. But the ear has its own kind of logic.

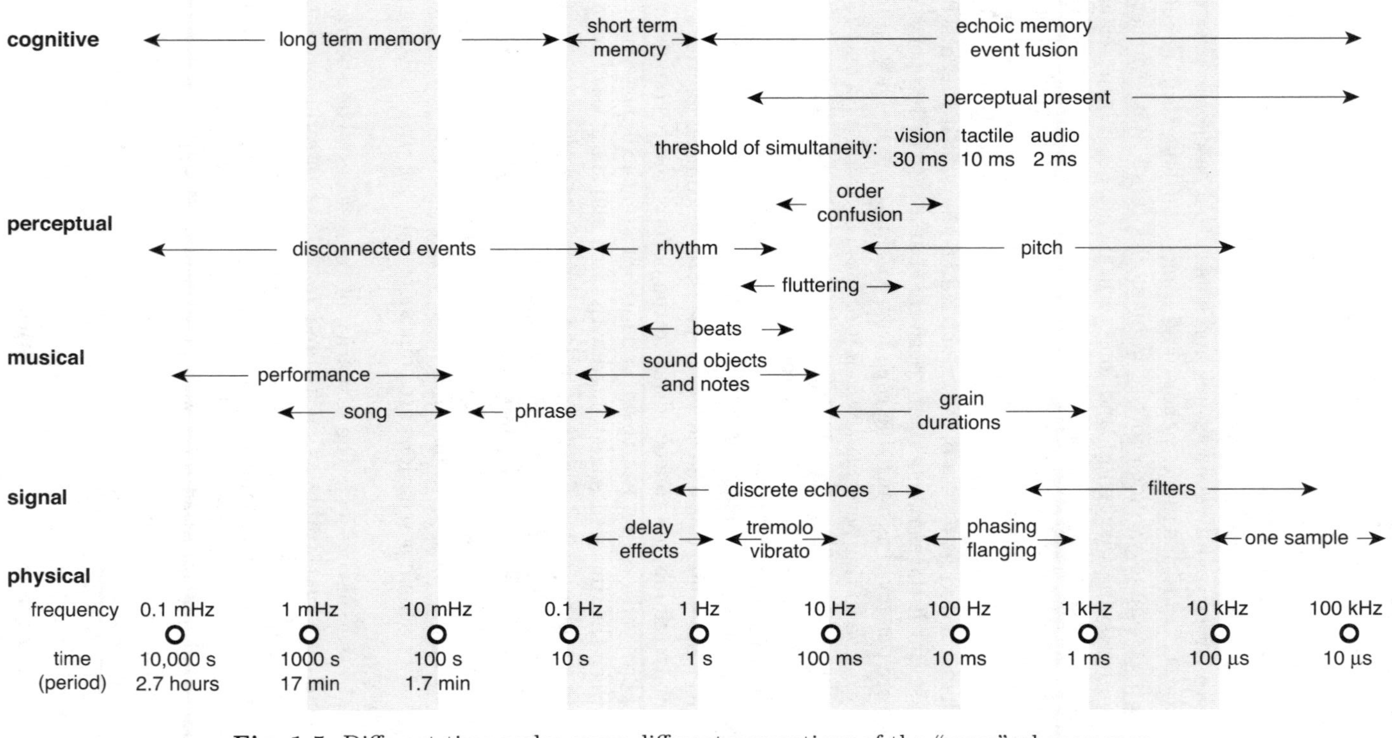

Fig. 1.5. Different time scales cause different perceptions of the “same” phenomenon

1.5 shows the approximate time scales at which various cognitive, perceptual, and musical events occur.

After the disparate sensory impressions are bound into coherent objects of perception, these objects are themselves grouped together based on similarity or proximity. Short term memory is where patterns such as words, phrases, melodies, and rhythms are gathered into perceptual streams. Long term memory is where larger cognitive structures and conceptual categories are stored; abstract ideas, forms, language, poems, and songs. But long term memory is not a passive receptacle where short term memories retire. Rather, there is a constant interplay between short and long term memories. Whenever an object is present in short term memory, it activates similar objects from within long term storage; these are then recirculated in parallel with the new events.

There are also differences in the perception of events at different time scales that mirror these differences in cognition. For example, if a series of short clicks is played at a rate of 3 per second they are heard as a series of short clicks. But if the same clicks are performed at a rate of 100 per second, then they are perceived as a buzzing tone with a definite pitch. Thus "pitch" is the name we give to this perception when it occurs between 20 Hz and 20 kHz, while we call it "rhythm" when the interval between clicks is longer, between about $\frac{1}{10}$ and 3 s. There is even a different vocabulary to describe the rates of these phenomena: pitch is described as being low or high; rhythm is described as being slow or fast. See Chap. 4 for sound examples and further discussion. At yet longer time intervals, the clicks are heard as disconnected events. Thus rhythmic patterns may be conceived (as in the orderly succession of day and night) or perceived (as in a heartbeat, a dance, or a musical passage).

In between the time scales associated with pitch and those associated with rhythm lies a region (called *fluttering* in Fig. 1.5) where sound is perceived in brief bursts. Rainsticks, bell trees, and ratchets, for example, produce sounds that occur faster than rhythm but slower than pitch. Similarly, drum rolls and rapid finger taps are too fast to be rhythmic but too slow to be pitched. Roughly speaking, pitched sounds occur on the same time scale as echoic memory and rhythmic perceptions are coincident with the time scales of short term memory.

Musical usage also reflects the disparity of time scales. The shortest isolated sounds are perceivable as clicks, and may have duration as short as fractions of a millisecond. These are called "grains" of sound. In order to have a clear sense of pitch, a sound must endure for at least about 100 ms, and this is enough time to evoke impressions of pitch, loudness, and timbre. Such sensations are typically fused together to form sound objects, which are commonly called "notes" if they are played by an instrument or sung by a voice. Groups of notes cluster into phrases, and phrases coalesce into songs, or more generally, into performances that may last up to a few hours.

Finally, Fig. 1.5 shows the time scales at which various kinds of signal processing occur; from the single sample (which may be from about 5 kHz to 200 kHz for audio), through filters (such as lowpass, bandpass, and highpass), various special effect processing such as flanging and phasing, and the rate at which vibrato and tremolo occur. These signal processing methods occur within the zone of event fusion and so effect the quality of a sound (its timbre, vibrato, spectral width, attack characteristics, etc.). Multi-tap delay line effects extend into the time scale dominated by short term memory and thus can change the perception of rhythmic events.

The above discussion has focused on how the scale of time interacts with our cognitive, perceptual, and musical makeup. A different, though related issue is how we perceive the flow of time. This depends on many factors: the emotional state of the observer, how the attention is directed, familiarity with the events, etc. In addition, the perception of time depends on the nature of the events that fill the time: repetition and a regular pulse help time to pass quickly while irregular noises or unchanging sounds tend to slow the perception of time. Issues of duration and time perception are explored more fully in Chap. 4.

1.3 Illusions of Sound Perception

There are a number of optical illusions that are frequently cited as demonstrating that our senses do not always accurately report the reality of the world around us. For instance, Fig. 1.6(a) shows a diagonal line that is interrupted by two parallel vertical lines. It appears that the two halves of the diagonal are misaligned. Part (b) shows the Necker cube (familiar to quilters as "tumbling blocks") which can be seen in two ways: either as jutting up and to the right or as receding back and to the left. Interestingly, it is usually possible to perceive this shape either way, but not both simultaneously. The figure in (c) appears to be a forked object. At the left it appears to have two prongs, while at the right it has three prongs. Most people "see" a triangular shape in part (d), even though the reality is that there are only three pac-man-like partial circles. Part (e) shows Penrose's "impossible tribar" which appears to be a triangular solid built from three 90° right angles. Even though geometry proves that the sum of all angles in a triangle must total 180°, it still *looks like* the tribar could exist in three-dimensional space. The ever-ascending stairway (f) from [B: 161] was made famous by the graphic artist M. C. Escher in his prints "Ascending and Descending" and in the enigmatic "Waterfall."

Less well known, but just as fundamental as such visual tricks, are everyday aspects of audio perception that are "illusions" in the sense that what we perceive is very different from the reality in the physical world surrounding us. For example, if I were to say "think of a steady, continuous, unchanging sound like the tone of an organ," you could most likely do so without difficulty. In reality, however, there is nothing "steady" or "continuous" or "unchanging" about an organ sound. The physical reality is that an organ sound (like any sound) is an undulating wave with alternating regions of high and low pressure,

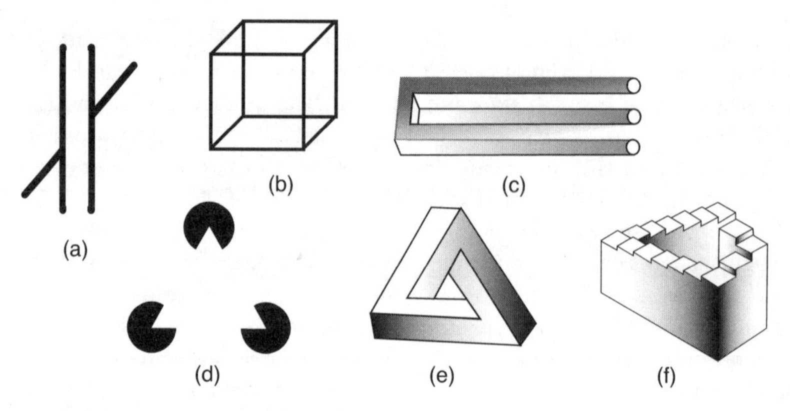

Fig. 1.6. Several visual illusions demonstrate that the visual system does not always perceive what is presented. In (a), the diagonal appears to be misaligned (it is not). (b) can be perceived in two completely different ways: as angled up and to the right, or as angled down and to the left. The "objects" that appear to be depicted in (c), (e) and (f) cannot exist. There is no triangle in (d), only three partial circles. The stairway in (f) appears to ascend (or descend) forever.

air molecules must be constantly wiggling back and forth. Were the pressure changes and the motions to cease, then so would the sound.[4]

Because the undulations occur at a rate too fast to perceive separately, the auditory system "blurs" them together and they achieve the illusion of steadiness. Thus sounds which occur closer together in time than the threshold of simultaneity (Fig. 1.5 places this at about 1 ms) are merged into a "single sound." This is parallel to the everyday illusion that television (and movies) show continuous action; in reality movies consist of a sequence of still photos which are shown at a rate faster than the threshold of simultaneity for vision, which is about 20 Hz. Closely related to the illusion of continuity is the illusion of simultaneity: sounds appear to occur at the same time instant even though in reality they do not. The ear[5] tends to bind events together when they occur close to each other in time and to clearly separate others that may be only slightly further apart.

Another common auditory "illusion" is based on the ear's propensity to categorize certain kinds of sounds. For example, it is easy to say the vowel "a" and to slowly (and "continuously") change it into the vowel "e." Thus there is a continuum of possible vowel sounds: "a" at one end, two thirds "a" and one third "e," one half "a" and one half "e," one third "a" and two thirds "e," continuing to a pure "e" at the other end. Yet no matter who is speaking, one never perceives the intermediate vowel sounds. The sound is automatically categorized by the auditory system into either "a" or "e," never

[4] The only truly steady sound is silence.

[5] When there is no danger of confusion, it is common to use "the ear" as shorthand for "the human auditory system."

into something in between. This is called categorical perception, and it has obvious importance in the comprehension of language.

While there are many similarities between visual and auditory perceptions, there are also significant differences. For example, in 1886 Mach [B: 136] demonstrated that spatial symmetry is directly perceptible to the eye whereas temporal symmetry is not directly perceptible to the ear. Unlike vision, the human ability to parse musical rhythms inherently involves the measurement of time intervals.

1.3.1 Illusions of Pitch

Loosely speaking, pitch is the perceptual analog of frequency. Acousticians define pitch formally as "that attribute of auditory sensation in terms of which sounds may be ordered on a scale extending from low to high." Sine waves, for which the frequency is clearly defined, have unambiguous pitches because everyone orders them the same way from low to high.[6] For non-sine wave sounds, such an ordering can be accomplished by comparing the sound with unknown pitch to sine waves of various frequencies. The pitch of the sinusoid that most closely matches the unknown sound is then said to be the pitch of that sound.

Pitch determinations are straightforward when working with musical instruments that have a clear fundamental frequency and harmonic overtones. When there is no discernible fundamental, however, the ear will often create one. Such *virtual* pitch (see [B: 225] and [B: 226]) occurs when the pitch of the sound is not the same as the pitch of any of its overtones. This is shown on the Auditory Demonstrations CD [D: 24], where the "Westminster Chimes" song is played using only upper harmonics. In one demonstration, the sounds have spectra like that shown in Fig. 1.7. This particular note has partials at 780, 1040, and 1300 Hz, which is clearly not a harmonic series. These partials are, however, closely related to a harmonic series with fundamental at 260 Hz, because the lowest partial is 260 times 3, the middle partial is 260 times 4, and the highest partial is 260 times 5. The ear recreates the missing fundamental, and this perception is strong enough to support the playing of melodies, even when the particular harmonics used to generate the sound change from note to note. Thus the ear can create pitches even when there is no stimulus at the frequency of the corresponding sinusoid. This is somewhat analogous to the "triangle" that is visible in Fig. 1.6(d).

Perhaps the most striking demonstration that pitch is a product of the mind and not of the physical world is Shepard's [B: 211] ever rising tone, which is an auditory analog of the ever-ascending staircase in Fig. 1.6(f). Sound example [S: 4] presents an organ-like timbre that is constructed as diagrammed in Fig. 1.8. The sound ascends chromatically up the scale: after

[6] With the caveat that some languages may use different words, for instance, "big" and "small" instead of "low" and "high."

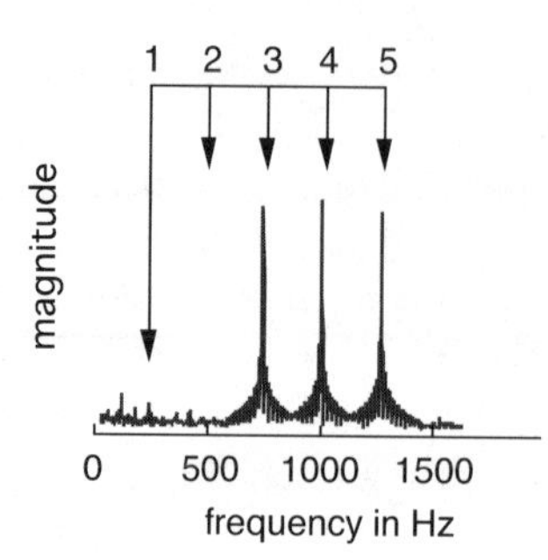

Fig. 1.7. Spectrum of a sound with prominent partials at 780, 1040, and 1300 Hz. These are marked by the arrows as the third, fourth, and fifth partials of a "missing" or "virtual" fundamental at 260 Hz. The ear perceives a note at 260 Hz, which is indicated by the extended arrow.

ascending one full octave, it has returned to its starting point and ascends again. The perception is that the tone rises forever (this version is about 5 minutes long) even though it never actually leaves a single octave!

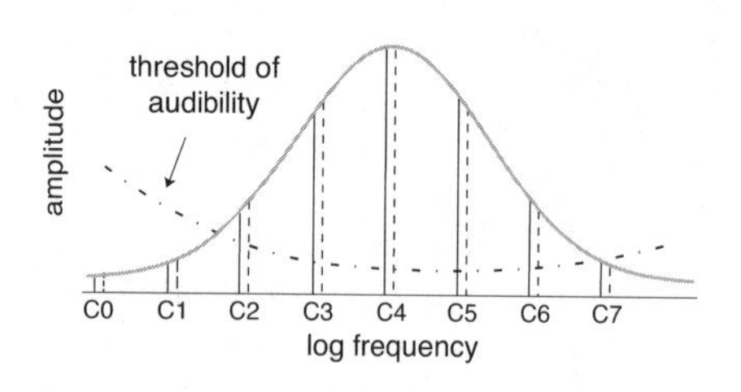

Fig. 1.8. The Shepard tone of sound example [S: 4] contains many octaves of the same note (indicated as C). Successive tones increase in frequency with amplitudes that increase (for low frequencies) and decrease (for high frequencies). The dotted lines indicate the $C\sharp$ tone. As the highest frequencies dissolve into inaudibility, the lowest frequencies become perceptible.

1.3.2 Why Illusions Happen

Already we have encountered illusions of continuity and simultaneity, illusions of categorization, and two different kinds of pitch illusions (the missing fundamental and the ever-rising sound). As will become clear in Chap. 4, there are many other kinds of auditory illusions, not all of them operating at very short time scales. Why is the ear so easily fooled into reporting things that do not exist (such as the missing fundamental), into failing to sense things that do exist (such as the vowel that is in reality halfway between "a" and "e"), and into perceiving things that cannot be (such as an ever-rising pitch)?

The ear's job (to engage in some introductory chapter anthropomorphization) is to make sense of the auditory world surrounding it. This is not an easy job, because sound consists of nothing more than ephemeral pressure waves embedded in a complex three-dimensional world. The sound wave that arrives at the ear is a conglomeration of sound from all surrounding events merged together. The ear must unmerge, untangle, and interpret these events. The ear must capture the essence of what is happening in the world at large, simultaneously emphasizing the significant features and removing the trivial.

Does that sound represent the distant rustling of a lion or the nearby bustling of a deer? The distinction may be of some importance.

Given this formidable task, the ear has developed a sophisticated multi-level strategy. In the first stage, collections of similar sense impressions are clustered into objects of perception called *auditory events*. This occurs on a very short time scale. At the second stage, auditory events that are similar in some way are themselves grouped together into larger chunks, to form patterns and categories that are most likely the result of learning and experience.

For example, the low level processing might decode a complex wave into an auditory event described by the phoneme "a." This represents a huge simplification because while there are effectively an infinite variety of possible waveshapes, there are only about 45 distinct phonemes.[7] At the next stage, successive phonemes are scanned and properly chunked together into words, which can then invoke various kinds of long term memory where something corresponding to meaning might be stored.

As another example, the low level processing might decode a complex waveform into an auditory event such as the performance of a musical "note" on a familiar instrument; the C of a flute. Again, this represents a simplification because there are only a few kinds of instruments while there are an infinite variety of waveforms. At the next stage, several such notes may be clustered to form a melodic or rhythmic pattern, again, condensing the information into simple and coherent clusters that can then be presented to long term memory and parsed for meaning.

Thus the ear's strategy involves simplification and categorization. A large amount of continuously variable data arrives; a (relatively) small amount of well categorized data leaves, to be forwarded to the higher processing centers. In the normal course of events, this strategy works extremely well. If, for instance, two sounds are similar (by beginning at the same time, by having a common envelope, by being modulated in a common way, by having a common period, by arriving from the same direction, etc.) then they are likely to be clustered into a single event. This makes sense because in the real world, having such similarities implies that they are likely to have arisen from the same source. This is the ear doing its job.

If we, as mischievous scientists, happen to separate out the cues associated with legitimate clustering and to manipulate them independently, then it should come as no surprise that we can "fool" the ear into perceiving "illusions." The pitch illusions are of exactly this kind. It would be a rare sound in the real world that would have multiple harmonically related partials yet have no energy at the frequency corresponding to their common period (such as occurs in Fig. 1.7). It would be an even rarer sound that spanned the complete audio range in such a way that the highest partials faded out of awareness exactly as the lowest partials entered.

[7] in English. This is not to suggest that the phonemes themselves are innate; rather, it is the processes that allow the ear to recognize and categorize phonemes.

1.3.3 Why Illusions Matter

Illusions show the limitations of our perceptual apparatus. Somewhat paradoxically, they are also helpful in distinguishing what is "really" in the world from what is "really" in our minds.

Consider two friends talking. It might appear that a tape recording of their conversation would contain all the information needed to understand the conversation. Indeed, you or I could listen to the recording, and, providing it was in a language we understood, reconstruct much of the meaning. But there is currently no computer that can do the same. Why? The answer is, at least in part, because the recording does not contain anywhere near "all" the information. There are two different levels at which it fails. First, the computer does not know English and lacks the cultural, social, and personal background that the two friends share. Second, it lacks the ability to parse and decode the audio signal into phonemes and then into words. Thus the computer fails at both the cognitive and the perceptual levels.

The same issues arise when attempting to automate the interpretation of a musical passage. What part of the music is in the signal, what part is in the perceptual apparatus of the listener, and what part is in the cognitive and/or cultural framework in which the music exists? Features of the music that operate at the cognitive level are unlikely to yield to automation because the required information is vast. Features that fundamentally involve perceptual processing may yield to computer analysis if an appropriate way to preprocess the signal in an analogous fashion can be found. Only features that are primarily "in the signal" are easy. Illusions can help distinguish which parts of our sense impressions correspond directly to features of the world, and which do not. As will become clear, the things we call "notes," "beats," "melodies," "rhythms," and "meter" are objects of cognition or perception and not primary sense impressions; they are "illusions" in the mind of the listener and not intrinsic properties of the musical signal.

1.4 Beat Tracking

One interesting aspect of musical rhythm is the "beat," the steady foot-tapping pulse that drives music forward and provides the temporal framework on which the composition rests. Is the beat directly present in the signal, is it a perceptual construct, or does it require high level cognitive processing?

Though it may be tempting to imagine that the beat really exists in the musical signal itself (because it is so conspicuous in our conception), it does not. For example, there may be a syncopated section where all the energy occurs "off" the beat. Or a song may momentarily stop and yet the beat continues even in the absence of sound. Something that can exist without sound cannot be in the signal!

Figure 1.9 shows a number of the physical and perceptual terms associated with musical rhythm. The waveform depicts a bit more than five seconds of Scott Joplin's *Maple Leaf Rag* (which may be heard in sound example [S: 5]). The beats are shown above, aligned with the waveform. While several beats are clearly visible in the waveform (the final three, for instance, involve obvious amplitude fluctuations), many are not. The stretch between 30.5 s and 33.5 s is devoid of obvious amplitude changes, yet the beat goes on.

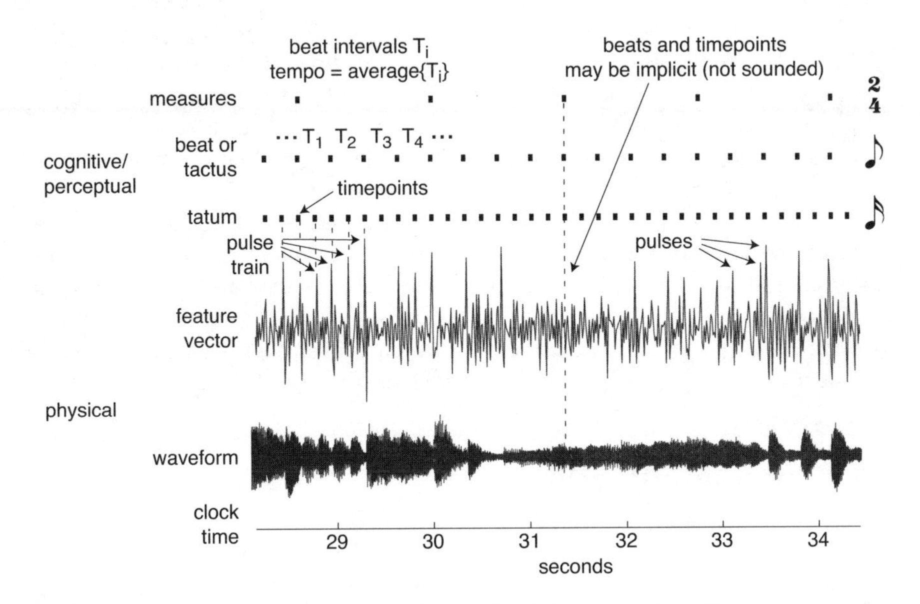

Fig. 1.9. A few seconds of Joplin's *Maple Leaf Rag* is used to illustrate a number of the terms associated with rhythm. The waveform is a direct representation of the physical pressure wave from which the feature vector is derived. Perceptual terms include the tatum ("temporal atom"), beat (or tactus), beat interval, and tempo. Cognitive terms include measures, time signatures, and musical notations which correlate with (but are distinct from) their perceptual counterparts (e.g., the tatum corresponds to the sixteenth note while the beat corresponds to the eighth note). Perceived pulses typically align with the tatum (and/or beat) though they need not in all circumstances.

Comparing the waveform to the line of dots that represent the beat shows why it can be difficult to recover the beat directly from the waveform. Feature vectors may be helpful as an intermediate step; they are derived from waveforms but are designed to emphasize significant features of the sound. For example, the feature vector shown in Fig. 1.9 was constructed by calculating the short term spectrum and measuring the change in the spectrum from one "instant" to the next. Large values (in either the positive or negative directions) indicate large changes. To the extent that the ear is sensitive to such spectral changes, the large values of the feature vector correspond to perceptible pulses. Clearly, choosing good feature vectors is an important but tricky business. This is discussed at great length in Chap. 4.

When such pulses occur at regular intervals, they tend to induce a perception of repetitiveness. For instance, the string of six pulses beginning at 28.5 s coincides with six successive timepoints. In this case, the pulses occur at a rate faster than it is comfortable to tap the foot (i.e., faster than the beat), and the grid of approximately equally spaced timepoints that is aligned with the pulses is called the tatum (the regular repetition of the most rapid temporal unit in the piece). Thus the beat and the tatum are similar; they are both regular, repetitive perceptions of a steady flow. They exist because of and persist despite the moment by moment sound of the piece which may or may not reinforce the regularity over any given duration. The difference is that the tatum is always the fastest such regular grid while the beat may be slower. Typical beat intervals are between about 300 and 700 ms.

The tatum is also typically the rate at which the fastest notes in a musical score are written; in this case, sixteenth notes. Because there are two tatum timepoints for each beat, the beat is therefore represented in the score by the eighth notes, and the measure by a $\frac{2}{4}$ time signature. These latter notions, involving the musical score, are clearly higher level cognitive constructions and such notations are discussed in Sect. 2.1.2.

1.5 Why Study Rhythm?

> *Analyzing and modeling the perception of musical rhythm provides insights into non-verbal knowledge representations, quantification of musicological theories, and intelligent tools for music performance and composition.* [B: 217]

Understanding the workings of the human mind is one of the great scientific frontiers of our time. One of the few paths into the brain is the auditory system, and discovering the boundaries between auditory cognition, perception, and the signals that arrive at our ears is a way to probe at the edges of our understanding. Building models that try to mimic particular human abilities is a great way to proceed: when the models are successful they lead to better algorithms and to new applications. When the models fail they point to places where deeper understanding is needed. Studying the rhythmic aspects of music is one piece of this larger puzzle.

Three important aspects of rhythmic phenomena are its nonverbal nature, its relationship with motor activity, and its relationship with time. Rhythmic knowledge is nonverbal, yet operates in a hierarchical, multi-tiered fashion analogous to language with "notes" instead of "phonemes" and "musical phrases" instead of "sentences." Rhythmic phenomena express a kind of meaning that is difficult to express in words – just as words express a kind of meaning that is difficult to express in rhythm.

Second, rhythmic activities are closely tied into the motor system, and there is an interplay between kinesthetic "meaning" and "memory" and other kinds of meaning and memory. From the work song to the dance floor, the synchronization of activities is a common theme in human interactions that can help to solidify group relationships.

Third, rhythmic activities are one of the few ways that humans interact with time. We sense light with our eyes and sound with our ears. But what organ senses the passage of time? There is none, yet we clearly do know that it is passing. Gibson [B: 73] concludes that time is an intellectual achievement, not a perceptual category. By observing how time appears to pass, Kramer [B: 117] explores the interactions between musical and absolute time, and shows how musical compositions can interrupt or reorder time as experienced. Indeed, Chap. 10 shows very concretely how such reorderings can be exploited as compositional elements. In arguing that music and time reveal each other, Langer [B: 122] states elegantly that music "makes time audible."

How do we learn about time? Children playing with blocks are learning about space and spatial relationships. Talking, singing, and listening to speech and music teach about time and temporal relationships. Jody Diamond's comments [B: 47] about gamelan music apply equally well to the study of rhythm in general:

> *The gamelan as a learning environment is well suited to some important educational goals: cooperative group interaction, accommodation of individual learning styles and strengths, development of self-confidence, creativity...*

Rhythm and Transforms focuses on a few of the simplest low level features of musical rhythms such as the beat, the pulse, and the short phrase, and attempts to create algorithms that can emulate the ability of listeners to identify these features. We take a strictly pragmatic viewpoint in trying to relate things we can measure to things we can perceive, and these correlations demonstrate neither cause nor effect. The models are essentially mathematical tricks that may be applied to sound waveforms, and the signal processing techniques emphasize properties inherent in the signal prior to perceptual processing.

Nonetheless, as the discussion throughout this chapter suggests, the models are often inspired by the operation of the perceptual mechanisms (or, more accurately, guesses as to how the perceptual mechanisms might operate). For example, Chaps. 5–7 explore mathematical models of periodicity detection. To make these applicable to musical signals, a kind of perceptual preprocessing is applied which extracts certain elementary features from the waveform. These derived quantities (like the feature vector of Fig. 1.9) feed the periodicity detection. Similarly, Chap. 7 describes an un-biological model of beat extraction from musical signals based on a Bayesian model. These function in concert with perceptually inspired features that are extracted from the musical signal.

Several new and exciting applications open up once the foot-tapping machine of Fig. 1.2 can reliably locate the beats and basic periodicities of a musical performance:

Musical Editing: Identification of beat boundaries allows easy cut-and-paste operations when editing musical signals.

An Intelligent Drum Machine: Typical drum machines are preprogrammed to play rhythms at predefined speeds and the performers must synchronize themselves to the machine. A better idea is to build a drum machine that can "listen" to the music and follow the beat laid down by the musicians.

External Synchronization: Beat identification enables automated synchronization of the music with light effects, video clips, or any kind of computer controlled system. This may be especially useful in the synchronization of audio to video in film scoring.

A Tool for Disc Jockeys: Any identified levels of metrical information (as fast as the tatum or as slow as the phrase) can be used to mark the boundaries of a rhythmic loop or to synchronize two or more audio tracks.

Music Transcription: Meter estimation is required for time quantization, an indispensable subtask of transcribing a musical performance into a musical score.

Beat-based Signal Processing: Beats provide natural boundaries in a musical signal, which can be used to align a variety of signal processing techniques with the music. For example, filters, delays, echoes, and vibratos (as well as other operations) may exploit beat boundaries in their processing. This is discussed in Chap. 9 and appropriate algorithms are derived.

Beat-based Musical Recomposition: Once the beat boundaries are located, composers can easily work with the beat intervals, an underexplored compositional level. Several surprising techniques are discussed and explored in Chap. 10.

Information Retrieval: The standard way to search for music (on the web, for instance) is to search metadata such as file names, `.mp3` ID tags, and keywords. It would be better to be able to search using melodic or rhythmic features, and techniques such as beat tracking may help to make this possible.

Score Following: In order for a computer program to follow a live performer and act as a responsive accompanist, it needs to sense and anticipate the location of musically significant points such as beat boundaries and measures.

Personal Conducting: Combining the beat tracking with an input device (such as a wand that could sense position and/or acceleration) and a method of slowing/speeding the sound (such as a phase vocoder, see Sect. 5.3.4), the listener can "conduct" the music at a desired tempo and with the desired expressive timing.

Speech Processing: Rhythm plays an important role in speech comprehension because it can help to segment connected speech into individual phrases and syllables.

Visualization Software: Designed to augment the musical experience by presenting appropriate visuals on a screen, visualization software is a popular adjunct to computer-based music players. Many of these relate the visuals to the music using the amplitude of the audio signal (so that, for instance, louder passages move faster), the shape of the waveform, or various transforms. It would clearly be preferable to also have them able to synchronize to the beat of the piece.

1.6 Overview of *Rhythm and Transforms*

There are three parts to *Rhythm and Transforms.* There are chapters about music theory, practice, and composition. There are chapters about the psychology and makeup of listeners, and there are chapters about the technologies involved in finding rhythms.

Music: Chapter 2 discusses some of the many ways people think about and notate rhythmic patterns. Chapter 3 surveys the musics of the world and shows many different ways of conceptualizing the use of rhythmic sound.

Perception: The primary difficulty with the automated detection of rhythms is that the beat is not directly present in the musical signal; it is in the mind of the listener. Hence it is necessary to understand and model the basic perceptual apparatus of the listener. Chapter 4 describes some of the basic perceptual laws that underlie rhythmic sound.

There are three approaches to the beat finding problem: transforms, adaptive oscillators, and statistical methods. Each makes a different set of assumptions about the nature of the problem, uses a different kind of mathematics, and has different strengths, weaknesses, and areas of applicability. Despite the diversity of the approaches, there are some common themes: the identification of the period and the phase of the rhythmic phenomena and the use of certain kinds of optimization strategies.

Transforms: The transforms of Chap. 5 model a signal as a collection of waveforms with special form. The Fourier transform presumes that the signal can be modeled as a sum of sinusoidal oscillations. Wavelet transforms operate under the assumption that the signal can be decomposed into a collection of scaled and stretched copies of a single mother wavelet. The periodicity transform presumes that the signal contains a strong periodic component and decomposes it under this assumption. When these assumptions hold, then there is a good chance that the methods work well when applied to the search for repetitive phenomena. When the assumptions fail, so do the methods.

Adaptive Oscillators: The dynamical system approach of Chap. 6 views a musical signal (or a feature vector derived from that signal) as a kind of clock. The system contains one or more oscillators, which are also a kind of clock. The trick is to find a way of coupling the music-clock to the oscillator-clock so that they synchronize. Once achieved, the beats can be read directly from the output of the synchronized oscillator. Many such coupled-oscillator systems are in common use: phase locked loops are dynamic oscillators that synchronize the carrier signal at a receiver to the carrier signal at a transmitter, the "seek" button on a radio engages an adaptive system that scans through a large number of possible stations and locks onto one that is powerful enough for clear reception, timing recovery is a standard trick used in cell phones to align the received bits into sensible packets, clever system design within the power grid automatically synchronizes the outputs of electrical generators (rotating machines that are again modeled as oscillators) even though they may be thousands of miles apart. Thus synchronization technologies are well developed in certain fields, and there is hope that insights from these may be useful in the rhythm finding problem.

Statistical Methods: The models of Chap. 7 relate various characteristics of a musical signal to the probability of occurrence of features of interest. For example, a repetitive pulse of energy at equidistant times is a characteristic of a signal that is likely to represent the presence of a beat; a collection of harmonically related overtones is a characteristic that likely represents the presence of a musical instrument playing a particular note. Once a probabilistic (or generative) model is chosen, techniques such as Kalman filters and Bayesian particle filtering can be used to estimate the parameters within the models, for instance, the times between successive beats.

Beat Tracking: Chapter 8 applies the three technologies for locating rhythmic patterns (transforms, adaptive oscillators, and statistical methods) to three levels of processing: to symbolic patterns where the underlying pulse is fixed (e.g., a musical score), to symbolic patterns where the underlying pulse may vary (e.g., MIDI data), and to time series data where the pulse may be both unknown and time varying (e.g., feature vectors derived from audio). The result is a tool that tracks the beat of a musical performance.

Beat-based Signal Processing: The beat timepoints are used in Chap. 9 as a way to intelligently segment the musical signal. Signal processing techniques can be applied on a beat-by-beat basis: beat-synchronized filters, delay lines, and special effects, beat-based spectral mappings with harmonic and/or inharmonic destinations, beat-synchronized transforms. This chapter introduces several new kinds of beat-oriented sound manipulations.

Beat-based Musical Recomposition: Chapter 10 shows how the beats of a single piece may be rearranged and reorganized to create new structures and rhythmic patterns including the creation of beat-based "variations on a theme." Beats from different pieces can be combined in a cross-performance synthesis.

Beat-based Rhythmic Analysis: Traditional musical analysis often focuses on the use of note-based musical scores. Since scores only exist for a small subset of the world's music, it is helpful to be able to analyze performances directly, to probe both the literal and the symbolic levels. Chapter 11 creates skeletal *rhythm scores* that capture some of the salient aspects of the rhythm. By conducting analyses in a beat-synchronous manner, it is possible to track changes in a number of psychoacoustically significant musical variables.

1.7 Sound Examples: Teasers

Rhythm and Transforms is accompanied by a CD-ROM that contains many sound examples in `.mp3` format that are playable in iTunes, Windows Media Player, Quicktime, or almost any other audio program. The sound examples[8] are an integral part of the book. You will miss many of the most important aspects of the presentation if you do not "listen along." This section presents a few highlights, tidbits of sound that suggest some of the results and sound manipulations that are possible using beat-oriented audio processing.

One of the primary examples used throughout *Rhythm and Transforms* for sound demonstrations is the *Maple Leaf Rag*, a ragtime masterpiece composed at the start of the twentieth century by Scott Joplin. The *Maple Leaf Rag* became the most popular piano tune of its era, selling over one million copies of the sheet music. A reproduction of the cover of the original sheet music is shown in Fig. 1.10 along with a portrait of Joplin. Joplin's music enjoyed a revival in the 1970s when the copyrights expired and his work entered the public domain. Because there are no legal complications, it is possible to freely augment, manipulate, expand, and mutilate the music. *Ragtime* literally means "time in tatters," and I would like to imagine that Joplin would not be offended by the temporal shredding and rhythmic splintering that follows.

The idea of tracking the beat can be heard in the *Maple Tap Rag* [S: 6] which superimposes a brief burst of white noise at each detected beat point. It is easy to hear that the process locates times when listeners might plausibly tap their feet. The technology needed to accomplish this task is discussed in Chaps. 5–7. Once the beat locations are found, there are many kinds of processing that can be done. It is possible to remove some of the beats, leading to the *Maple Leaf Waltz* [S: 131]. It is possible to employ signal processing techniques in a beat-synchronous manner as in the *Beat Gated Rag* [S: 85], the *Make It Brief Rag* [S: 142], and the *Magic Leaf Rag* [S: 141]. It is possible to remove all the tonal material, leaving only the transients as in the *Maple Noise Rag* [S: 91] or leaving only atonal material in each beat as in the *Atonal*

[8] Sound examples are designated by [S:] and described in detail in the list of examples starting on p. 295.

Fig. 1.10. The *Maple Leaf Rag* can be heard in sound example [S: 5]. The first four measures of the musical score can be found in Fig. 2.3 on p. 27. The complete musical score is available on the CD [B: 107], which also contains a standard MIDI file sequenced by W. Trachtman.

Leaf Rag #2 [S: 99]. It is possible to map all of the harmonics of every note to a desired location: *Sixty-Five Maples* [S: 104] maps every overtone to a harmonic of 65 Hz while the *Pentatonic Rag* [S: 111] maps all overtones to scale steps of the five-tone equal tempered (5-tet) scale. The *Make Believe Rag* [S: 115] alternates among a number of different n-tets and the different mappings play a role analogous to chord changes, even though it is the underlying tuning/temperament that is changing. *Rag Bags #1* and *#2* [S: 155] create hybrid sound collages that merge 27 different renditions of the *Maple Leaf Rag*.

Julie's Waltz by Mark Schatz [S: 8] provides a detailed case study in Chap. 11 of how beat-based *feature scores* can display detailed information about aspects of a musical performance (such as timing and timbre) that are missing from a standard musical score. The little-known song *Soul* [S: 7] by the (now defunct) Ithaca-based band *Blip* is also used extensively to demonstrate the various techniques of sound manipulation in a vocal rock context. Successful beat tracking of *Soul* is demonstrated in Table A.1(10) on p. 289 along with many others. The *Soul Waltzes* [S: 132] show that hard driving rhythms need not be confined to $\frac{4}{4}$ time signatures. *Atonal Soul* [S: 100], *Noisy Souls* [S: 93], and *Frozen Souls* [S: 120] demonstrate the removal of all tonal material, the elaboration of the noise component, and textural changes due to spectral freezing. The effects of the processing on Beil's voice are often remarkable: sometimes silly and sometimes frightening. These examples give only a taste of the possibilities.

2

Visualizing and Conceptualizing Rhythm

There are many different ways to think about and notate rhythmic patterns. Visualizing and Conceptualizing Rhythm *introduces the notations, tablatures, conventions, and illustrations that will be used throughout* Rhythm and Transforms. *The distinction between symbolic and literal notations is emphasized.*

Rhythmic notations represent time via a spatial metaphor. There are two approaches to the notation of rhythmic activities: symbolic and literal. Symbolic approaches accentuate high level information about a sound while literal representations allow the sound to be recreated. A good analogy is with written (symbolic) vs. spoken (literal) language. Text presents a concise representation of speech but cannot specify every nuance and vocal gesture of a native speaker. Similarly, symbolic representations of music present a concise description of the sound, but cannot specify every nuance and musical gesture that a musician would naturally include in a performance. Standard musical notation, drum tablatures, and MIDI transcriptions are all examples of symbolic notations. Literal notations allow a (near) perfect reproduction of a performance. Somewhat paradoxically, by preserving all the information about a performance, literal representations make it difficult to focus on particular aspects of the sound that may be aurally significant. For example, while a symbolic notation may show the fundamental timepoints on which the music rests, these may not be discernible in a literal representation. Similarly, symbolic representations of the pitches of musical events are easy to comprehend, yet the pitch may not be easy to extract from a literal representation such as a `.wav` file, a spectrogram, or a granular representation.

A third class might be called abstract notations, where the score itself is intended to be a work of (visual) art. Several modern composers have created idiosyncratic notations for particular purposes in individual pieces, including Cage's indeterminate notations, Crumb's emulation of Medieval symbolic designs, Penderecki's ideograms, and Lutoslawski's mobiles. These are certainly interesting from an artistic perspective and may be quite instructive in terms of the composer's intentions for a particular piece. They are not, however, general notations that can be applied to represent a large class of sound; rather, they are designed for a unique purpose. A variety of visual metaphors are described in Sect. 2.3.

This chapter surveys the varieties of notations that have been used over the centuries to represent rhythmic phenomena, beginning with symbolic methods and then exploring various kinds of literal notations.

2.1 Symbolic Notations

From a near infinite number of possible timbres, rhythms, pitches, and sonic gestures, symbolic notations extract a small number of features to emphasize in pictorial, numeric, or geometric form. In many cases, time is viewed as passing at regular intervals and this passage is depicted by equal divisions of space. In other cases, time (and/or duration) is itself represented symbolically.

2.1.1 Lyrical Notation

One of the earliest forms of rhythmic notation were markings used to annotate chants, though metrical symbols for syllables were well known to the ancient Greeks. As codified in the anonymous *Discantus Positio Vulgaris* in the early 13th century, these were built on the distinction between long, strong, accented syllables notated with a dash –, and short, weak, unaccented syllables labeled ⌣. Many of these terms, such as those in the prosodic notation of Table 2.1 are still used in the analysis of stress patterns in poetry.

Table 2.1. The five common elements (feet) of the metrical structure of English verse

name	stress pattern	symbol	examples
trochee	long–short	– ⌣	singing, pizza, rigid
iamb	short–long	⌣ –	appear, remark, event
dactyl	long–short–short	– ⌣⌣	tenderly, bitterly, specimen
anapest	short–short–long	⌣⌣ –	in the night, on the road, up a tree
amphibrach	short–long–short	⌣ – ⌣	acoustic, familiar, Sethares

There are two parts to *scansion*, the rhythmic analysis of poetic meter. First, partition the phrase into syllables and identify the most common metric units from Table 2.1. Second, name the lines according to the number of feet. The words monometer, dimeter, trimeter, tetrameter, pentameter, hexameter, heptameter, and octameter, describe the names for one through eight feet per line respectively. For example, Shakespeare's witches sing

– ⌣ – ⌣ – ⌣ – ⌣
Double Double toil and trouble

in trochee with four feet per line, hence in trochaic tetrameter. The lines

⌣ ⌣ – ⌣⌣ – ⌣ ⌣ – ⌣ ⌣

For the moon never beams without bringing me

– ⌣ ⌣ – ⌣ ⌣ – ⌣ ⌣ –

Dreams of the Beautiful Annabel Lee

from Edgar Allan Poe's *Annabel Lee* are in regular anapestic heptameter, while Christopher Marlowe's line about Helen of Troy (from *Dr. Faustus*)

⌣ – ⌣ – ⌣ – ⌣ – ⌣ –

Was this the face that launched a thousand ships

is iambic pentameter.

This kind of meter is an arrangement of primitive elements (such as those in Table 2.1) into groups based on stress patterns, accented syllables, and the number of syllables per line and defines the rhythmic structure of poetic stanzas. Though such notation can be useful in following the metrical flow of a poem, it is imprecise because it does not describe actual temporal relationships such as the relative durations of the – and ⌣s, nor how long pauses should last.

In principle, the rhythm of chant is the rhythm of ordinary spoken Latin since the intent of the recital is to make the meaning as clear as possible. In practice, it may be desirable to standardize the chants so that important words are emphasized. Guido D'Arezzo, a Benedictine monk working in the eleventh century, invented the first form of chant notation that was able to notate both pitch and rhythm using a four lined staff. This early musical notation (using symbols such as the longa, the brevis, the maxima and the semibrevis as shown in Fig. 2.1) allowed a notation where certain vowels could be elongated and others abbreviated. Especially when wishing to notate nonlyrical music, greater precision is needed. D'Arezzo's system eventually evolved into the modern system of musical notation.

Fig. 2.1. This small segment of a chant from an illuminated manuscript looks remarkably modern with lyrics positioned beneath symbols that indicate both pitch and duration. But there is no clef and no time signature, and there are only four lines per staff.

Lyrically inspired notations also play a role in many oral traditions. For example, Indian *tabla* players traditionally use a collection of *bols* (syllables for "mouth drumming," see Sect. 3.8) to help learn and communicate complex drumming patterns. During the Renaissance, articulation techniques such as the tonguing of recorders and cornets and the bowing of violin were conceptualized syllabically. Ganassi's *Fontegara* [B: 69] shows three major kinds of articulations: *taka*, *tara*, and *lara*, which embody hard-hard, hard-soft, and soft-soft syllables. Such (sub)vocalizations are particularly useful when a recorder attempts to mimic the voice. Lyrical notations have also had impact on modern theories of rhythmic perception such as that of Cooper and Myers [B: 35], which use the five prosodic rhythmic groups of Table 2.1 as a basis for metrical hierarchy. This is discussed further in Sect. 3.2.

2.1.2 Musical Notation

Modern musical notation is, at heart, a set of instructions given by a composer (or arranger) to a performer for the purpose of describing how a piece should be performed. Since performers typically create sound by playing notes on an instrument (or using their voice), it is natural that the notation should be written in "notes." Standard notation contains two parts, a staff that represents the pitches and a method of specifying the duration of each note. The discussion here focuses on the rhythmic portion of the notation.

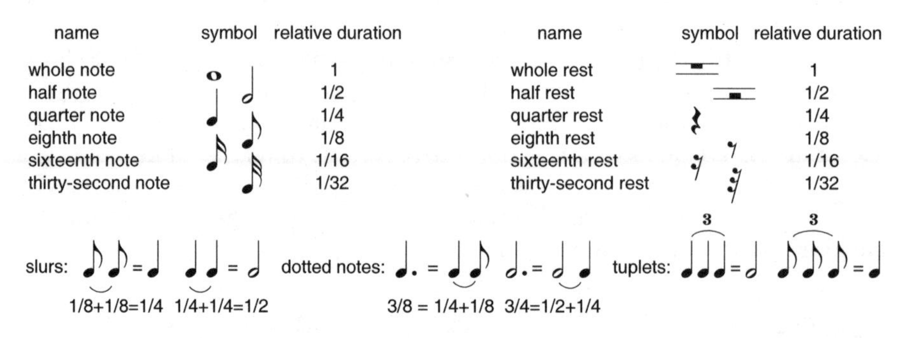

Fig. 2.2. There are six common kinds of notes and six kinds of rests (silences), each with a different duration. The whole tone has a duration of one time unit and all others are specified as fractions of that time. Thus if the whole tone represents one second, the quarter note represents 250 ms. In addition, if a note is followed by a small dot or period, its duration is increased by half. Thus, if the above quarter note were dotted, it would represent a duration of 375 ms. Two notes connected by a slur ⌣ are played as a single note with duration equal to the sum of the two notes. Tuplets allow the specification of durations that are not factors of 2; in one example a half note is divided into three equal parts while in the other a quarter note is divided into three equal parts.

Figure 2.2 shows the various kinds of note symbols and their relative durations. By definition, the whole tone represents a time duration of one time unit, and the others are scaled accordingly. Besides the notes themselves, standard notation also specifies a time signature that looks something like a fraction, for example $\frac{4}{4}$ and $\frac{6}{8}$. The bottom number specifies the note value that defines the beat (4 specifies a quarter note beat while 8 specifies an eighth note beat). The top number states how many beats are in each measure.[1] Thus in $\frac{4}{4}$ time, the length of a measure is equal to the time span of a whole note (since this has the same duration as four quarter notes). In practice, this can be divided in any possible way: as four quarter notes, as four eighth notes plus a single half note, as eight eighth notes, etc. Similarly, in $\frac{6}{8}$ time, there are six eighth note beats in each measure. Again, any combination is possible: six eighth notes, two quarter notes plus two eighth notes, one quarter note plus four eighth notes, etc.

To be concrete, Fig. 2.3 shows the first four measures of Scott Joplin's *Maple Leaf Rag*. The time signature is $\frac{2}{4}$. A quarter note receives one beat and there are two beats per measure. The bass in the lower staff moves along with complete regularity, each measure containing four eighth notes (notes are played simultaneously when they are stacked vertically). In the treble (top) staff, the division is more complex. The first measure contains five sixteenth notes, one eighth note, and one sixteenth rest (totalling two beats). The second measure, in contrast, is divided into a quarter note and four sixteenth notes.[2] Overall, the piece moves at the rate of the sixteenth note, and this forms the tatum of the piece. The complete musical score for the *Maple Leaf Rag* is available on the CD in the files folder [B: 107]. The piece itself can be heard in [S: 5].

Fig. 2.3. The first four measures of the *Maple Leaf Rag* by Scott Joplin. The full score, drawn by J. Paterson, may be found on the CD [B: 107].

In principle, the same musical passage can be written with different time signatures. The *Maple Leaf Rag*, for instance, can be notated in $\frac{4}{8}$ since two quarter notes per measure is the same as four eighth notes per measure. It could also be notated in $\frac{4}{4}$ by changing all the sixteenth notes to eighth notes,

[1] Measures are typically separated by vertical lines.

[2] One of the sixteenth notes is tied to the quarter note (by the slur) so that there are actually only four notes played in the measure. The three sixteenth notes are followed by a single note with duration $\frac{1}{4} + \frac{1}{16}$. The complete measure is again two beats.

all the eighth notes to quarter notes, etc., and requiring four beats per measure instead of two. Indeed, some MIDI transcriptions represent the meter this way.

Another example is the rhythm represented in Fig. 2.4(a). This rhythm is popular throughout much of Africa; it is typically notated in $\frac{6}{8}$ as shown. Though it could logically be written in $\frac{3}{4}$, it is not. The time signature $\frac{3}{4}$ is reserved for dances that have a "three" feel to them, like a waltz. The difference between such equal-fraction time signatures is stylistic and infused with history.

2.1.3 Necklace Notation

Representing temporal cycles as spatial circles is an old idea: Safî al-Din al-Urmawî, the 13th century theoretician from Baghdad, represents both musical and natural rhythms in a circular notation in the *Book of Cycles* [B: 3]. Time moves around the circle (usually in a clockwise direction) and events are depicted along the periphery. Since the "end" of the circle is also the "beginning," this emphasizes the repetition inherent in rhythmic patterns.

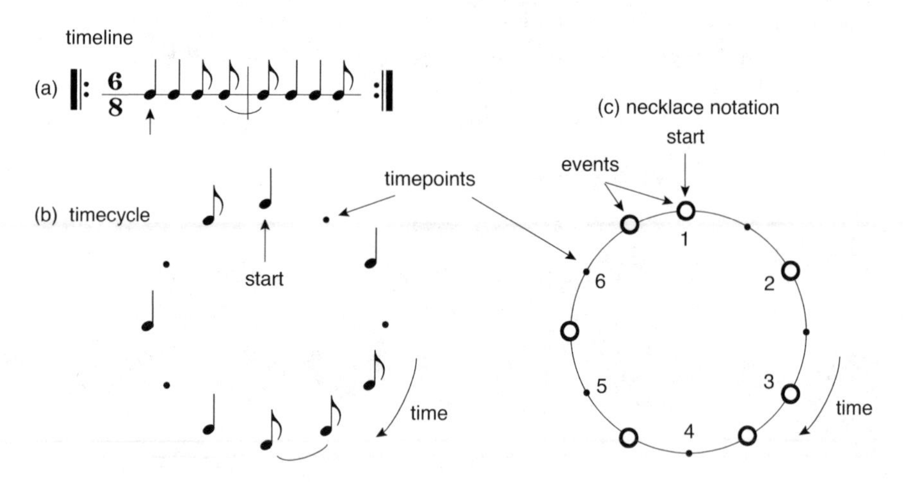

Fig. 2.4. The rhythmic pattern (see also Fig. 1.3 and sound example [S: 2]) is represented in musical notation (a) and then translated into the timecycle notation (b) where the repetition is implicit in the circular structure. The necklace notation in (c) replaces the redundant note symbols with simpler donut-shaped events, and the beats are labeled inside the circle. In both (b) and (c), the twelve timepoints define the tatum, a regular grid of time on which all events lie. The tatum must be inferred from the $\frac{6}{8}$ time signature in the musical notation.

Anku [B: 4] argues that African music is perceived in a circular (rather than linear) fashion that makes the necklace notation particularly appropriate. Performances consist of a steady ostinato against which the master drum performs a series of rhythmic manipulations. The ostinato, visualized as a background of concentric circular rhythms each with its own orientation, reveal staggered entries that sound against (or along with) the regular beat. The master drummer "improvises" by choosing patterns from a collection of rhythmic phrases commonly associated with the specific musical style. "It is in these complex structural manipulations (against a background of a steady ostinato referent) that Africa finds its finest rhythmic qualities" [B: 4].

The necklace notation is useful in showing how seemingly "different" rhythms are related. Part (a) of Fig. 2.5 shows traditional rhythms of the Ewe (from Ghana), the Yoruba (from Nigeria) and the Bemba (from Central Africa). All three are variants of the "standard rhythm pattern" described by King [B: 111]. In (b), the Yoruba pattern (called the *Kànàngó*) is shown along with the accompanying *Aguda*, which can be rotated against the primary rhythm for variety.

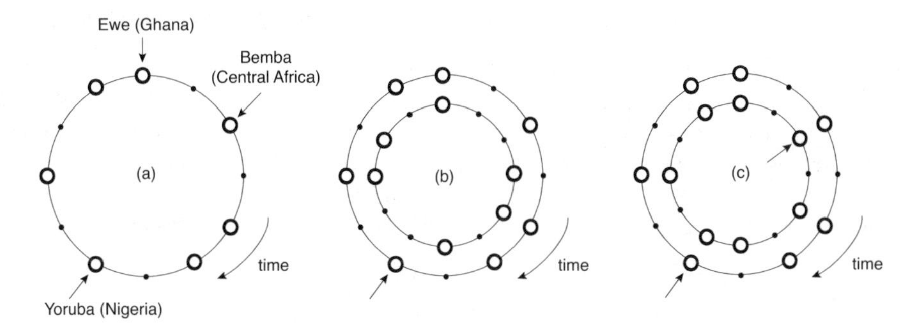

Fig. 2.5. Three traditional rhythms of Africa are variants of King's "standard pattern." They are the same as the rhythmic motif of Figs. 1.3 and 2.4 but interpreted (and perceived) as having different starting points. (b) demonstrates how two drums may play with each other, and in (c) the same pattern appears in both the inner and outer necklace, though rotated in time. These are demonstrated in [S: 9].

In (c), the same *Kànàngó* rhythm is played on the drum and sung, but the two are out of phase with each other. This technique of playing a rhythmic pattern along with a time delayed version of itself can provide "a strong sense of driving forward through the time continuum" [B: 140]. This technique is called a "gap" (*zure*) in Japanese folk music. Similarly, Sachs [B: 187] comments on the use of a rhythmical shift between a singer's melody and the accompaniment that may last for long stretches as the instrument keeps slightly ahead of the singer "by an eighth note or less."

2.1.4 Numerical Notations

Perhaps the simplest form of rhythmic notation begins with a time grid in which each location represents a possible note event. If a note is present at that point in time, the location is labeled "1" while if no event occurs it is labeled "0." The points of the underlying time grid are assumed to be fixed and known. For example, one cycle of the Ewe rhythm of Fig. 2.5(a) is represented as

1 0 1 0 1 1 0 1 0 1 0 1

in the binary notation. One cycle of the two simultaneous rhythms in Fig. 2.5(b) (starting at the arrow) is

1 0 1 0 1 1 0 1 0 1 1 0
0 0 1 1 0 1 0 0 1 1 0 1 .

Some authors use a variant in which different numbers may appear in the timeslots, depending on some attribute of the instrument or of the event: its volume, the type of stroke, or its duration. For example, Brown [B: 21] weights the amplitude of the note events by the duration, which may be useful when attempting automatic meter detection. In this scheme, the Ewe rhythm is

2 0 2 0 1 2 0 2 0 2 0 1.

More sophisticated weighting schemes exploit the results of psychoacoustic experiments on accents. Povel and Okkerman [B: 175] study sequences composed of identical tones. Individual tones tend to be perceptually marked (or accented) if they are

(i) relatively isolated
(ii) the second tone of a cluster of two
(iii) at the start or end of a run (containing three or more elements)

For example, consider the rhythm shown in Fig. 2.6, which compares Brown's duration-weighted scheme with Povel's accent-weighted scheme. Along with the binary notation, these are examples of what Jones [B: 106] calls patterns *in* time, sequences that are defined element by element within the flow of time.

Another kind of numerical notation translates each musical note into a number specifying the duration as an integer multiple of the underlying tatum. For the rhythm in Fig. 2.6 this is the eighth note, and the "duration notation" appears just below the musical notation. Jones [B: 106] calls these patterns *of* time, because each element represents the temporal extent of an event.

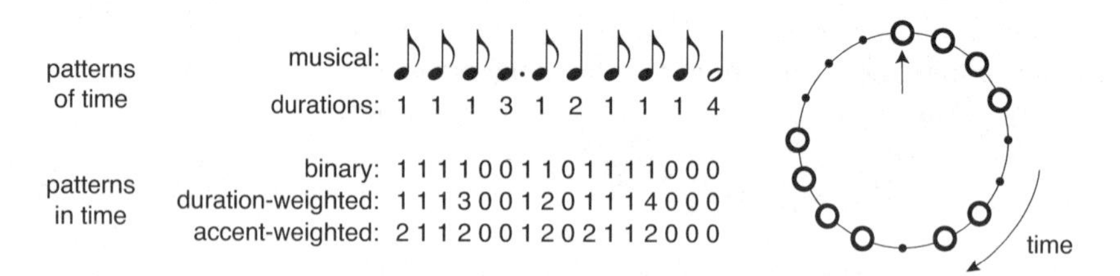

Fig. 2.6. Several different numerical notations for the same rhythm, which is shown in musical and necklace notations and is performed in [S: 10]

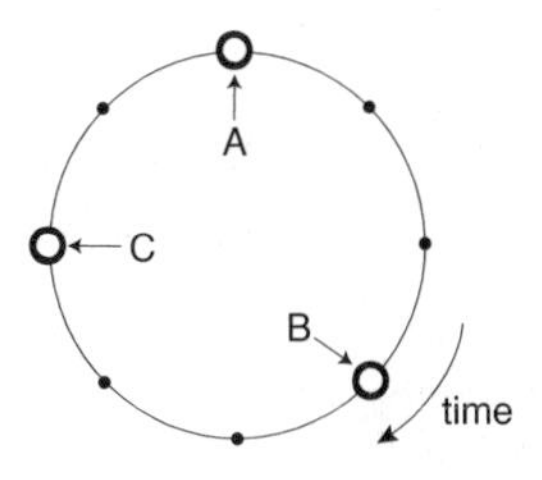

Fig. 2.7. The three functions representing the three rhythms A, B, and C are identified as elements of the same equivalence class under the shift operator

2.1.5 Functional Notation

The binary notation can also be translated into the form of mathematical functions. This is probably the most abstract of the notations, yet it allows the mathematician to demonstrate certain properties of repeating (i.e., rhythmic) patterns. Suppose first that there is an underlying beat (or tatum) on which the rhythm is based and let the integers $\mathcal{Z}$ label the timepoints in the tatum. Following Hall [B: 87], define the function $f : \mathcal{Z} \to \{0, 1\}$ by

$$f(k) = \begin{cases} 1 \text{ if is there is a note onset at } k \\ 0 \text{ otherwise} \end{cases} .$$

The function f represents a periodic rhythm if there is a p such that $f(k) = f(k + p)$ for all $k \in \mathcal{Z}$. The smallest such p is called the *period* of the rhythm. A *rhythm cycle* is defined to be an equivalence class of all p-periodic functions f modulo the shift operator s, which is defined as $(s \cdot f)(k) = f(k - 1)$.

This shift property captures the idea that the rhythm is not just a vector of timepoints, but one that repeats. The equivalence property implies that the 8-periodic functions corresponding to sequences such as $A = \{10010010\}$, $B = \{01001001\}$, and $C = \{10100100\}$ all represent the "same" underlying cyclical pattern, as is demonstrated in Fig. 2.7.

Using such functional notations, it is possible to investigate questions about the number of possible different rhythm patterns, to talk concretely about both symmetric and asymmetric rhythm patterns, and to explore tiling canons (those that "fill" all possible timepoints using a single rhythm cycle with offset starting points, see [B: 87]).

2.1.6 Drum/Percussion Tablature

Drum and percussion tablature is a graphical form of the binary representation in which each possible location on a two-dimensional grid is filled to represent the presence of an event or left empty to indicate silence. Time is presented on the horizontal axis along with verbal directions for counting the beat. The instruments used are defined by the various rows of the grid. For example, Fig. 2.8 shows a two measure phrase played on a full drum kit.

The drum and percussion tablature shows when and how to strike the instruments and there are archives of such tablature available on the web [W: 13]. Percussion grids such as in Fig. 2.8 are also a popular interface for programming drum machines such as Roland's TR-707.

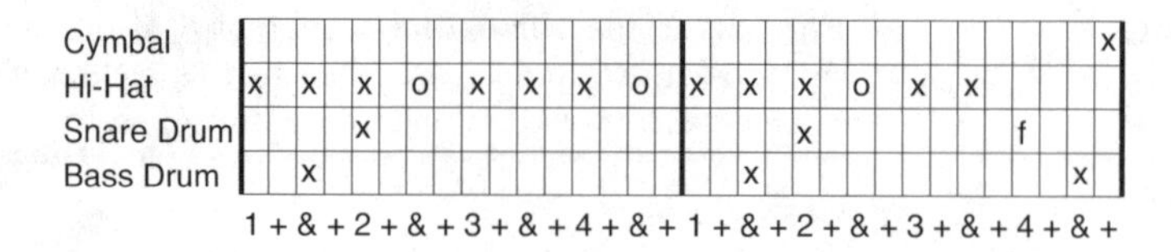

Fig. 2.8. Drum tablature lists the various percussion instruments on the left. The time grid shows when each drum should be hit and the numbering below shows how to count the time. In this case, each of the two measures is counted one-ah-and-ah two-ah-and-ah three-ah-and-ah four-ah-and-ah. The type of stroke is also indicated: 'x' means a normal hit, 'o' means an open hi-hat, and 'f' means a flam on the snare. Typically, there is a legend describing the symbols used on any given tablature. This example is performed in [S: 11].

2.1.7 Schillinger's Notation

Joseph Schillinger [B: 189] (1895–1943) proposed a graphical technique for picturing musical composition that he hoped would one day replace musical notation. In this notation, a grid of squares represents time moving in the horizontal direction; one square for each timepoint. The vertical dimension is pitch, typically labeled in semitones. Figure 2.9, for instance, shows the start of Bach's *Two-Part Invention No. 8* [S: 12]. The contour of the melody line is immediately apparent from the graphical representation, and Schillinger considered this an important step in helping to create a scientific approach to melody. Schillinger showed how to modify the musical work graphically through variation of its geometrical properties, and how to compose directly in the graphical domain. Other factors than pitch may also be recorded analogously, for example, a curve might show the loudness at each grid point, the amount of vibrato, or various aspects of tone quality.

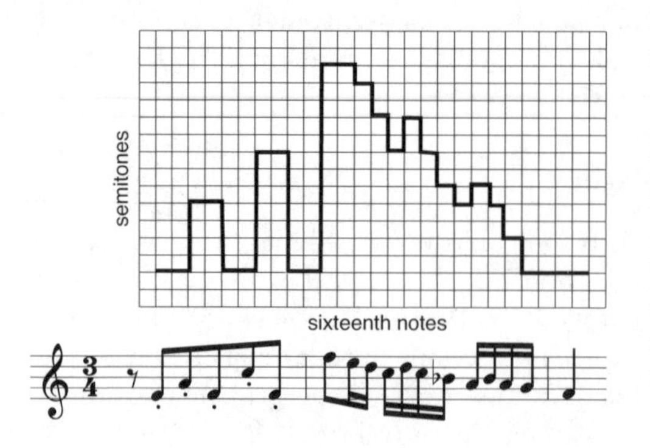

Fig. 2.9. The first two measures of Bach's *Two-Part Invention No. 8* are shown in musical notation and in Schillinger's graphical notation. In the grid, each horizontal square represents one time unit (in this case, a sixteenth note) and each vertical square represents one semitone. The contour of the melody is immediately apparent from the contour of the curve.

Schillinger's theory of rhythm merges this graphical notation with the idea of interference patterns. When two sine waves of different frequencies are added together, they alternate between constructive and destructive interference. This alternation is perceived as beats if the differences between the frequencies is in the range of about 0.5 to 10 Hz. Schillinger schematizes this idea using square waves in which each change in level represents a new note. When two (or more) waves are combined, new notes occur at every change in level. This allows creation of a large variety of rhythmic patterns from simple source material, and the *Encyclopedia of Rhythms* [B: 190] provides a "massive collection of rhythm patterns."

Two examples are shown in Fig. 2.10. In part (a), a square wave with two changes per unit time is combined with a square wave with three changes per unit time. Overall, changes (new notes) occur at four of the six possible times. The resulting rhythm is also given in standard musical notation, and in the necklace notation. Similarly, part (b) shows a four-against-three pattern, which results in a more complex polyrhythm. Schillinger's system is a method of generating polyrhythms (see also Sect. 3.9), that is, a method of combining multiple steady pulse trains, each with its own period. When played simultaneously, the polyrhythm is sounded.

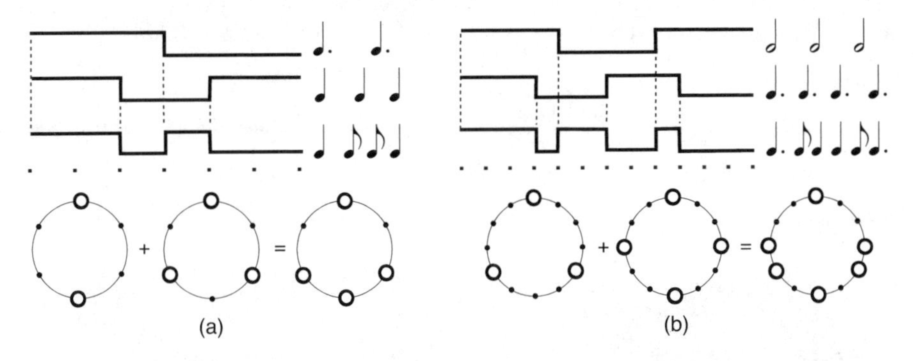

Fig. 2.10. (a) One wave repeats every three time units and the other repeats every two time units. Combining these leads to a three-against-two polyrhythm. (b) One wave repeats every four time units and the other repeats every three time units. Combining these leads to a four-against-three polyrhythm. Both are shown in Schillinger's graphical notation, in musical notation, and in the necklace notation. Dots represent the steady pulse. These and other polyrhythms are demonstrated in [S: 27].

2.1.8 MIDI Notation

In 1982, a number of synthesizer manufacturers agreed to a common specification for the transfer of digital information between electronic musical instruments. While originally conceived as a way to synchronize the performance of multiple synthesizers and keyboards, the MIDI (Musical Instrument Digital Interface) specification is flexible enough to allow computers to interact with electronic keyboards in many ways. Since then, MIDI has been adopted throughout the electronic music industry.

An outgrowth of the MIDI protocol is the Standard MIDI File (SMF) format [W: 49], commonly indicated by the file extension `.mid`, which allows computer programs to store and recall MIDI performances. While not originally intended as a method of notation, the SMF has become a common way of transferring and storing musical information and hence a standard way of representing musical scores and keyboard performances.

Raw MIDI data is difficult to parse directly. For example, the list of numbers in Table 2.2 represents a single note (number 53 = $F2$, the F just below middle C) played on channel 1 (MIDI allows 16 simultaneous channels of information). The note is struck with a "velocity" of 100 (out of a maximum of 127). Time passes (represented by the variable length "delta time") and then the note is turned off. Fortunately, the computer can be used to aid in the organization and visualization of the MIDI data.

Table 2.2. The right table shows the MIDI event list for the five note pattern in Fig. 2.11. The left table shows the raw MIDI data for the first note alone.

MIDI Data	Comment
144	note on channel 1
53	note number 53 (F2) – bell
100	note velocity
129	delta time
112	delta time continued
128	note off channel 1
53	note number 53 (F2)
64	note off velocity

Time	Note	Velocity	Duration
1|1|000	F2	↓100 ↑64	1|000
1|2|000	F2	↓64 ↑64	1|000
1|3|000	F2	↓64 ↑64	0|240
1|3|240	F2	↓64 ↑64	1|000
1|4|240	F2	↓64 ↑64	0|240

For example, the right hand side of Table 2.2 shows an "event list," a way of organizing the MIDI data. The first row in this table represents the same information as the previous column of numbers. Each of the other rows represents a subsequent note event, and the five notes are shown in timeline notation in Fig. 2.11. They can be heard in [S: 13]. Observe that the SMF data incorporates information about the metric structure (the time signature) of the piece. For example, the first note occurs at measure 1, beat 1, tick 0. By default, there are four beats in each measure and each beat has 480 ticks. These defaults can be changed in the file header. The duration of the first note is given on the right as 1|000, which specifies the duration as one beat). Similarly, the third and fifth notes have a duration of 0 beats and 240 ticks (1/2 of a beat). This representation makes it easy for software to change the tempo of a MIDI performance by simply redefining the time span of the underlying tick.

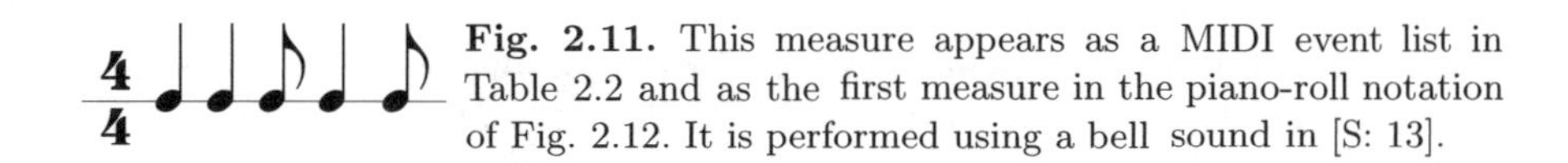

Fig. 2.11. This measure appears as a MIDI event list in Table 2.2 and as the first measure in the piano-roll notation of Fig. 2.12. It is performed using a bell sound in [S: 13].

Even more useful than event lists, however, are visualizations of MIDI data such as the piano-roll of Fig. 2.12. In this representation, the vertical axis represents MIDI note numbers (such as $F2 = 53$ above) and the corresponding notes of the piano keyboard are shown graphically on both sides of the figure. Time moves along the horizontal axis, marked in beats and measures. When representing percussion, each row corresponds to a different instrument (instead of a different note). In the general MIDI drum specification, for instance, the row corresponding to $C1$ is the bass drum with MIDI note number 36, $D1 = 38$ is the snare, and $F\sharp 1 = 42$, $G\sharp 1 = 44$, and $A\sharp 1 = 46$ are various kinds of hi-hat cymbals. These are labeled in the figure along with the corresponding MIDI note number. The drum pattern is played in [S: 14].

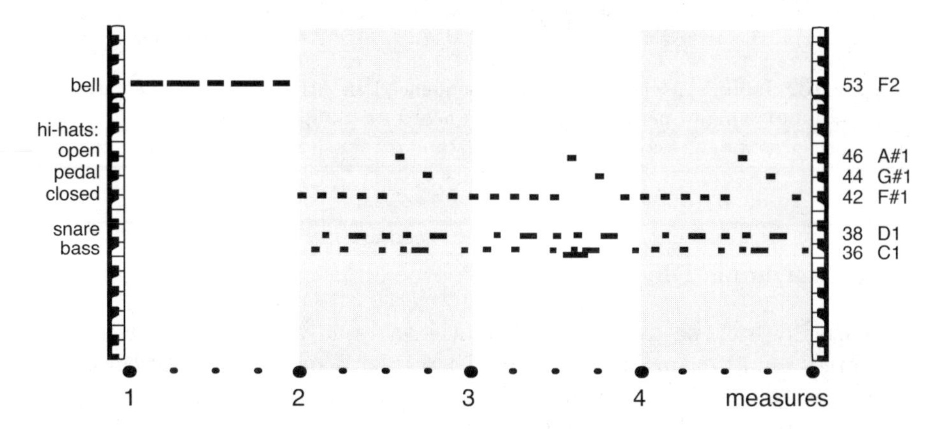

Fig. 2.12. A standard MIDI drum track is shown in piano roll notation. Each of the four measures is divided into four equal beats. The first measure represents the same five notes (performed with a bell sound in [S: 13]) as are given in Table 2.2 and shown in the timeline of Fig. 2.11. The final four measures represent a common "rock" oriented drum pattern, and can be heard in [S: 14].

More common than using the piano-roll for notating percussion is to notate complete musical performances. The first four measures of the *Maple Leaf Rag* by Scott Joplin are displayed in MIDI piano-roll notation in Fig. 2.13. This can be readily compared to the musical score from Fig. 2.3 and is performed in [S: 5]. Thus standard MIDI files, like all the notations discussed so far, are a symbolic representation of the music rather than a direct representation of the sound. When using software to play standard MIDI files, it is easy to change the "sound" of the piece by assigning a different synthesizer "patch," for example, to play the *Maple Leaf Rag* with a violin sound or a xylophone sound instead of the default piano. If you are not familiar with this technique, you may want to experiment with the MIDI file of the *Maple Leaf Rag*, which is available on the CD [B: 107], or download some standard MIDI files from the Classical Music Archives [W: 9].

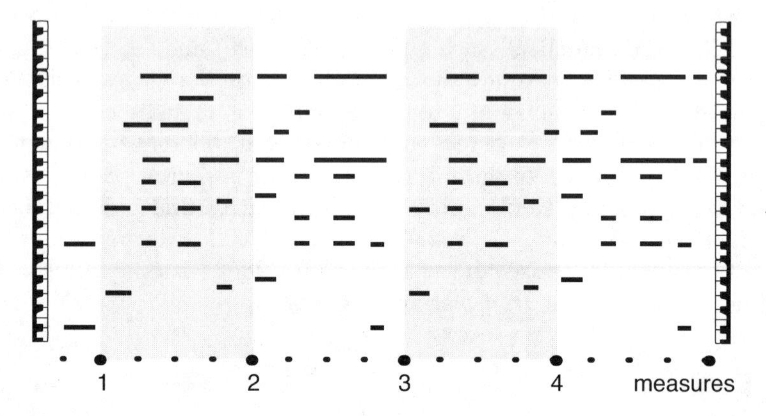

Fig. 2.13. Joplin's *Maple Leaf Rag* is sequenced in MIDI form on the CD (see [B: 107]) and a piano performance can be heard at [S: 5]. Shown here are the first four measures for easy comparison to the standard musical notation in Fig. 2.3.

2.1.9 Harmonic Rhythm

When a rhythmic pattern is played sufficiently rapidly, it becomes a tone. For example, sound example [S: 33] plays a regular sequence of identical clicks separated in time by N milliseconds. When N is large ($N = 500$), it is perceived as a steady rhythm at a rate of two clicks per second; when N is small ($N = 10$) it is perceived as a tone with a pitch that corresponds to a frequency of $\frac{1}{10 \text{ ms per pulse}} = 100$ Hz.

Suppose that the rhythm consists of two pulse trains sounding simultaneously. Cowell [B: 37] draws a parallel between the ratio of the pulse rates and the interval between the pitches of the resulting tones. Suppose that the pulses in the first rhythm are separated by N_1 ms and those in the second rhythm are separated by N_2. Each will have a pitch corresponding to the inverse of the pulse rate, and so the interval (ratio) between the two tones is $\frac{N_2}{N_1}$. For example, with $N_1 = 10$ and $N_2 = 20$ ms, the two pitches are in a 2:1 ratio. Thus they are an octave apart. If N_2 were 15 ms, then the ratio would be 3:2, corresponding to a musical fifth.

This is shown in Fig. 2.14, which lists the first six harmonics of a C note. The ratio formed by successive harmonics and the corresponding musical intervals appear in the next columns. If the period of the fundamental is one time unit, then the time occupied by each of the harmonics is shown graphically. Finally, the ratios are expressed in terms of musical (rhythmic) notation.

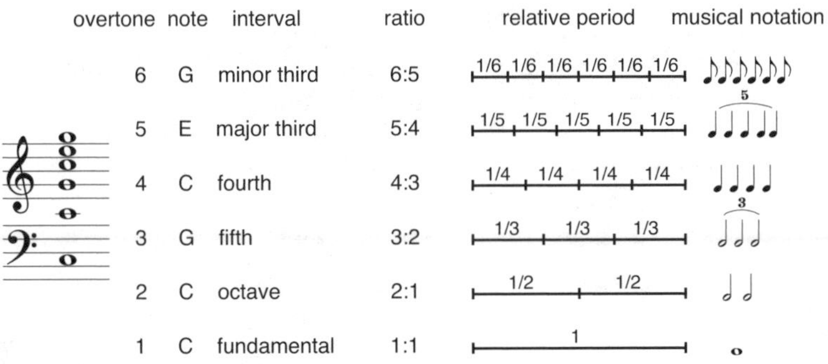

overtone	note	interval	ratio	relative period	musical notation
6	G	minor third	6:5	1/6, 1/6, 1/6, 1/6, 1/6, 1/6	
5	E	major third	5:4	1/5, 1/5, 1/5, 1/5, 1/5	5
4	C	fourth	4:3	1/4, 1/4, 1/4, 1/4	
3	G	fifth	3:2	1/3, 1/3, 1/3	3
2	C	octave	2:1	1/2, 1/2	
1	C	fundamental	1:1	1	

Fig. 2.14. A complex musical tone with fundamental C has six harmonics. The intervals formed by successive overtones and the corresponding ratios are shown. These ratios correspond to the period of vibration of the overtone, which are shown schematically. These periods are translated into musical notation in the final column.

This provides an analogy between the harmonic overtones of musical sounds and the simple-integer relationship found in certain rhythmic patterns. For example, the perfect fifth is represented by a frequency ratio of 3:2, hence it is analogous to a pair of rhythmic pulses with periods in the ratio 3:2. This is the same three-against-two polyrhythm as in Fig. 2.10(a). The polyrhythm in part (b) is in a 4:3 ratio and thus corresponds to a perfect fourth. Similarly, the major and minor thirds correspond to 5:4 and 6:5 polyrhythms. Not shown are the 5:3 (major sixth) and 8:5 (minor sixth) rhythms.

Using this kind of argument, Jay [W: 20] states that harmony and rhythm are different aspects of the same phenomenon occurring at radically different speeds. If a harmony is lowered several octaves, pitches become pulses, and the more consonant harmonic intervals become regularly repeating rhythms. The greater the consonance, the simpler the rhythm. On the other hand, rhythm is converted into harmony by raising it five to ten octaves. Thus harmony is fast rhythm; rhythm is slow harmony.

Stockhausen [B: 224] takes an analogous view in "*. . . How Time Passes. . .*" where he argues that pitch and rhythm can be considered to be the same phenomenon differing only in time scale. Stockhausen makes several attempts to create a scale of durations that mirrors the twelve notes of the chromatic scale. These include:

(i) a series of durations that are multiples of a single time unit: T, $2T$, . . . , $12T$

(ii) subdividing a unit of time T into twelve fractions T, $\frac{T}{2}$, $\frac{T}{3}$, ..., $\frac{T}{12}$
(iii) dividing time into twelve durations that are spaced logarithmically between T and $2T$: T, αT, $\alpha^2 T$, $\alpha^3 T$, ..., $2T$, where $\alpha = \sqrt[12]{2}$

It is not easy to construct a scale of durations that makes sense both logically and perceptually. The idea of a fundamental relationship between harmony and rhythm is enticing and the logical connections are clear. Unfortunately, our perceptual apparatus operates radically differently at very slow (rhythmic) and very fast (pitched) rates. See Sect. 3.9 for sound examples and further discussion.

2.1.10 Dance Notation

> *Labanotation* serves the art of dance much as music notation serves the art of music. Ann Hutchinson [B: 100].

Music is but one human rhythmic activity, and there are a wide variety of notations and conventions that describe dance, juggling, mime, sports (such as gymnastics, ice skating, and karate), and physical movements (such as physical therapy and body language). In [B: 99], Ann Hutchinson reviews a number of dance notations, beginning with the 15th Century Municipal Archives of Cervera, Spain, where various steps are notated by collections of horizontal and vertical lines and a series of abbreviations. Raoul Feuillet's "track drawings" show a floor plan annotated with foot and arm movements. Theleur's method uses stick figures to show leg, body, and arm motions. Saint-Léon's stenochorégraphie places symbolic stick figures (showing motion) on a musical staff (indicating time). Margaret Morris observed that "all human movements take place around an imaginary central axis," and used abstract symbols placed on a pair of three-lined staves to express various movements: gestures of the head and arms on the upper staff, activities of the legs and feet on the lower staff, and movement of the body in between.

Perhaps the most common dance notation today is *Labanotation*, named after the Hungarian Rudolf von Laban (1879–1958). Laban conceived of his "kinetographi," which combines the Greek words *kinētikos* (to move) and *graphos* (writing), as an attempt to analyze and record every aspect of human movement [W: 24]. Like musical notation, Labanotation is based on a series of regular beats, which are typically displayed running vertically from bottom to top. A vertical center line divides two columns that represent the left and right sides of the dancer. Motions occurring to the left (or right) side of the body are thus sensibly displayed on the left (or right) side of the center line. Symbols are placed in the columns: rectangles are altered in shape to indicate direction, color and/or shading are used to show level, and length is used to indicate duration. Special symbols are used to represent the joints (shoulders, knees, fingers), for surfaces of the body (palm, face, and chest), and special signs can be used for a wide variety of actions such as touching, sliding, stamping, clapping, etc. A readily accessible introduction to Labanotation can be found at the Dance Notation Bureau [W: 12].

For example, Fig. 2.15 shows two measures of a tango in Labanotation, as recorded by Andreas Maag [W: 25]. There are two columns because there are two dancers, and the notation is augmented with verbal descriptions of unusual or characteristic motions. Different dance styles have different sets of specialized symbols.

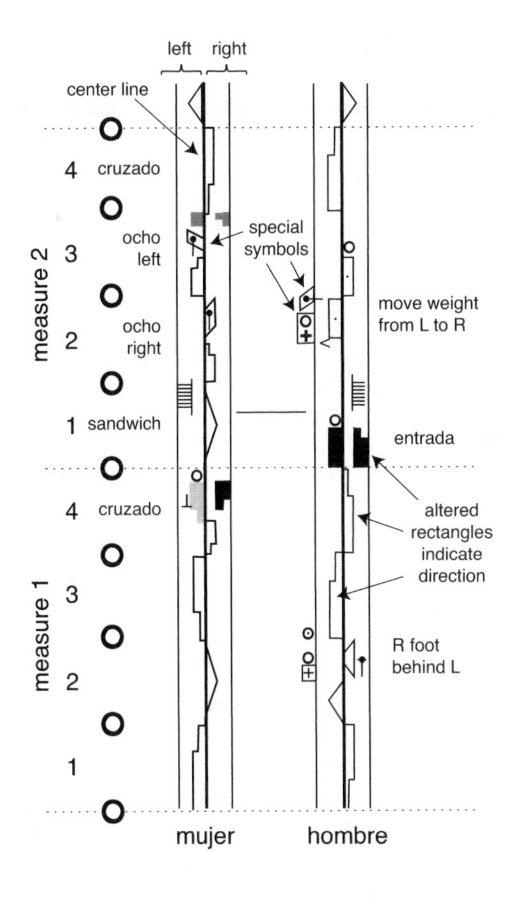

Fig. 2.15. Two measures of a tango notated in Labanotation by Andreas Maag [W: 25]. The two dancers perform somewhat different actions and their motions are notated in separate columns that are synchronized (vertically) in time. The annotation reads, in part: Normal salida (2 steps) then step to change feet (R.F. behind the L.F., do not turn the pelvis); 2 steps. Entrada (footstop) and sandwich (R.F. of the hombre between the feet of the mujer). Bow, hombre R.F. touches mujer feet (but not on top). The hombre leads a turn and interrupts it right away (beat 7) with R.F. Pulls R.F. back and leads ocho to finish. The figure is redrawn and annotated with permission. The circles representing the beat locations are added to emphasize the similarities with beat-based musical notations.

2.1.11 Juggling Notation

Juggling is the action of repeatedly tossing a number of objects (typically balls) into the air and catching them again. It requires skill and agility because no hand holds more than one ball at a time and there are typically more balls than hands. Like music, the actions must be performed rhythmically. Like dance, the actions must respect gravity.

Over the years, jugglers have developed a variety of ways to notate their art. Some notations are more flexible (can describe a large variety of different patterns) and some are more compact. One recent innovation is the "beat-based juggling notation" of Luke Burrage (described at the Internet Juggling Database [W: 19]), which notates what each hand is doing at every beat in time. Beever's comprehensive *Guide to Juggling Patterns* [B: 10] written for "jugglers, mathematicians and other curious people," gives a rich overview of juggling techniques and notations ("like sheet music for jugglers"), and demonstrates a variety of related mathematical results.

Perhaps the most popular notation is the siteswap notation (see [B: 227], [W: 48]), which represents throws by integers that specify the number of beats in the future when the object is thrown again. Because there are many constraints (no hand holds more than one ball, balls must return to a hand after a short time in flight) these integer patterns represent juggling patterns. Beever [B: 10] builds up a logical description of the process that includes the time, site (e.g., left or right hand, elbow, knee), and position (to the left, right or right side of the body) of the throw, the position and site of the catch, and the airtime of the flight. From these, and under nominal assumptions, he derives the siteswap base which provides a shorthand notation for general use.

For example, Fig. 2.16 shows the schematic "ladder" notation of the cascade, a standard three-object juggling pattern. Time moves vertically from the bottom to the top. Beats occurring at roughly equal timepoints are represented by the small circles. The location of each of the three balls is indicated by its horizontal position: held in the left hand, in the air, or held in the right hand. At each beat timepoint, one ball is being thrown, one is being caught, and one is in flight. Since there are three balls and since each ball is thrown every three beats, this pattern is represented succinctly as "3." Except for trivial changes (such as starting with a different hand), this completely specifies the cascade.

By changing the number of balls and the order of arrivals and departures, a large variety of different juggling patterns can be represented. For example, the right hand part of Fig. 2.16 shows a four-ball juggling pattern notated "534" in the siteswap notation. In this pattern, one ball moves regularly back and forth between the two hands every three beats. The other three balls follow a more complex pattern. Observe that odd numbers are always caught by the opposite hand and even numbers are caught by the tossing hand. Hence the five beat throw moves from one hand to the other while every four beat toss is caught by the throwing hand. The throwing pattern of the three balls repeats every 18 beats. The website at [W: 48] has a useful and entertaining visualization tool that automatically simulates any legal siteswap pattern.

The beats (successive timepoints) in juggling are analogous to beats in music since they provide a rhythmic frame on which all activity is based. All of the well known juggling notations are symbolic since they specify only the most important events (the throwing and catching of the balls) and do not specify the exact hand positions, the activities of the hands (other than at

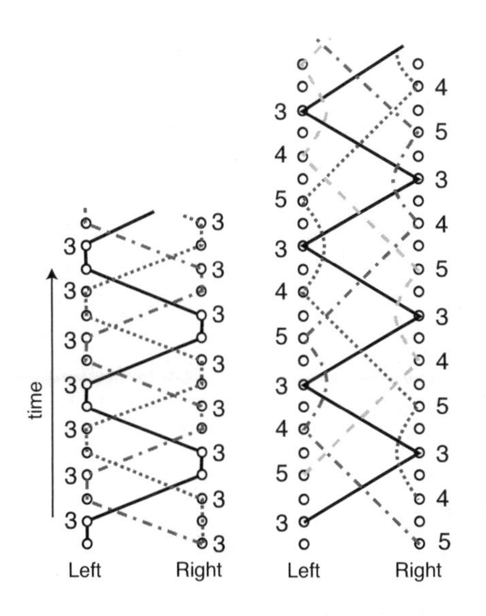

Fig. 2.16. Each rung in the "ladder notation" (described in [B: 10]) represents one (roughly equal) timepoint. The balls are represented by the different lines and the location of each ball is specified (in the left hand, in the right hand, or in the air) at each timepoint. The numbers indicate how many beats must pass before the ball can be thrown again. In the left ladder, the standard three-ball cascade, this is always three beats and this pattern is "3" in the siteswap notation. The "534" on the right is a four ball juggling pattern with one ball that changes hands every three beats (like the cascade) and three balls that follow a more complex pattern. "345" and "453" represent the same pattern with different starting hands.

the instants of catching and throwing) nor the actual trajectories of the balls in flight. Nonetheless, there is enough information in the notation to allow jugglers to learn and invent new patterns.

2.2 Literal Notations

The great strength of symbolic notations for music is that they show the structure of a piece at a high level. The symbols (the notes, rests, MIDI events, tablatures, etc.) represent the underlying structure of a piece by providing instructions that can be readily translated by people or machine into performances of the work. The weakness of symbolic representations is that they do not specify many salient factors such as the timbre of the instruments or the exact timing of the events. In short, they are not a literal record of the sound, but a reminder of what the sound is like.

Literal notations allow full reproduction of a performance even though important aspects of the sound may become lost among a flood of (near) irrelevant data. For example, while pitch is clearly an important aspect of musical performance, waveform representations (such as that in Fig. 2.17) do not display pitch in an obvious way. Similarly, the fundamental timepoints on which a rhythmic passage is built may not be clearly marked. Literal notations are, by their nature, recordings of particular performances and not representations of the underlying composition. This distinction is discussed further in Sect. 12.3.

2.2.1 Waveforms

The most common literal method of musical representation is direct storage of the sound wave. Common analog storage mechanisms are cassette tapes and LP records. Common digital storage techniques sample the waveform and then store a representation of the samples on magnetic tape, in optical form on a CD, or in the memory of a computer. All of these technologies record the variations in a sound pressure wave as it reaches a microphone. Larger numbers (greater deviations) indicate higher pressures, and sound is perceived when the fluctuations in the waves are between (about) 20 and 20,000 cycles each second. When stored digitally, this requires thousands of numbers per second (88,200 numbers for each second of stereo music). The most common way to view the data is to plot the values vs. time, as is done in Fig. 2.17.

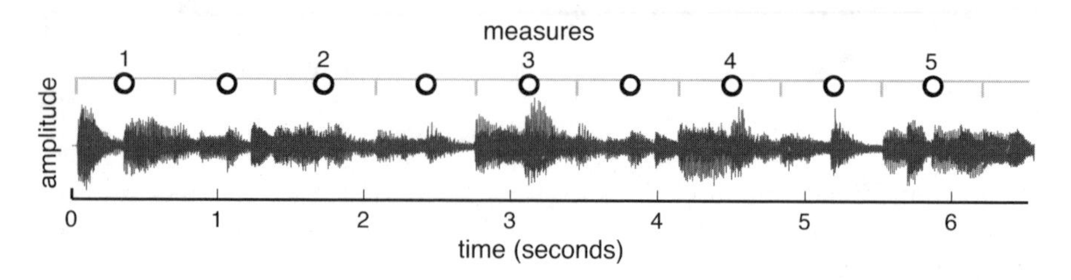

Fig. 2.17. This is a plot of the raw sound wave data for the first six seconds of Joplin's *Maple Leaf Rag*. The larger blobs represent louder passages. The measures and beats are also notated, though these are not directly present in the data, they have been added to help orient the reader and to enable comparison with the musical score Fig. 2.3 and the MIDI notation Fig. 2.13. It is not easy to see individual notes in this representation.

Playing back a sound file such as a `.wav` or a `.mp3` allows a listening experience that is much like that experienced near the microphone at which the recording was made. The timbre of the instruments and the timings are reproduced in (nearly) unaltered form. The performance is fully specified by the sound file. What is not obvious, however, are the higher level abstractions: the notes, the beats, the melodies. Indeed, it is very difficult to determine such high level information directly from a waveform. This problem in musical processing is analogous to the well known problem of creating a computer program that can understand connected speech.

Waveform representations typically represent time linearly, but they may also use other geometrical constructions. Analogous to the necklace notation for symbolic cyclical patterns, the waveform can be plotted around a circle. Figure 2.18, for instance, shows the first two measures of the *Maple Leaf Rag*. The progress of time is specified in beats, which are indicated by the small circles. Such a representation is particularly appropriate when the music repeats regularly at the specified interval.

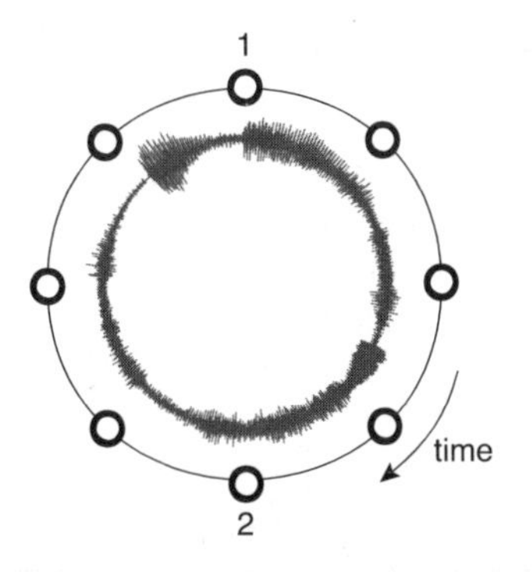

Fig. 2.18. The first two measures of the *Maple Leaf Rag* are displayed in a polar (circular) plot. Time moves in a clockwise direction and the beats are indicated by the small circles.

It is common to use the waveform representation as a method of storing music (this is what CDs do) but it is not common to compose directly with waveforms. In one approach, the composer chooses a collection of individual time points and amplitudes, and then invokes a computer program to algorithmically interpolate the intermediate samples [B: 242]. The algorithms may generate the intervening data using a nonlinear dynamical system, via some kind of random process, or via a process of hierarchical construction. Wavetable synthesis directly generates short snippets of waves that are used as building blocks for larger sound structures. By creative looping and moving the loop points, it is possible to create ever-changing waveforms from a small number of specified tables. This synthesis technique was used in the PPG synthesizers and in the more modern Waldorf wavetable synthesizers.

2.2.2 Spectrograms

The spectrum looks inside a sound and shows how it can be decomposed into (or built up from) a collection of sinusoids. For example, guitar strings are flexible and lightweight, and they are held firmly in place at both ends under considerable tension. When plucked, the string vibrates in a far more complex and interesting way than the simple sine wave oscillations of a tuning fork or an electronic tuner. Figure 2.19 shows the first $\frac{3}{4}$ second of the open G string of my Martin acoustic guitar. Observe that the waveform is initially very complex, bouncing up and down rapidly. As time passes, the oscillations die away and the gyrations simplify. Although it may appear that almost anything could be happening, the string can vibrate freely only at certain frequencies because of its physical constraints.

For sustained oscillations, a complete half cycle of the wave must fit exactly inside the length of the string; otherwise, the string would have to move up and down where it is rigidly attached to the bridge (or nut) of the guitar. This is a tug of war the string inevitably loses, because the bridge and nut are far more massive than the string. Thus, all oscillations except those at certain privileged frequencies are rapidly attenuated. This is why the spectrum shows large peaks at the fundamental and at the integer harmonics: these are the

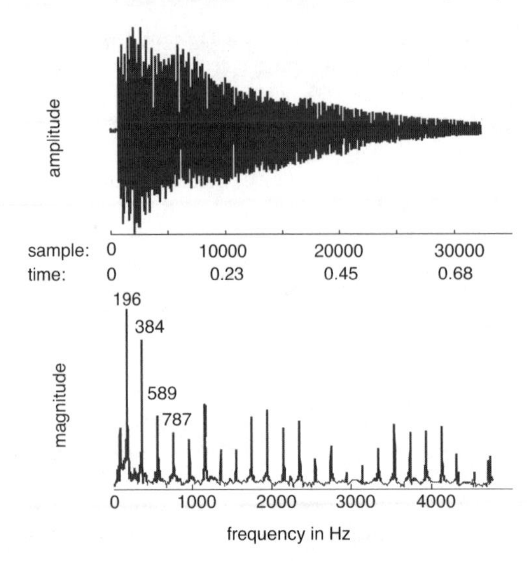

Fig. 2.19. Waveform of a guitar pluck and its spectrum. The top figure shows the first $\frac{3}{4}$ second (32,000 samples) of the pluck of a G string of an acoustic guitar. The spectrum shows the fundamental at 196 Hz and evenly spaced overtones at 384, 589, 787, etc. These are called the harmonics of the sound; they occur near simple integer multiples of the fundamental since $384 \approx 2 \times 196$, $589 \approx 3 \times 196$, and $787 \approx 4 \times 196$. More than twenty harmonics are clearly distinguishable.

frequencies that "fit" inside the length of the string. Spectra such as Fig. 2.19 are typically calculated in a computer using the Discrete Fourier Transform (DFT) and its numerically slicker cousin, the Fast Fourier Transform (FFT).

Spectrum plots are useful when studying the composition of isolated sounds (such as a guitar pluck) and more generally to sounds that do not change over time. But they are not typically useful for complex evolving sounds such as a rendition of the *Maple Leaf Rag*. In this case, it is common to use a sequence of short spectral snapshots called a spectrogram, which shows how the spectrum changes over time. Figure 2.20 shows the waveform and spectrogram of the start of the *Maple Leaf Rag*.

Spectrograms are built by partitioning a sound into small segments of time called frames and then applying the FFT to each frame. All the FFTs are placed side-by-side; large values are represented by dark coloration, small values are shaded lightly. Thus time is represented on the horizontal axis, frequency is represented on the vertical axis, and energy (or loudness) is represented by darkness of the shading. In addition it is possible to create spectrograms dynamically and in other geometrical arrangements. See, for instance, [W: 52].

Since a spectrogram is a visual representation of a sound, it is possible to synthesize sound from an image, and to manipulate the sound graphically. The "image synthesizer" in Eric Wenger's *Metasynth* [W: 31] transforms images into sound by translating each pixel into a short fragment of sound based on its horizontal location (mapped to time), vertical location (mapped to frequency), and color (mapped to stereo placement). Both the vertical and horizontal scales are flexible; frequencies can be specified in almost any scale and time can move at any rate. Using default parameters, the stretched girl

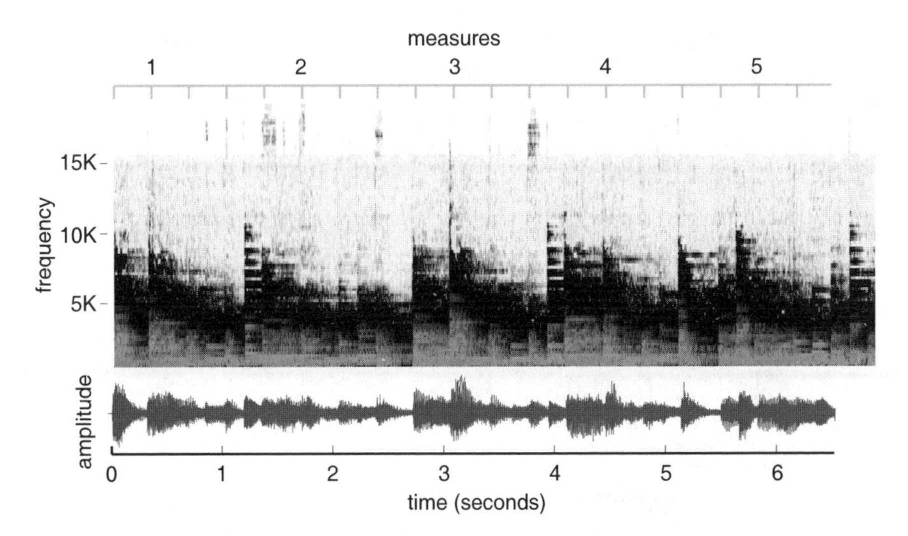

Fig. 2.20. Waveform and spectrogram of the first 6.5 seconds of the *Maple Leaf Rag*

in Fig. 2.21 has a duration of about 24 seconds. She may be heard in sound example [S: 15].

The pictorial representation encourages manipulation of the picture using the kinds of tools familiar from graphics and drawing programs: cutting and pasting, selecting ranges, inverting colors, and drawing with various shaped "pens." *Metasynth* also contains a number of uniquely music-oriented tools: pens that draw the shape of a harmonic series, grid lines that mark off the time axis in a rhythmic pattern, tools that sharpen attacks by emphasizing edges, spray brushes that splatter tiny fragments of sound across the pallette, and tools that add a haze after each sound (reverb) or before each sound (pre-verb).

Another innovation in *Metasynth* is the use of images as filters. The filter palette uses the same mapping (horizontal for time, vertical for frequency) as the image synthesizer, but is not heard directly. Rather, the filter image is applied point-by-point to the sound image. This allows the creation of arbitrarily complex filterings that can be different at each time instant. Figure 2.22, for

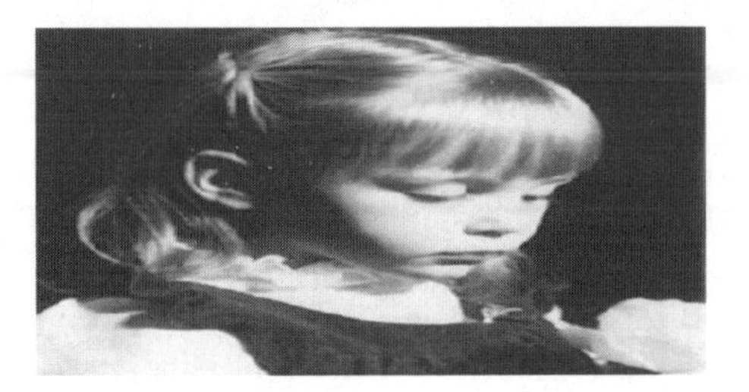

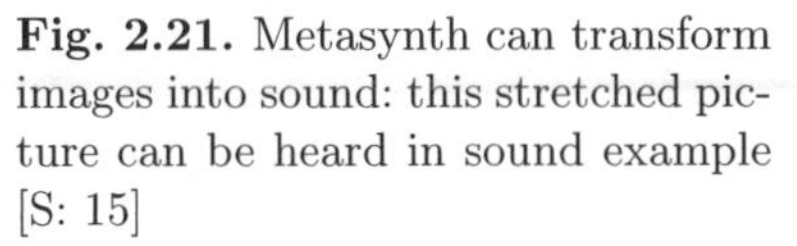
Fig. 2.21. Metasynth can transform images into sound: this stretched picture can be heard in sound example [S: 15]

instance, shows two image filters in (a) and (c). These are applied to the first few seconds of the *Maple Leaf Rag* and result in the spectrograms shown in (b) and (d). The vertical stripes of (a) suggest a stuttering effect, and this is borne out the sound examples [S: 16]. Similarly, the graceful sweeping arch of (c) suggests a lowpass effect that slowly changes to highpass and back.

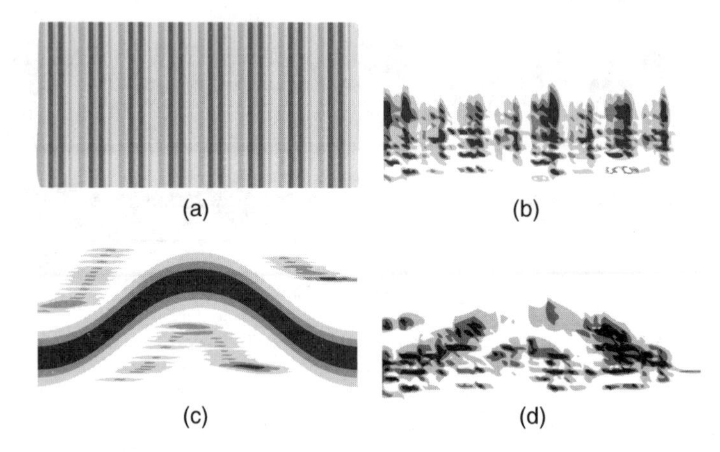

Fig. 2.22. *Metasynth* uses one image to filter another. The two filters in (a) and (c) are applied to the first few seconds of the *Maple Leaf Rag*. The resulting spectrograms are shown in (b) and (d), and may be heard in sound example [S: 16].

Such graphically oriented sound manipulation represents a change in paradigm from conventional methods of creating and modifying sounds. In the standard method, musical instruments are used as the sound source, and these are orchestrated using symbolic notations such as a musical score. In contrast, *Metasynth* generates its sound directly from a picture, directly in its literal notation. The strength of the system is that the composer is supplied with a large variety of perceptually meaningful tools for changing and rearranging the basic sound material. The high level abstractions of the musical score include the pitch of notes and the regular time unit defined by the measure. The high level abstractions in spectrogram "scores" include the frequency content of a signal as displayed in the image and time-grids that can be arbitrarily specified.

2.2.3 Granular Representations

Any signal can be be represented by a collection of (possibly overlapping) acoustic elements called grains (see Fig. 2.23). Though each individual grain sounds like a brief click, aligned masses of grains can represent familiar sounds and make it easy to specify certain kinds of complex sound clouds and auditory textures. This decomposition of sound into quanta of acoustical energy was first proposed by Gabor [B: 68] in 1947, but was popularized in the 1970s by Xenakis [B: 247] and Roads [B: 181], who were among the first to exploit the power of computer-based synthesis using a granular technique.

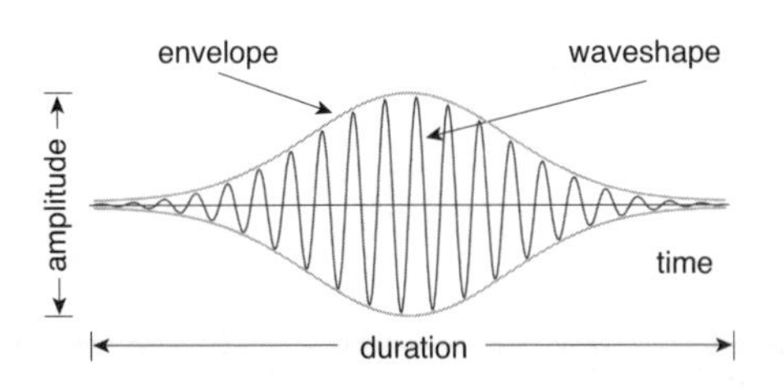

Fig. 2.23. Sound grains are characterized by their duration, amplitude, envelope, and waveshape. If the envelope is a Gaussian curve and the waveshape is a sinusoid (as shown), then it is a "Gabor grain." Other kinds of grains use different waveshapes, envelopes, and durations.

Each grain is characterized by its duration, envelope, amplitude, and waveshape. In Gabor's original work, the envelope had a Gaussian (bell curve) shape (because this provides an optimal trade-off between spread in frequency and spread in time) and the waveshape was sinusoidal. Since the durations are very short, individual grains sound like short clicks, though longer duration grains may give a pitch sensation as well. In the two examples [S: 17], both individual grains and small clouds of grains are clearly audible. [S: 18] is built from a variety of different grain shapes synchronized to a shuffle pattern.

Grains become useful when many are clustered together. Xenakis pictures each grain as a small dot, and places the dots onto a "screen" whose axes specify the frequency and intensity of the grain. A collection of screens (called a "book") is then played back over time (like the individual still frames of a movie) in order to create complex evolving sound textures. Figure 2.24 illustrates.

For example, in Xenakis' piece *Concret PH*, which premiered at the Brussels World's Fair in 1958, the grains were drawn from recordings of a crackling wood fire that were cut into one-second fragments. They were then recombined using tape splicing techniques. In *Analogique B*, Xenakis created grains from the output of an analog sine wave generator by cutting the tape recording into short sections. The grains were then combined by placing them on screens with probabilities drawn from an exponential distribution. The "degree of order" was controlled by random numbers drawn from a first-order Markov chain.

Because so many grains are needed, specifying all the information is a nontrivial task. Roads [B: 181] (and others [B: 41], [B: 231]) have developed sophisticated ways to use stochastic procedures (random numbers) to choose from among the most salient variables. In these schemes, the composer chooses a collection of parameters that specify the probabilities that certain kinds of grains occur at certain times. The computer then generates the bulk of the actual data and realizes the composition. Typical parameters that might be controlled are:

(i) the density of the grains (number of grains per second)

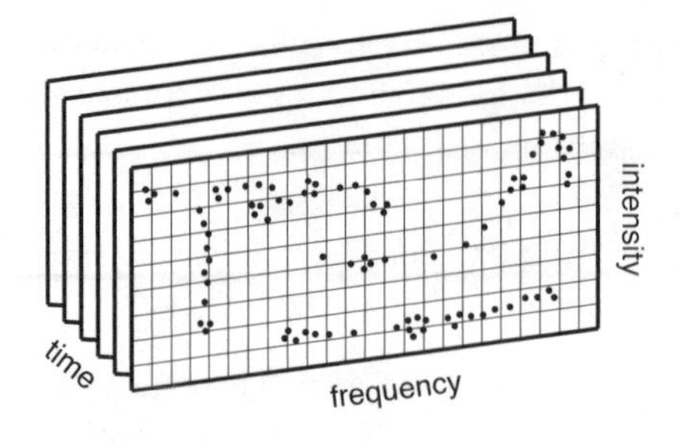

Fig. 2.24. Each grain of sound is represented as a single dot sprayed out onto a grid or screen with differing frequencies and intensities. Collections of screens (a "book of screens") defines the evolution of a complex sound through time.

(ii) the durations of the grains
(iii) the waveshapes within the grains (sinusoid, square wave, noise, impulsive, sampled from a given source, etc.)
(iv) intensity/amplitude of each grain
(v) temporal spacing between grains
(vi) grain envelopes
(vii) frequencies of the grains (e.g., limits on the highest and lowest frequencies)
(viii) frequencies: scattered or aligned
(ix) timing (whether the grains are synchronous or asynchronous)
(x) placement in the stereo field

This represents another conception of musical composition. Rather than working with "notes" and "instruments," the composer specifies a method, or an algorithm for choosing parameters within that method. Rather than conceiving the piece as fixed in a musical score which requires musicians for performance, it is fixed in a computer program which requires appropriate software and hardware for its realization. Granular synthesis is a method of composition in a literal notation (that of sound grains) rather than a method that operates at the symbolic level. Perhaps the most important part of any such method is that the control possibilities provided to the composer must make perceptual sense. They must impose some kind of (relatively) high level order on the literal representation so that it is comprehensible. The list of parameters (i)–(x) above can thus be viewed as an attempt to allow the composer direct control of certain high level variables. These primitives are quite different from the standard idea of a musical "note," with its pitch, volume and timbre, but it is not hard to gain an intuitive feel for parameters such as density, opacity, and transparency.

For example, a cloud containing many 100 ms grains might be perceived as continuous and solid, whereas if it were created from 1 ms grains the cloud would be thin and transparent. If the waveshapes are sinusoids then the sound mass might appear sparkling and clear, while if the waveshapes are jagged and noisy the cloud would occupy more of the spectrum. One of the strengths of sound clouds is that they can evolve over time: in amplitude, density, internal tempo, harmonicity, noise, spectrum, etc. The waveshape might be chosen a priori, generated by an algorithmic process, or perhaps harvested from a sound file.

Computer programs that implement granular synthesis include Road's `Cloud Generator`, Erbe's `Cornbucket` (used in conjunction with `Csound`), Bencina's `Audiomulch`, and McCartney's `SuperCollider`. See [B: 181] and [B: 54]. A variety of experiments with grainlets, pulsars, and glissons are documented in Roads *Microsound*. `Kyma` [W: 23] is a mixed hardware/software package that implements granular synthesis along with a variety of other synthesis techniques.

There are a number similarities between granular techniques and the spectrogram approach of the previous section. Indeed, if the windows used in partitioning the signal are chosen to be the same as the envelope of the grain, and if the waveshape is a sinusoid, then the two representations are logically equivalent. However, there are differences. Grains tend to be of very short duration, while the window length in the FFT-based methods is typically large enough to allow representation of low frequencies (limitations of the spectral methods are discussed further in Chap. 5). The granular technique allows any waveshape within its envelope, while the spectrogram requires the use of sinusoids. Moreover, as we have seen, the spectrogram and the granular communities have developed different kinds of high level abstractions that help to make composition and sound manipulation more transparent for the composer.

As will become apparent in Sect. 5.4, there are also close similarities between granular methods and wavelet transforms. Wavelets also have short "grains," though they are not called this. One difference is that wavelets are not all the same duration; low frequency wavelets are longer and high frequency wavelets are shorter. This helps to make wavelet representations more efficient.

2.3 Visual and Physical Metaphors for Rhythm

Rhythm refers to orderly recurrence in any domain. Visual rhythms may involve alternations of light and dark, of up and down, of colored patterns, or of symbols. Tactile rhythms may involve alternations between strong and weak or may lie in the motion of our bodies. Architectural rhythms may be built from repeated structural or decorative elements such as windows, columns, and arches. Interestingly, we do not appear to be able to sense rhythmic phenomenon directly in either the olfactory or the gustatory senses [B: 65].

For example, Figs. 2.25 and 2.26 show two artistic conceptions of rhythm inspired by Matisse's *Two Dancers* and Mondrian's *Rhythm of Black Lines*. These provide an artistic representation of two fundamental aspects of rhythm: motion and repetition. Like Matisse's famous *Two Dancers*, Fig. 2.25 shows a collection of disembodied shapes that can be seen as representing two people dancing. The darker figure (with disembodied head) is poised to catch his partner after flinging her into the air. Or does this show a flying warrior

Fig. 2.25. Rhythm as motion: like Matisse's *Dancers* paintings, this tries to capture the idea of motion. An instability (the person in the air) drives these figures towards resolution. "Dance," Matisse said, means "life and rhythm."

pouncing on his beheaded victim just as the prey begins to topple? There are several illusions associated with these shapes: why they appear to be people when in reality they are collections of blobs, why they appear to be in motion when in reality the blobs are utterly fixed on the page. Whether dancers or warriors (and other interpretations are possible), the essence of the motionless figure is motion!

Like Mondrian's well known *Rhythm of Black Lines*, Fig. 2.26 shows a collection of crosshatched black lines that repeat at regular intervals. The lines are punctuated by shaded regions and small variations in the lengths and connections of the lines. This is a visual representation of the kinds of *repetition with variation* that is so common in musical phenomenon. A completely regular structure might represent the unchanging beat or the metrical level; the idiosyncratic lengthenings and shortenings correspond to the variations that make the work interesting and exciting.

One of the best known examples of proportion and balance in classical architecture is the Parthenon, the crown of the Acropolis. Built as a temple to Athena around 440 BC, its massive columns, ornamental friezes (showing processions of men and Gods), metopes (representing battles between good and evil, between the Greeks and their enemies) and pediments (with dozens of statues representing stories of various Gods, especially Athena) are elegant, detailed, and exemplify the flowing kinds of elements associated with architectural rhythm. Many of its features are in the proportion of the golden section.

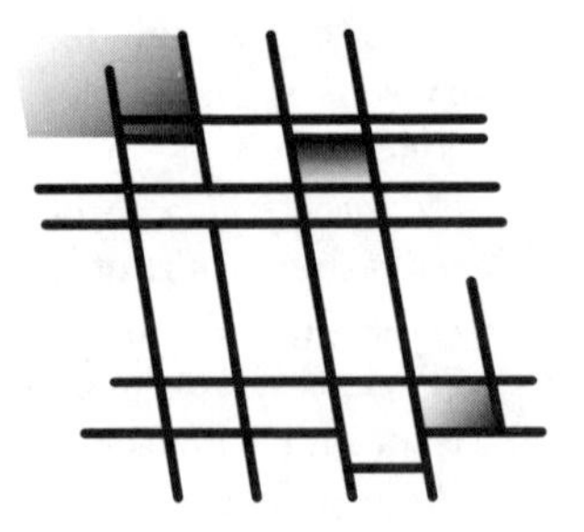

Fig. 2.26. Visual rhythms unfold in space while auditory rhythms evolve over time. Repetition plays a key element in both visual patterns and temporal rhythms.

Figure 2.27 shows the facade, the floor plan, and a small piece of the frieze, all of which demonstrate a mastery of variety within repetition. Goethe refers to such repetitive architectural structures as "frozen music," though a modern view of architecture as "people moving through light filled spaces"[3] suggests a closer analogy with musical improvisation.

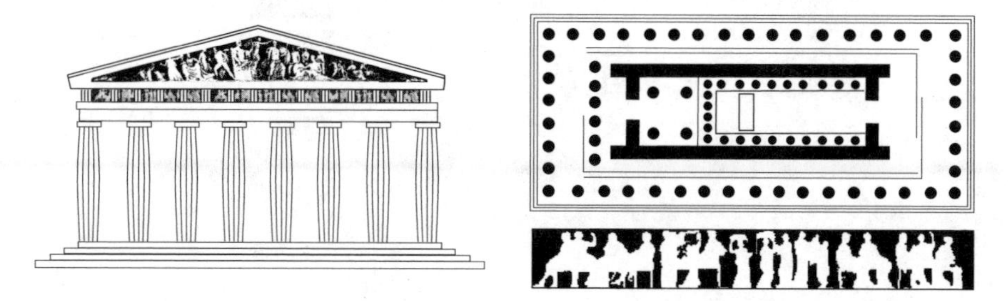

Fig. 2.27. The Parthenon shows the kinds of proportion and balance achieved in classical architecture. The facade, floor plan, and a section of the frieze sometimes called the Elgin marbles (currently in the British museum) all show clear rhythmic features.

Among the oldest metaphors for rhythm are the motions of the bodies of the solar system such as the daily path of the sun, the monthly waxing and waning of the moon and the cyclical behavior of the planets. Many plants have rhythmic behaviors and rhythmic structures: the daily opening and closing of a flower, the yearly cycles of a tree shedding and regrowing its leaves, the bamboo plant with its long stalk punctuated at semi-regular intervals by knots. A necklace of beads provides an analogy in which the beads represent the regular beats while the circularity represents the repetitive metrical structure. Each sand grain dropping through a sand clock is metaphorical of a tiny grain of sound, and the regularity of sand grains passing through the neck represent the regularity of the musical experience. Zuckerkandl [B: 250] suggests that a (water) wave provides a powerful analog of rhythm: the repetitive feeling of relaxation, rising tension, approach to a climax, and then the final ebb.

Rhythm and Transforms deals primarily with the steady recurrence of audible impressions, with rhythmical sounds that represent the organization of time into parts that are perceptible to the ear. Though visual representations and physical metaphors are often useful (especially in a format such as a paperbound book), the subject matter is sound. Accordingly, the sound examples on the CD represent the most important part of the argument. Visualizations provide analogies or metaphors; sound examples are the real thing.

[3] Roger Tucker, private correspondence.

3

Varieties of Rhythmic Experience

Surveying the musics of the world shows many different ways of conceptualizing the use of rhythmic sound.

Rhythmic music occupies a large part of the traditions of Asia, Africa, and India, and may play an even larger role in these traditions than in the West. In his overview of World music, Sachs [B: 187] states:

> What harmony means to the West, the almost breathlike change from tension to relaxation, is in the East provided by the rhythm. In avoiding the deadly inertia of evenness, rhythm helps an otherwise autonomous melody to breath in and out – just as harmony does in the West.

That rhythm is a key element of Indian music is suggested by the makeup of temple and imperial orchestras. Sachs cites two examples: the Rajarajeśvara temple at Tanjore, which in AD 1051 had 72 drummers (of 157 musicians), and Emperor Akbar's band, which in the 16th century had 23 wind instruments and 42 drums.

Even in traditions without large imperial bands, rhythm provides the glue that holds music together. And there are a surprising number of ways that it can be sticky: the Western conception of a metrical hierarchy, the African notion of the recurring timeline, the "inner melody" of the gamelan, the circular conception of the Indian *tala*, the "groove" of modern funk. This chapter presents concrete examples of a variety of different kinds of rhythmic thought, including standard Western practice in Sects. 3.2–3.3, the timecycles of Africa and its Latin descendants in Sects. 3.4–3.7, the *tala* of the Indian subcontinent in Sect. 3.8, polyrhythms in Sect. 3.9 and polymeters in Sect. 3.13, the cyclic conceptions of Indonesian gamelan music in Sect. 3.10, and modern dance styles such as funk (Sect. 3.11) and hip-hop (Sect. 3.12). In all of these musics and in all known dance music, there is an underlying beat. Sometimes the music proceeds in synchrony with the beat, sometimes it moves against the beat, and sometimes the beat provides a temporal framework on which more complex units are hung.

Though these conceptions represent a large part of the World's music, they are not exhaustive. Indeed, the next section discusses music that does not utilize a foot-tapping beat. Instead, the rhythms are conceptualized as being based on poetry or on metaphors with rhythmic breathing.

3.1 Fluid, Unmeasured, and Beatless

Music among the ancient Greeks was primarily a vocal art. In the *Principles and Elements of Harmonics*, Aristoxenes equates the organization of music with the organization of impassioned speech. He classifies rhythmic patterns of poetry into a series of long and short syllables, much as in Sect. 2.1.1. Plato states in the *Republic* that rhythm and melody are regulated by language. In the *Ethos of Rhythmic Modes*, rhythms are classified by their effects on the soul. Thus rhythms have intrinsic qualities: some are stable, others are emotional, some are vulgar and others induce calm. Of course, we do not really know exactly what music sounded like in Plato's day.

Fortunately, parts of the modern worlds repertoire may be best thought of as extensions of poetry, and these may provide a reasonable analogy with the concept of music as the Greeks might have conceived it. In these lyrically inspired traditions, melodies and rhythms are typically used as an aid for memory and not (primarily) as an end in themselves. For example, many Maori traditional songs are strongly word-oriented and many have no meter. Malm [B: 140] calls this *heterometric*, meaning that the accent (and hence metric structure) continually shifts to maintain alignment with the text. The Orthodox church has a long tradition of chanting in which text takes priority over melody. The most important aspect of the chant is to retain intelligibility of the words; emphasis and stress must agree with those of the lyrics, and a regular rhythmic form is only of secondary concern. Similarly, the word *avaz* is used to describe an unmeasured rhythmic style that revolves around classical Persian poetry.

The *gagaku*, literally "elegant music," is a tradition of the Japanese Imperial Court more than 1100 years old. The music makes use of rhythms based on the principle of elastic or flexible breathing. In these styles the melody moves from beat to beat as if the beats were the three parts of breathing: a deep inhalation, a momentary pause while holding the breath, and then a slow exhalation. The music moves slowly with a focus on the detailed expression of tonal quality.

In contrast, the bulk of this chapter explores the wide variety of world music which is built upon and organized around a steady pulsation.

3.2 Meter

Rhythm typically occurs at several different levels simultaneously. For example, the beat defines the basic "foot-tapping" time scale on which rhythmic and dance music are constructed. The tatum (recall Fig. 1.9 on p. 15) represents the fastest pulsation present in the music. Both beat and tatum are examples of *regular successions*, grids of equal time durations. The beat and the tatum form a (two level) metrical hierarchy because the beats coincide with the timepoints of the tatum.

Meter concerns larger groupings at multiple levels. Western music theory and practice views these levels as a hierarchy of regular successions. Just as audible events that form the tatum are clustered into beats, beats can be grouped into measures, measures collected into phrases, and so on up through the highest levels of organization. Meter is typically characterized as a periodic alternation between strong accented beats and weak unstressed beats. A beat is accented when it is aligned with events from levels above and/or below it in the metric hierarchy. A beat is unstressed when it does not coincide.

What does this mean in a practical sense? How might such hierarchies arise? Consider the single unaccompanied melody shown schematically by the black dots in Fig. 3.1. Time is marked out at the most rapid level by the tatum which occurs at twice the rate of the beats labeled 1-2-3-4. The tatum arises from the melody because it is defined by a grid of the fastest note events, which occur in the groups marked II, V, and VII.

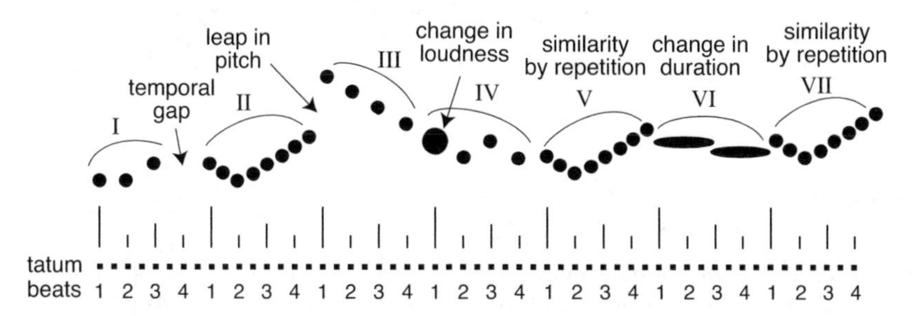

Fig. 3.1. The higher (measure) level is formed by a clustering of the melody at the beat level. Many features can cause clustering: separation in time, separation in pitch, changes in loudness or duration, and the perception of repetitive figures. One possible realization of this melody can be heard in [S: 19].

But the melody also defines groupings at a higher level. Several clusters of notes are shown, each cluster separated from its neighbors by some feature. The grouping labeled I is separated from grouping II because there is a pause. Group II is separated from III because the melody jumps to a higher pitch. Group III is separated from IV because the first note is louder (indicated by the larger black dot). Groups V and VI are set off from their neighbors because they are recognizable as repetitions of II. Group VI is distinct because its durations are longer (represented by the elongated dots).

Essentially any feature that can be used to group or cluster sounds (the psychoacoustics of clustering will be discussed further in Chap. 4) can be used to partition the melody into groups. Because the groups themselves form a regular succession of equal length durations, they form the next level of the metric hierarchy, in this case, the measures. Thus the accents are aligned with the cues that allow the grouping to take place and help to define the higher level of the hierarchy.

This can be carried out almost any number of times, creating ever larger sonic structures. Figure 3.2 shows the next level, where four-measure phrases are clustered by their internal similarities and distinguished by their differences. The first phrase A is separated from B by transposition of pitch. B is distinguished from C by the more rapid pace near the boundary. Of course, any feature of the sound that causes segmentation can help to distinguish phrase boundaries: common techniques involve changes of instrumentation or timbre, changes of tonal center (chord), changes of loudness, etc. Since all the phrases are of equal duration, they also form a regular succession, the next highest level in the hierarchy. And so on...

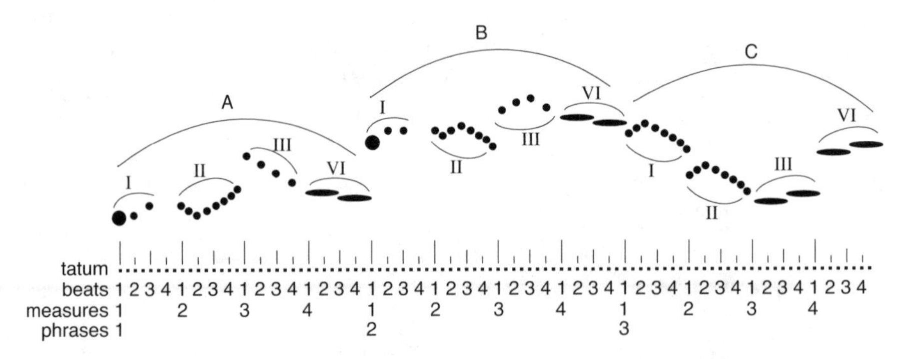

Fig. 3.2. The phrase level groupings are formed by clustering the measures in the same ways that the measures are formed by clustering the beats. Any feature that causes clustering can help segment the piece into ever higher levels.

Accents vs. Groups

Do accents cause perceptual groupings, or do perceptual groupings cause accents? Various writers have taken both sides of this chicken-and-egg issue.

Cooper and Meyer [B: 35] focus on internal perceptual features to define accent as "a stimuli (in a series of stimuli) which is marked for consciousness in some way." Cooper and Meyer observe that each of the five prosodic rhythmic groups (iamb, anapest, trochee, dactyl, and amphibrach, recall Table 2.1 on p. 24) consist of one or two unaccented beats grouped with one accented beat. They then show how these basic rhythmic groups can be clustered together into ever-higher levels to form a metrical hierarchy. In this view, the basic rhythmic groups form a set of recurrent accents that allow meter to measure the passage of time. For Cooper and Meyer, then, the rhythmic groupings cause the accents. Others, such as Berry [B: 14], use meter to mean the partitioning of music's time by accent. Benjamin [B: 11] wrestles with this issue at length.

Any such discussion hinges on exactly what is meant by accent. An early definition is due to Hornbostel [B: 97]: accent is "any quality that differentiates one tone from another." This is what Lerdahl and Jackendoff [B: 128] call a *phenomenal accent*: "any event at the musical surface that gives emphasis or stress to the musical flow." Included in such accents are features of a performance such as attacks of pitch events, sudden changes in timbre or dynamics, long notes, leaps in pitch, harmonic changes, and so on. They distinguish two other kinds of accent. *Rhythmic accents* are points of stability in a melodic or harmonic phrase such as occurs in a cadence. *Metric accents* denote beat locations (timepoints) that occupy strong metrical positions, for instance, the start of a measure. Thus Lerdahl and Jackendoff distinguish grouping (an organization of auditory intervals) from meter (an organization of durationless time points).

3.3 Additive vs. Divisive

The various classes of rhythms represent different ways of conceptualizing rhythm and they correspond to different ways of perceiving the rhythm. *Additive rhythms* begin with short segments that are added together, while *divisive rhythms* begin with a whole and successively divide it into smaller pieces. Thus additive rhythms are constructed and understood from the "bottom up," while divisive rhythms are constructed and understood from the "top down" [B: 131].

Western notions of measure and meter are often thought of as divisive. The measure is the primary unit which is then subdivided in various ways into various patterns and motifs, typically either in groups of three, four, six or eight. In this framework, it is common to think of meter as a grouping mechanism (i.e., the number of beats in a measure) but also as a fixed hierarchy of accents (for example, in 4/4, the first beat is accented the most followed by the third, lastly the fourth).

During the Renaissance, being able to clearly articulate multiple divisions of a passage was considered crucial to musicianship. Ganassi's *Fontegara* [B: 69] instructs recorder players in the "art of playing divisions," and observes that divisions may be either simple or compound in three aspects: in time, rhythm, or melody.

(i) *Simple in time:* each time interval (such as a measure) is divided into n equal parts
(ii) *Compound in time:* n may change as the piece develops
(iii) *Simple in rhythm:* within each division, all note values are the same (e.g., all quarter notes or all eighth notes)

(iv) *Compound in rhythm:* different note values occur within a single interval
(v) *Simple in melody:* the melodic contour is the same for each interval
(vi) *Compound in melody:* the melodic contour may vary from interval to interval

Several of these are illustrated in Fig. 3.3. Different performers may choose to subdivide a melody differently and the rhythmic complexities associated with multiple simultaneous subdivisions are a driving force in these musical styles.

Fig. 3.3. The basic melody (a) is divided in several ways: (b) is simple in time, rhythm, and melody, (c) is simple in rhythm and melody, but compound in time, while (d) is simple in time, and compound in melody and rhythm. Sound example [S: 20] plays (a), (b), (c) and (d) in succession.

In the additive perspective, there is a regular grid of short equidistant segments of time called the *tatum.* Notes are felt against the tatum by adding together the total number of tatum elements in the duration of the note. Consider, for example, the rhythm shown in Fig. 2.4 on p. 28. The tatum consists of the twelve timepoints dotting the circle. The rhythm is expressed additively as the collection of notes with durations 2+2+1+2+2+2+1. The tatum in the *Maple Leaf Rag* of Fig. 2.3 (on p. 27) occurs at the speed of a sixteenth note, at a rate of eight per measure. This forms a regular temporal grid on which all notes in the piece are sounded, though not all the grid positions are used.

The divisive perspective emphasizes the relationship between the notes of the rhythm and the perceptual beats. In the case of the rhythm of Fig. 2.4, the beats occur every two timepoints, as labeled inside the circle. The relationship is that the first three notes of the rhythm occur "on" the beat (i.e., synchronized with the beat timepoints) while the remaining notes occur "off" the beat (i.e., at timepoints between the beat locations). Thus the "same" rhythm can be thought of as either additive or divisive.

3.4 Timelines

Anku [B: 4] argues that there is a "recurrent rhythm" that underlies acts of both performance and listening in African music. This is typically associated with the role of the bell pattern found in many West and Central African drum ensembles. However, this "controlling structural concept is not always externalized along with the music." In other words, it may sometimes be sounded and sometimes be implicit. Waterman [B: 237] calls this the "subjective beat," which is supplied by the listener and around which the musician elaborates. Similarly, Chernoff [B: 30] quotes Abraham Adzenyah, a Fanti drum master, who "always keeps in mind a 'hidden rhythm' within his improvisations."

Some of these basic bell patterns have already appeared in Figs. 2.4 and 2.5. Figure 3.4 elaborates on the basic 12-timepoint bell pattern by showing the supporting drum patterns for the *Slow Agbekor* as reported by Chernoff. Each individual pattern is quite simple. What gives the rhythm its complexity and drive is the way that the parts fit together. Indeed, African musicians and listeners do not locate themselves in the music by counting from the main beat. Rather, they find their entrances in relation to the other instruments, that is, in relation to the cross-rhythmic fabric of the complete cycle.

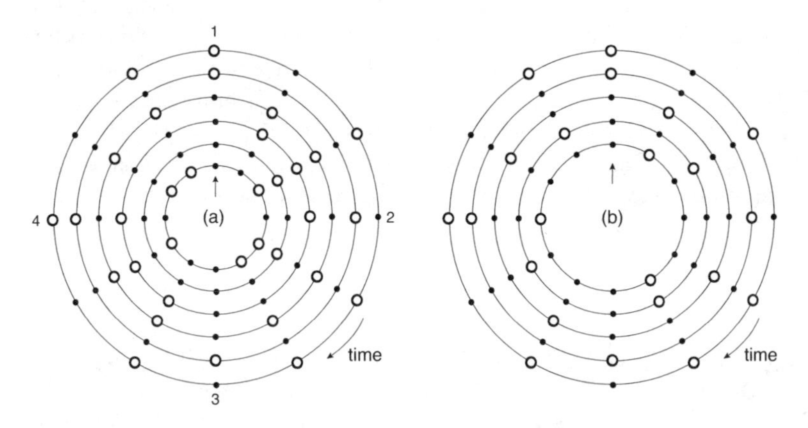

Fig. 3.4. (a) Supporting drums for the *Slow Agbekor* include the bell (outer circle), the rattle, the *kagan*, the *kidi*, the *totogi* and the *kroboto* on the inner circles. (b) The second, third, and fourth circles show three possible cross-rhythms. The different points of unity between the cross-rhythms and the *agbekor* bell pattern suggest different accents, and different possible ways of hearing the bell. See [S: 21].

One fascinating aspect of rhythmic patterns such as Figs. 3.4 (and 3.5 below) is that points on the cycle that seem to require the most emphasis (such as the "one" and the "three" in (a)) are often marked least by the instruments. The four shakes of the rattle are the basic beats to which spectators clap hands and to which dancers move. Part (b) (adapted from [B: 30]) shows how various accents (shown in the figure by the completely regular patterns on the various inner circles) can establish a number of different cross-rhythms. When playing different rhythmic patterns simultaneously, each rhythm helps to determine the way that the others are perceived. Chernoff comments:

> The establishment of multiple cross-rhythms as a background in almost all African music is what permits a stable base to seem fluid. Stable rhythmic patterns are broken up and seemingly rearranged by the shifting accents and emphases of other patterns.

Like many of the rhythmic styles throughout Africa, Gahu drum music (of the Ewe of Ghana) is typically performed by a group of musicians. The percussionists play distinct and contrasting patterns that, when sounded simultaneously, interweave to form a complex whole. Rather than concentrating on their own individual drum line, performers "hear" and synchronize to the whole pattern. One rhythm means nothing without the others.

There are several instruments that typically define the time cycle in drum Gahu. The *gankogui* is a pair of hand-held iron bells with two contrasting pitches. The *gankogui* pattern is shown in the outer circle of Fig. 3.5. The rhythm may be conceptualized in the additive framework with durations $3 + 4 + 4 + 2 + 3$. When hearing this divisively, the tension between the notes and the four beats (three of which fall off the beat) helps to drive the rhythm forward. The *axatse* is a kind of rattle made with a large number of beads woven in a fishnet design around a hollowed out gourd. It emphasizes the implicit polyrhythm between the *gankogui* and the beat as it plays the rhythm in the middle circle of Fig. 3.5. The *kaganu* is a narrow drum that is open at the bottom which plays the rapid off-beats shown in the inner circle, increasing the density of the pattern.

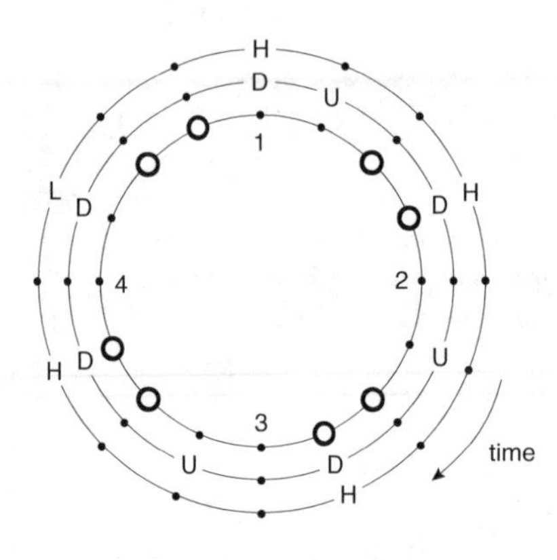

Fig. 3.5. The basic time of drum Gahu. The outer circle shows the high (H) and low (L) pitches of the *gankogui* bells. The *axatse* rattle (middle circle) makes different sounds on its downstroke (D) and its upstrokes (U). The *kaganu* drum plays the off-beat pattern in the inner circle. The four (implicit) beats are marked 1-2-3-4. Sound example [S: 22] demonstrates.

If a simple regular beat can be heard in different ways, then it should not be surprising that an intricate sound mass like the drum Gahu may be heard in different ways. Locke details five possible rhythmic modes, one for each of the possible starting points. For example, the second, third and fourth strokes all lie one timepoint behind the beat. If these become accentuated, they may cause the beat to "turn around" and rotate one timepoint counterclockwise. Though the notes and durations remain the same, the musical function changes; the upbeats in one interpretation are the downbeats in another. This is an example of how a continually repeating cyclical phrase can be mentally reordered.

In performance, the *gankogui*, *axatse*, and *kaganu* act as the basic time keepers while a variety of other drums implement a call and response that exploit a large number of techniques of rhythmic variation [B: 130] including:

(i) dynamics: varying the intensity or loudness of strokes
(ii) bending time: slight purposeful deviations in timing
(iii) displacement: shifting the placement of a motif in time
(iv) ornamentation: grace and/or ghost notes that may occur slightly before or slightly after a given stroke
(v) repetition: repeating a given phrase or motif
(vi) segmentation: partitioning long phrases into shorter ones
(vii) augmentation: combining smaller phrases into larger ones
(viii) doubling: hitting two strokes where normally one would occur
(ix) pausing: resting when normally there would be a stroke
(x) substitution: exchanging one kind of stroke for another

Many scholars who write about African music observe that musical training, and in particular rhythmic training, typically begins at an early age. Agawu [B: 1] transcribes a Northern Ewe children's game-song which accompanies a two measure clap pattern. The interesting feature of the four-bar melody, as schematized in Fig. 3.6, is that it can begin on either the first or the second bar of the clapping pattern. By varying the length of the last note in the melody, or by resting longer or shorter times, the performers may switch from one mode to another. Even here in a simple children's game there is a tension between the steady background repetition (the regularity of the clapping) and the perceptual changes in metric and rhythmic weight that characterize the content of the music. Blacking [B: 16] reports a similar phenomenon among the Venda of the Transvaal.

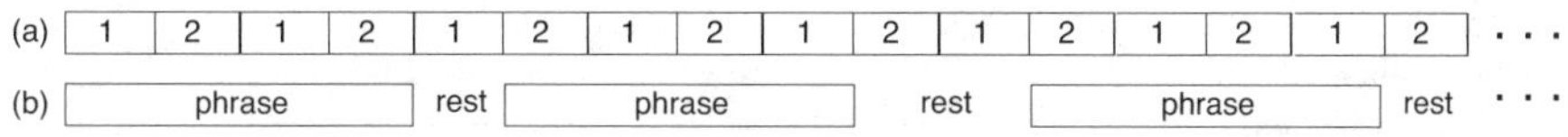

(a) 1 2 1 2 1 2 1 2 1 2 1 2 1 2 1 2 . . .
(b) phrase rest phrase rest phrase rest . . .

Fig. 3.6. The clapping in a children's game repeats in steady two bar phrases as schematized in (a). The melody is sung in four bar phrases as shown in (b), followed by a long held note (rest) that can be either one or two bars long. The relationship between the clapped pattern and the singing changes, since sometimes the singing begins on the first measure of the clapped pattern and sometimes on the second.

Agawu [B: 1] (like Hornbostel [B: 97] and Nketia [B: 155] before him) draws a distinction between "free" and "strict" rhythms. Free refers to music lacking a clear sense of periodicity and meter, and is typically associated with songs that reproduce the rhythms of speech. Strict rhythm, in contrast, "refers to the presence of a tactus, a palpable metric structure, and a resultant periodicity."

3.5 The Clave

The popularization of latin dances such as the salsa, the rumba, the mambo, the merengue, and the cha cha chá (among others) has brought Western musicians into direct contact with a different way of conceptualizing rhythm. The focus of the salsa, for instance, is on a rhythmic unit called the *clave* that forms a backbone around which the music is structured. In simplest form, the clave is a two measure pattern in $\frac{4}{4}$ time, containing three pulses in the first measure and two in the second. The clave forms a point of reference in the sense that singers, horns, pianos (and other instruments) either play "along" with the clave or "against" it. Figure 3.7(a) shows the clave rhythm in the necklace notation. The beats of the two measures are indicated inside the circle.

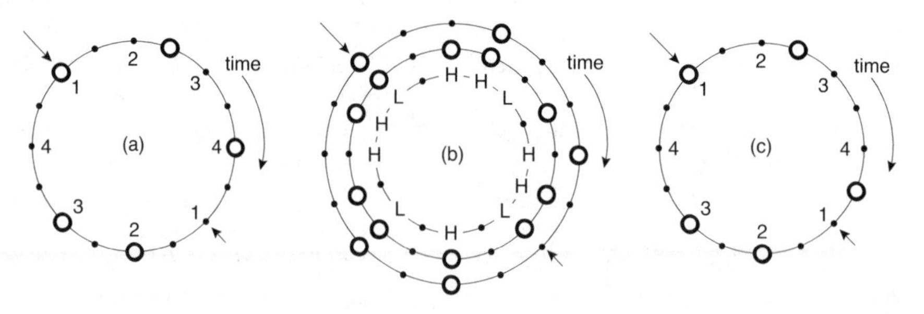

Fig. 3.7. The son clave rhythm is arranged in necklace notation; the 3-2 clave begins at the larger arrow while the 2-3 clave begins at the smaller arrow. (a) The beats of the two $\frac{4}{4}$ measures are indicated inside the circle along with the 16 timepoints that represent the tatum. (b) repeats the basic clave in the outer circle and shows how various other rhythmic parts complement, augment, and can substitute for the straight clave pattern. The middle circle shows the *cáscara.* The inner circle shows a bell pattern with low (L) and high (H) bells. (c) shows the *guanguancó* (rumba) clave. These are performed in [S: 23].

The clave is a two measure pattern, but it can "start" on either measure. If the melody begins on the first measure then it is called a 3-2 clave (because the three strikes occur in the first measure while the two strikes occur in the second measure) [B: 233]. Similarly, if the melody starts on the second measure, then it is called the 2-3 clave. Sheller comments [B: 71], "Once the song begins and the clave starts, the clave never changes. But the 'one' may change." Thus a melody may begin in the first measure in one part of a song and it may begin in the second measure in the another part of the song. Geometrically, the melody may rotate against the fixed clave.

The clave rhythm is sometimes played on a pair of rounded wooden sticks (called claves). Sometimes, one of the other rhythm instruments (such as the *cáscara* or bell of Fig. 3.7(b)) plays a complementary pattern. And sometimes, the clave is not directly sounded at all. Whether it is played or not, the clave is embedded in the music; it is an essential ingredient for the musician, for a dancer, and for the listener. Here again is a phenomenon where the ear can perceive pattern that may not be objectively present in the music.

3.6 Samba

Brazilian culture is a remarkable combination of African, Native American, and Iberian influences, and the music is no exception. Samba refers to both a dance and to the music that accompanies the dance. The first recorded samba was Ernesto dos Santos' *Pelo telefone* in 1917, and by the time of the New York World's Fair in 1939, the samba had become a popular ballroom-style dance. The Bossa Nova (New Wave), a combination of samba and cool jazz, arrived in America in the 1960s with the *The Girl From Ipanema*, a song by Antonio Carlos Jobim that has become a classic. Modern Samba is a dance and music style that can take many forms, from energetic Carnival dances to relaxed song-sambas played on mellow acoustic guitars. The samba-schools of Rio de Janeiro have helped to institutionalize samba as the national dance of Brazil, and the schools are credited with bringing Afro-Brazilian musical aesthetics to the cultural forefront. They present their music to the public in huge parades at Carnival.

Sambas are in lively double meter, and are characterized by a number of interlocking, syncopated lines. Two examples are given in Fig. 3.8. In both cases, the primary rhythm is played by a hand drum, which can be performed using a variety of articulations. The *surdo* is a large double headed drum played with a felt covered mallet and the open hand. The *ganza* is a metal shaker and the tamborim is a small frame drum played with a stick. The *guiro* is a notched or ridged gourd scraper that alternates its up and down strokes at each timepoint. The *agogo* is a pair of metal bells much like the *gankogui* bells of Sect. 3.4.

3.7 Vodou Drumming

Haitian sacred drumming is another style of music that grew in the New World from African roots. The music is governed by principles of interaction such as the call and response structure and the emergence of complex patterns from multiple interlocking rhythmic lines. Haitian music is rooted in the slave experience, and it is tied very closely to a variety of deities [B: 241]. Besides strictly musical duties, drummers are key agents in spirit possession

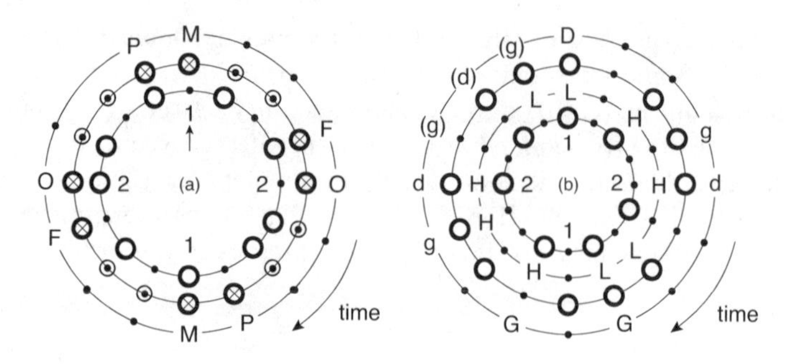

Fig. 3.8. Two versions of the samba. (a) is adapted from [W: 40]. Instrumentation includes the *surdo* (a hand drum) in the outer circle that can be struck in several ways: muffled with a stick M, softly with the fingers F, with the open hand O, and with the palm P. The *ganza* in the middle is played on every timepoint, but is accented in synchrony with the *surdo*. The *tamborina* plays the off beat accents. (b) is adapted from [B: 82]. The hand drum can be struck in the center with the left or right (D, G) palms, or with the fingers of the left or right hands (d, g). The shaker plays a regular pattern of three out of every four timepoints, while the *agogo* alternates its high H and low L bells. The inner circle is played on wooden clave sticks. Both are demonstrated in [S: 24].

ceremonies; they must know how to respond during possession, and they must not become possessed themselves.

The *ogan*, a kind of iron bell, plays a central role in the music of Haiti in much the way that the clave forms a foundation in Latin music and the *gankogui* grounds the Gahu timelines. Wilcken comments, "Even though the *ogan* is not always physically present in the Vodou drum ensemble, musicians and dancers feel its rhythmic pattern." Only members of the priesthood play the *ason*, a rattle that is typically used to set the tempo. The battery (drum ensemble) consists of three conically shaped drums: the largest is the master drummer's *maman*, the *segon* is played with one hand and one curved stick, while the smaller *boula* is typically played with two straight sticks.

Like other music with African origins, the rhythmic patterns move cyclically through time. Figure 3.9 and sound example [S: 25] demonstrate a skeletal version of the rhythmic framework of two pieces from the traditional Vodou rituals, the *Parigol* and the *Zepol*. The rhythms are not dissimilar to those found in the African timelines of the Yoruba (recall Fig. 2.5), which makes ethnographic sense since many of the early inhabitants of Haiti came from the region around Nigeria.

In both pieces, the *ogan* plays the basic timeline and the *ason* directs with purely off-beat strokes. The patterns set up a feeling of three-against-two between (for instance) the *segon* and the *maman* and the *ogan*. This opposition is strengthened when the singing is introduced, since it is most strongly felt in terms of binary divisions. The songs may be notated in $\frac{4}{4}$

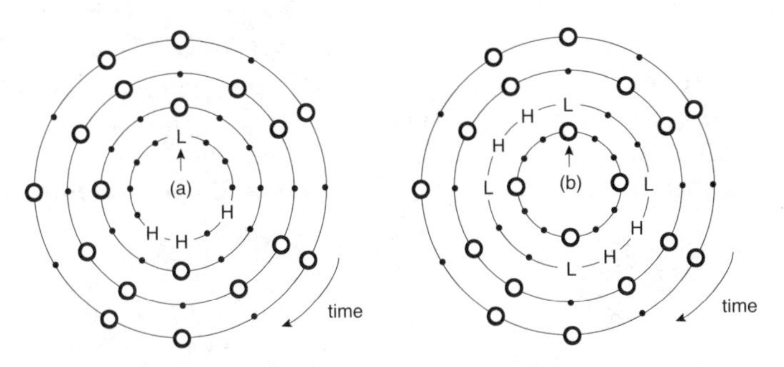

Fig. 3.9. Two pieces in the Haitian Vodou drumming repertoire contain rhythms not dissimilar from those found in the African timelines of Fig. 2.5. (a) The *Parigol*: drum parts are shown for the *ogan* (outer circle), the *ason*, the *boula*, and the *maman* (inner circle). (b) The *Zepol*: patterns are shown for the *ogan* (outer circle), the *ason*, the *boula*, and the *segon* (inner cricle). The L and H represent two different articulations (low and high) of the drum. See [S: 25].

despite being sung against the $\frac{12}{8}$ patterns above. A complete analysis of a set of ritual pieces is given in [B: 241].

The drummers use a variety of hand and stick techniques to add variation, changing both the envelope and the spectrum of the sound. The hand can either strike the drumhead or slide across it. When striking, the hand can be either cupped or flat, and the stroke can be either stopped (ending the strike on the surface of the drum) or unstopped (to bounce back from the drumhead). Similarly, when playing with sticks, the strokes can be either perpendicular to the skin or slanted, and the strokes may be either stopped or unstopped. Any of the strokes may hit towards the center or towards the rim of the drumhead, and it is not uncommon to strike the side of the drum with sticks for added variety.

3.8 Tala

The heart of the Indian musical theory of rhythm is the cyclical measure of time called the *tala*, a rhythmic experience arranged so that there is a feeling of return. The theory is particularly well developed by percussionists who typically play a pair of small drums called the *tablas*. A *tala* is easily pictured as an arrangement of equally spaced points around a circle, which may have anywhere from 3 to 128 beats, though the most common are between about 7 and 24. The cycles are typically subdivided into small groups of beats and are given names to aid in memorization [B: 72].

For example, in the *jhaptal tala* shown in Fig. 3.10(a), the ten beats of each cycle are subdivided into four segments 2+3+2+3. The cyclical nature of the *tala* gives special significance to the first beat; the *sam* is a point of convergence between the drummer and the other musicians, when the cycle ends as well as when it starts. The *tali* (clap) mark the start of the subdivisions and the *khali* (wave) indicate that the grouping is unstressed, providing a point of contrast. Listeners and musicians will sometimes keep track of the *tala* by clapping on the *sam* and *tali*, and waving their hand to the side on the *khali*. Observe that the groupings marked out by the *tali* and *khali* need not be isometric.

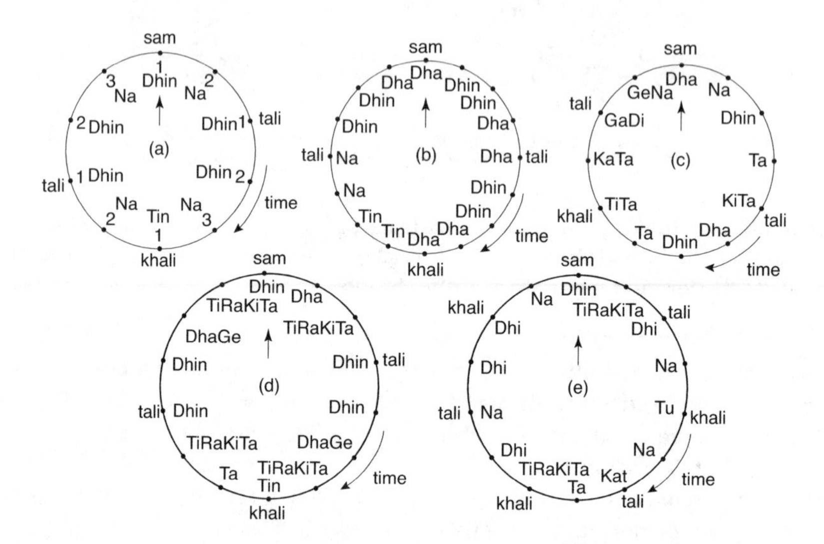

Fig. 3.10. (a) The Hindustani *jhaptal tala* from Malm [B: 140] is 10-beat cycle. The remaining *tala* are from [B: 36]: (b) the 16-beat *tintal theka* may be one of the oldest *tala*, (c) the 12-beat *choutal* is one of the most popular, (d) the 14-beat *jhumra tal* divides the cycle into $3 + 4 + 3 + 4$ and (e) the *ada choutal* divides its 14-beat cycle into seven equal segments, each with two beats. See [S: 26].

The syllables associated with each beat (*Dhin, Dha, Na,* etc.) are onomatopoeic mnemonics called *bols* that aid in the memorization of the *tala*. The *bols* represent particular strokes on the *tabla*. For example, the *Ga* is a left handed open stroke where the middle and ring fingers bounce from the surface of the drum. The *Na* (and the *Ta*) are rim strokes produced by hitting the index finger of the right hand against the rim while holding the last two fingers lightly against the edge. The *Ga* and *Ta*, when played together, form the *Dha*. The *Ka* is a left handed open slap. The *Tin* is a soft and delicate form of *Na*, and the combination of *Tin* and *Ga* forms the *Dhin*. Courtney [B: 36] provides a thorough catalog intended to help the beginning *tabla* player learn the traditional drum strokes and rhythmic patterns. Northern Indian (Hindustani) and Southern Indian (Karnatic) musics are different and the nomenclature for the *tala* and *bols* may vary significantly by region [B: 110].

Figure 3.10 shows several different *tala* with varying numbers of beats per cycle and varying numbers of subdivisions. In (c) there are several double-syllable beats such as *KaTa* and *GeNa*. These indicate that the beat is to be evenly subdivided into two strokes. Similarly, the *TiRaKiTa* of (d) and (e) divides the beat into four equal parts. Thus the Indian *tala* are additive at the level corresponding to the measure (the various *tali* and *khali* segments) and may be divisive at the level of the individual beat (recall the discussion in Sect. 3.3).

There are many ways that *tala* may be varied in performance. *Bols* may be substituted for each other. Multiple *bols* such as *DhaNa* may replace a single *Na* stroke to increase the density. *Bols* may be eliminated and replaced with silence. For example, the silence (indicated with –) immediately following the first beat in Fig. 3.10(b) helps to emphasize the *sam*.

Indian percussionists are well aware that perceptual shifts are sometimes needed when playing the "same" *tala* at widely different tempos. Some pieces may have a tempo so slow that it is difficult to conceptualize the rhythm. Strokes, numbers, or syllables can be mentally inserted to raise the tempo. At high speeds it may be necessary to conceptually shift the tempo downwards by factors of two until a more moderate tempo is achieved.

Malm comments on the differences between Indian and Western musicians:

> It would seem that the Western fascination with harmonic structures and the South Asian enchantment with melodic and rhythmic systems propelled these two grand traditions in very different directions. Western musicians need to sing solfège and recognize chord progressions by ear in order to feel and understand their past tradition; Indian musicians need to... beat and wave the divisions of the basic *tala* in order to be a true part of their musical world. Whereas the Western professional becomes aware of subtle variations in harmonic structures, the Indian becomes equally sensitive to the rhythms that tend to have become characteristic of each *tala*.

From the Indian's point of view, the "science of *tala*" is a medium for expressing rhythm in a logical and systematic way [B: 72]. Each *tala* represents a unique circular rhythmic structure that helps make time comprehensible through musical expression. According to Sen [B: 193], "music without *tala* is like a face without a nose."

3.9 Polyrhythms

Polyrhythms are the simultaneous sounding of two (or more) pulse trains where the tempos are not integer multiples of each other. The simplest such pattern is two-against-three, which appears in necklace notation in Fig. 3.11(a). This is performed at a tempo of 1.2 s per cycle in sound example [S: 27](i).

In principle, there are two tempos present. Yet when confronted with such patterns, listeners do not tend to sense ambiguity, rather, they quickly focus

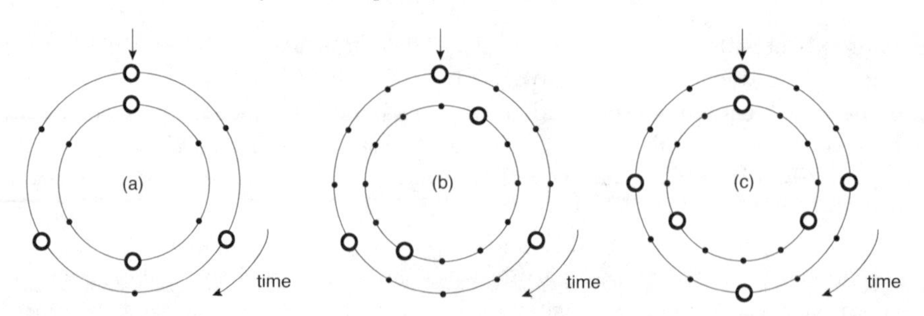

Fig. 3.11. Three simple polyrhythms are shown. In (a) the standard three-against-two rhythm is shown, and may be heard in sound example [S: 27](i). Observe that the rhythm is resolvable at the simplest level into six equal timepoints. In (b), the inner rhythm is rotated by 30 degrees, requiring twelve timepoints to represent all possible time intervals in each cycle. This variant is presented as [S: 27](v). (c) diagrams the four-against-three polyrhythm which can be heard in [S: 27](vi). Other more complex polyrhythms are performed in [S: 27](vii)–(x).

attention on one or the other tempo and hear that as the implied beat. Perhaps even more striking is that these polyrhythms tend not to be perceived as two separate "lines" at all. Rather, they are heard as a single chunk of rhythm. Thus three-against-two is more clearly described as "long-short-short-long" or – ⌣⌣ – in the lyrical notation of Sect. 2.1.1, than in terms of its constituent parts.

Blacking [B: 16] argues that an anthropological approach to musical systems (one that focuses on the behaviors of people as they perform and listen to music) makes more sense than an analysis based ultimately on the resulting sound. Any music is based on the human abilities to discover and interrelate ordered patterns in sound and there are often several possible structural explanations of any musical passage. Blacking provides several examples where the same surface structure (the 3-2 polyrhythm of Fig. 3.11(a)) can be produced by either one, two, or three performers among the Venda of the Transvaal. Though the sound is the same, the meaning of the passages is very different in each social context.

Blacking compares the rhythmic complexity of African music to the harmonic complexity of Western music

> The polyphony of early European music is in principle not unlike the polyrhythm of much African music; in both cases, performance depends on a number of people holding separate parts within a framework of metric unity, but the principle is applied "vertically" to melodies in polyphony and "horizontally" to rhythmic figures in polyrhythm.

Cowell [B: 37] draws an analogy between the ratios of rhythmical beats and the ratios of musical tones, as discussed in Sect. 2.1.9. For example, the ratio between the fundamental frequencies of two notes that are separated by a musical fifth (say C and G) is $\frac{3}{2}$. In terms of the periods, the lower tone repeats twice in the same time span that the upper note repeats three times, analogous to the $\frac{3}{2}$ polyrhythms in Fig. 3.11(a). Similarly, other familiar musical intervals correspond to various polyrhythmic timings: the major third to a time ratio of $\frac{5}{4}$ and the minor third to a time ratio of $\frac{6}{5}$. This idea of equivalence at different time scales is explored in sound examples [S: 27](i)–(iv) where the three-against-two polyrhythm is performed at different speeds: at 12 ms, 120 ms, 1.2 s, and 12 s per cycle. At 120 ms, the individual drum strokes of the drum merge and the rhythmic percept is lost. The sound becomes a rapid trilling, the chirping of a motorized cricket. At 12 ms, the drum character has disappeared completely and is replaced by a pair of pitches separated by a fifth, just as Cowell posits. When the cycle is stretched to 12 s, all perceptual connection between the drum hits is lost, and there is only a series of disconnected events.

Of course, the analogy cannot be pushed too far. As Cowell points out, there are few melodies that repeat the same note over a whole work, while there are many pieces in which the same metric structure repeats throughout. Perhaps more fundamental is that the perception of events at a 12 ms repetition rate is quite different from the perception of the "same" events at a 1.2 s repetition rate. It is this perceptual discontinuity that disrupts the unification of the rhythmic and tonal levels.

Many musicians and composers have explored the use of polyrhythms. Jacky Ligon [B: 129] comments about a group of frame-drummers where "everyone is playing in a number of simple patterns, all in different time-signatures... the "dum" strokes appear to be passed around between the drummers as the meters cause continuous new alignments of the strokes... it's a wonderful experience." My own polyrhythmic attempt can be heard in *Persistence of Time* [S: 28]. The tonal material, which uses an adaptive tuning to manipulate the pitches of the sounds, is documented elsewhere [B: 196], but it is laid on top of a strict three-against-two rhythmic bed.

3.10 Inner Melody and the Gamelan

The last several sections demonstrate that at least part of the human experience of rhythm is a psychological ordering of sensations that may not be present in the sound itself: the beat may not be played, the clave may be implied, the timeline may not be expressed in sound. Sorrell [B: 221] makes the parallel argument that Indonesian gamelan music is based on the concept of an "inner melody" which forms the common basis of all parts in a performance, yet which may never be literally played by any instrument. If, asks Sorrell, the inner melody...

> is not played as it is conceived, where is it? The answer is: in the minds of the musicians... the 'inner melody' is not played by any single

> instrument but is a kind of intuitive melodic core which influences the movement and direction of the whole ensemble. . . [the] inner melody is the melody that is sung by musicians in their hearts.

The gamelan "orchestras" of Java and Bali consist of a large family of metallophones that play exotically tuned phrases that repeat over and over, with variations that slowly evolve throughout pieces of near symphonic length. Sorrell gives several examples of the *balungan*, the melodic core around which a gamelan piece is structured. For example, the outer ring of Fig. 3.12(a) shows the four note balungan 2 – 7 – 2 – 3, where the numbers refer to notes of the *pelog* scale (a seven-tone scale that is unique to each gamelan). A typical realization of the piece plays this skeletal melody at differing rates and with different ornamentations and embellishments as in [S: 29].

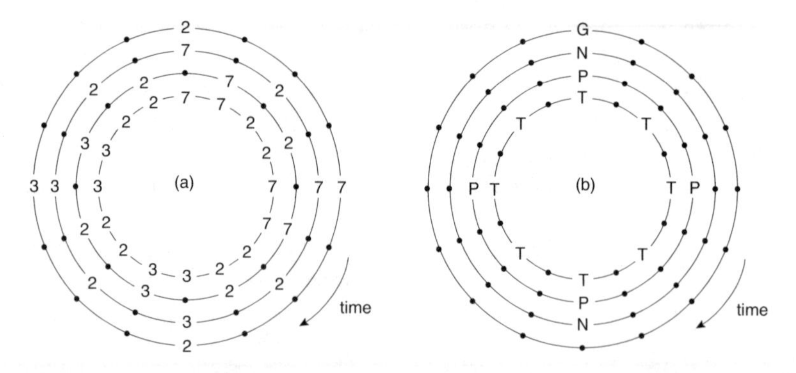

Fig. 3.12. Adapted from Sorrell [B: 221], part (a) shows a possible rendition of the *balungan* in the outer ring using the three inner instruments (which are played by the *peking*, the *demung* and the *slenthem* respectively). Part (b) schematizes the *irama*, the gamelan concept of tempo relationships based on ever faster binary divisions of the time cycle.

The *irama*, an important organizational principle in gamelan music, is shown schematically in Fig. 3.12(b), which is adapted from Malm [B: 140]. For instance, the gong may sound once per cycle. The *kenong* (labeled N) may sound twice per cycle, equally dividing the gong strokes. The *kempul* (labeled P) may sound four times per cycle, equally dividing the *kenong* strikes. The *kethuk* may then sound eight times per cycle, evenly dividing the strokes of the *kempul*. Thus the complete cycle is divided again and again into two even parts, until the fastest instruments may play at dizzying speed.

One of the first Europeans to document the extraordinary sounds of the gamelan was Jaap Kunst [B: 121], who coined the phrase *colotomic motion* to describe the style of music in which sections are marked by the entrance (and exiting) of various instruments. In gamelan performance, instruments generally enter and leave at the start or end of cycles which are typically punctuated by strokes of the largest gongs.

3.11 Funk

One of the quintessential "funk" performers is James Brown, whose songs are "all about grooves." Growing out of the "Rhythm & Blues" of the 1950s, Brown and his rhythm section sculpted a number of popular and influential songs over several decades. While each individual instrument in the rhythm section may only contain a few notes or a few hits, it is the interplay between the instruments that sets the "groove."

Figure 3.13 shows the basic rhythmic patterns in two of Brown's early songs, *Out of Sight* and *Papa's Got A Brand New Bag* [D: 8], both of which follow a standard twelve bar blues pattern. Both are grounded by Melvin Parker's drumming, which combines a regular eighth note hi-hat with a characteristic bass and snare. Tightly coupled with the drums is Sam Thomas's bass guitar, which is the most active and varied part in the rhythm section. The guitar parts (played by Les Buie and Jimmy Nelson) accent each hit of the snare. In addition, the horn section provides active and driving rhythmic accents and stabs throughout. These patterns repeat throughout the songs except for the "turn-around," the last two measures of each twelve-bar phrase. A single cycle of each is excerpted in [S: 30].

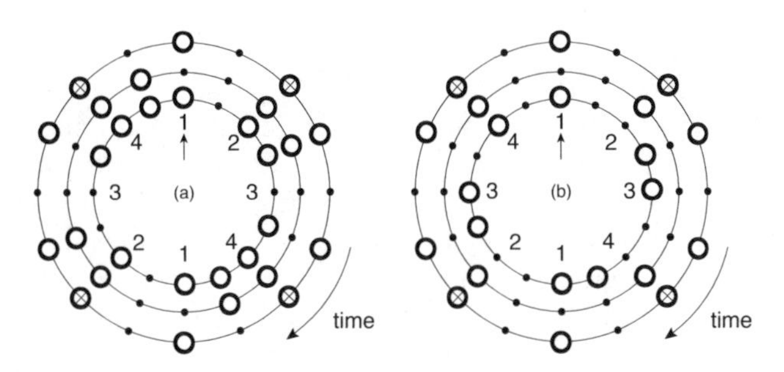

Fig. 3.13. Basic rhythm patterns in (a) *Out of Sight* and (b) *Papa's Got A Brand New Bag.* The drum kit is shown in the outer circle with the bass drum ◯ and snare ⊗. The hi-hat (not shown) is hit at each timepoint. The bass guitar (inner circle) plays the most active line with the most variation. The guitar plays chord "stabs" at regular intervals, accenting the two and four of the snare drum. The analysis is adapted from [B: 213].

The funk style can be interpreted as a melange of two musical practices. In the cyclical performance suggested by the timecycles of Fig. 3.13, in the use of simple interlocking elements to create a complex motive musical force, and in the use of two-bar phrases with shifting accents, the music resembles the traditional timelines. Where the traditional music is orchestrated with bells and drums, funk is orchestrated with basses, drum kits, horn stabs, and the occasional guitar.

The second influence appears in the harmonic motion driven by the twelve bar blues pattern:

$$E\flat 7\ E\flat 7 \mid E\flat 7\ E\flat 7 \mid A\flat 7\ A\flat 7 \mid E\flat 7\ E\flat 7 \mid B\flat 7\ A\flat 7 \mid E\flat 7\ E\flat 7$$

Since each cycle in Fig. 3.13 represents two measures, the blues structure imposes a larger cycle: after every five cycles, there is a two-bar break.

3.12 Hip-Hop

The recent emergence of hip-hop (and its lyrical cousin "rap") presents an interesting mix of highly inflected vocalizations superimposed on a bed of tightly metric rhythms. Much of the academic interest in this music relates to the (sub)cultural contexts in which the artists work, though analysts such as Krims [B: 119] have begun looking more carefully at the structure of the music. Krims distinguishes three vocal styles: the "sung" style close to popular music, the "percussive-effusive" style (where the voice is used to accent the musical texture and/or subdivide regular metrical units) and the "speech-effusive" style in which the rhythm of the voice mimics that of spoken language.

In stark contrast to the elaborate and nuanced timing of the vocals, the instrumental parts are often created using electronic percussion ("drum machines"), "scratching" (playing short snippets, usually from vinyl records), and "looping" (the process of layering many short phrases of pre-recorded samples). Common to all these techniques is the idea of combining pre-existing sounds (loops, samples from CDs, synthetic patterns from drum machines) into a rhythmic bed. Though any given element of the rhythm may be simple, by layering different rhythmic motifs, unpitched noises, exploiting electronic timbral distortions, combining sequences in different meters and different keys, the overall effect can be hypnotically complex. Krims calls this the "hip-hop sublime."

Figure 3.14 schematizes the sublime in a rap song by Ice Cube. The instrumentation includes several synthesizer parts, two basses (bass guitar and a synthesized bass) and drums. This pattern repeats between measures 1–56 of the song, and returns during later verses. Consider the bass guitar, which is one of the most prominent lines (shown on the outer circle). Observe that this is the same pattern (except for a rotation) as is played by the *gankagui* bell of Fig. 3.5. The sounds represented on the inner circles provide interlocking rhythmic motifs that help provide rhythmic force. The hip-hop sublime accomplishes some of the same musical tasks as the basic timelines of the earlier styles; where the traditional timeline is orchestrated with bells and drums, the sublime is orchestrated with samples and loops.

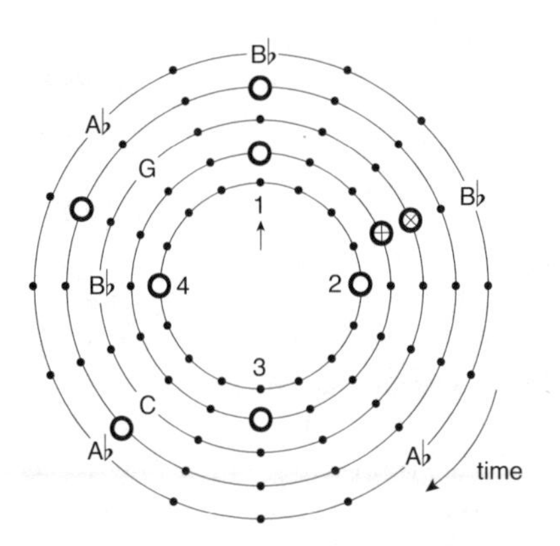

Fig. 3.14. The "hip-hop sublime" occupying measures 1–56 (and others scattered throughout) of *The Nigga Ya Love to Hate* by Ice Cube [D: 27]. Shown are the bass guitar (outer circle), bass drum, bass synth (with a chord stab indicated by ⊗), synth whoosh (plus cowbell ⊕), and snare drum (inner circle). The analysis is adapted from [B: 119] and a single cycle is excerpted in [S: 31].

In traditional African drumming, the timeline is used as a rhythmic base on which the master drummer improvises and on which dancers (and others) base their participation. The sublime provides a rhythmic base on which the rapper improvises with stylized vocal nuances and practiced rhymes. Table 3.1 emphasizes the analogy between the rapper and the master drummer by diagramming a single verse of a rap song and showing where (in the $\frac{4}{4}$ measures) the singer accents syllables.

Though there is no melody in the classical sense, the lyrics are placed rhythmically within the sublime as Table 3.1 shows. The first three lines begin on the second beat, and there are numerous regularities of accent and stress. The end-rhyme finalizes the couplet and then leads into a repeat of the line "Premier's on the breaks." This is closer to the singing style than to the speech-effusive style. The master drummers of Africa are often cited for drumming so as to imitate speech. It is an interesting irony that in the new style, the rapper has adopted the role of the drummer.

Table 3.1. The opening four lines of KRS-One's [D: 29] *MCs Act Like They Don't Know*. Analysis adapted from [B: 119].

1				2				3				4				
1	2	3	4	1	2	3	4	1	2	3	4	1	2	3	4	
	x		x	x		x	x	x		x						Clap your hands, everybody
	x	x		x				x	x		x					if you've got what it takes
	x	x		x	x		x			x	x		x		x	'Cause I'm K-R-S, and I'm on the
x		x	x	x				x			x	x				mike, and Premier's on the breaks!

In certain syncopated styles, the beat is not heard but implied. Nelson [B: 154] comments: "For African-American musicians who improvise when a rhythmic beat is already established, their challenge is to express knowledge of the beat by *not* playing it... a show of motility or skill by playing around the beat instead of playing on the beat... to illustrate mastery over time by playing against time."

3.13 Simultaneous Tempos

With polyrhythms, while each line is playing a different speed, all the lines are related to each other and typically the complete rhythm is heard as a new form rather than as a superposition of independent elements. Sachs [B: 187] points out that in some North American Indian music the drum accompaniment is played at a tempo that is independent of the tempo at which the voice sings. He gives several examples, including a Chippewa song where the singer proceeded in a steady rate of 168 beats per minute while the drummer played a more stately 104. In one extreme case the drum and voice are at nearly the same tempo. At the beginning,

> the drumbeat is slightly behind the voice, but it gains gradually until for one or two measures the drum and voice are together; the drum continues to gain, and during the remainder of the record it is struck slightly before the sounding of the corresponding tone by the voice. [B: 42]

This concept of drumming can be understood two ways. First, it may be that the two meters are perceived as separate streams, and hence are intentionally unrelated (other than by providing a background against which the other is performed). Another possibility is that this is a way of introducing variety despite the repetition of parts... at each repeat the individual parts are the same, but their relationship changes over time. This is schematized in Fig. 3.15.

Fig. 3.15. Two steady pulses, each at a slightly different tempo, are sounded simultaneously. The top pulse train precesses against the bottom. For a perceptual discussion of this phenomenon, see Fig. 4.16 on p. 101.

3.14 Synthesis

The discussion in Sects. 1.3 and 1.4 argued that at least part of the human experience of rhythm is a mental phenomenon or ordering that is not explicit in the sound. The beat, for instance, may or may not be physically present, even when it is readily perceptible. Similarly, in the dance music of Latin America, the clave is often implicit in a song rather than directly sounded in the performance. The timelines of Africa serve a similar function by framing the use of time, yet they may be unheard in the music. The inner melody of the Indonesian gamelan emphasizes the importance of the unsounded mental phenomenon. As Western musicians use meter, Indians use *tala*, Africans use the timeline. Each provides an internalized frame on which repetition and variation hang, for both listeners and performers. Thus rhythm is an emergent property, a product of consciousness. Many of the most important rhythmic structures are present only in the minds ear. Though they may be perceived quite clearly, they do not exist objectively in the sound. All music has meaning that goes beyond the sound.

This presents in stark form one of the major problems confronting any nonhuman or "artificial" intelligence that attempts to parse a signal created by humans for other humans. Only part of the information needed to understand the meaning of the signal is actually included in the signal; the rest is hidden in the social context of the communication, in the human perceptual apparatus, or in some common web of meaning enabled by human biological commonalities. Perhaps this is why computer-based speech recognition has achieved only limited success even after concerted effort by the scientific community for over half a century. It cannot be said that we have failed at this task for lack of effort! Similarly, efforts to automatically parse an audio performance and transform it into standard musical notation have not been fully successful. There are many tasks that humans can solve effortlessly that foil even the best efforts of the computational community.

What is universal in the human conceptualization and realization of musical rhythm? Surveying the musicological and ethnomusicological work summarized in this chapter leads to the conclusion that the answer primarily involves mental constructs. The listener is an active participant in a musical performance, often supplying an internal rhythm, beat, clave, timeline, melody, etc., that makes the rhythm intelligible. Indeed, this is why listeners may find music from other cultures confusing; they lack the appropriate mental framework with which to hear the meaning that the performer can so easily convey to the intended audience.

These cautionary notes show that from a computational point of view, it is important to distinguish the portion of the human rhythmic experience that is contained in the audio signal from the portion that is not. Those parts that are available in the waveform will be the focus of the remainder of *Rhythm and Transforms*. Those parts that are not resolvable without added cultural or biological information will be left for another day. Clearly, the "musical universals" discussed above are *not* part of the signal.

In order to focus on the possibilities for signal analysis, observe that almost all of the rhythmic notations, perceptions, and conceptualizations involve an underlying steady pulsation. This often occurs at a level below (faster than) the nominal beat. This level appears in the timecycles of the necklace notation as the equidistant timepoints arranged around the circles. It often appears in the musical notation in the time signature, though it may also occur at a more rapid rate that is a factor of two multiple of this basic rate. In the drum, binary, and functional notations, it appears as the fundamental time rate represented by the symbols. The many examples from around the world suggest that the kinds of music which are based on steady pulsation are a common part of our musical heritage.

This does fall considerably short of universality, however. Indeed, music that is fundamentally based on poetry, chant, or plain language tends to contain very complex stress patterns that may not conform to an underlying pulsation. Music that is built around "flexible breath" (such as the Japanese *gagaku* and Buddhist chant) or music that is primarily poetic (such as liturgical chants, the Ancient Greek music discussed by Plato, and the speech-like free rhythms of Africa) will not be good candidates for automated rhythmic analysis. We must, therefore, approach such problems cautiously.

4

Auditory Perception

The auditory system is not simple. Underlying the awareness of rhythmic sounds are basic perceptual laws that govern the recognition of auditory boundaries, events, and successions. Research into the mechanisms of perception sheds light on the physical cues that inspire rhythmic patterns in the mind of the listener. These cues help distinguish features of the sound that are properties of the signal from those that are properties of the perceiving mind. The beat is not in the signal; it is in your mind.

Auditory perception is the ability to make sense of sonic information, to organize successive sounds into coherent patterns. A sound begins as a physical disturbance in the air which propagates to the ear. It is then transduced into complex neural stimuli that are analyzed, selected and categorized into events with meaningful properties. The events enter short term memory where they can be related to other events and clustered into larger units and groups. Figure 4.1 shows the interactions between the various elements of the mind as it tries to understand a sound.

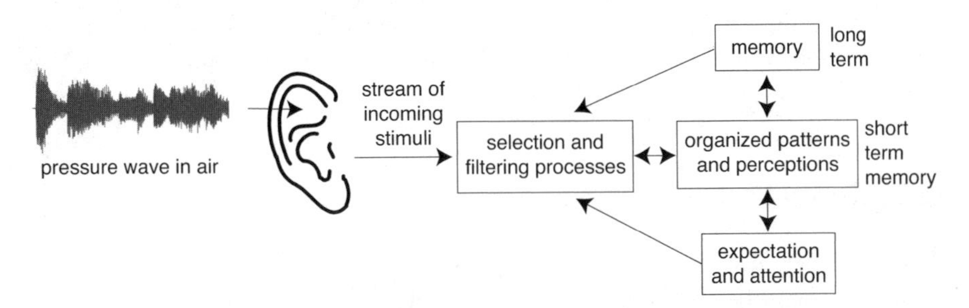

Fig. 4.1. Perception of sound is not a simple process; it begins with a physical waveform and may end with a high level cognitive insight (for example, understanding the meaning of a sound). There are constant interactions between long term memory, attention and expectation, and the kinds of patterns formed. There are also constant interactions between memory, attention, expectation, and the ways that the raw information is selected and filtered. The time span over which the short term memory organizes perceptions is called the *perceptual present*.

The selection and filtering operation takes a large amount of continuously variable data and condenses it into a (relatively) small number of events from a (relatively) small number of categories. Short term memory, where the patterns and perceptions are organized, does not operate using only input from lower level processing. Indeed, short term memory continually interacts with longer term memory and with consciousness (which appears in the figure as "attention" and "expectation"). Expectations and memories can influence the formation of patterns. Moreover, memory and expectation continually interact with the basic filtering and selection operations, and thus may influence what kinds of events are forwarded to short term memory. For example, someone expecting to hear a friend's voice is more likely to perceive that voice in a crowd. Someone anticipating a certain kind of musical phrase is more likely to perceive it, whether it occurs in reality or not.

Observe that the arrows leading from the memory and attention processes to the filtering and selection processes are unidirectional. Neither the conscious mind nor the long term memory directly access the raw sensory data. Indeed, we do not perceive a puff of air followed rapidly by a near sinusoidal tone; we hear someone playing a flute. We do not perceive a steady stream of harmonic tones passing through a time varying filter and punctuated with a series of glottal stops; we perceive spoken words.

Thus perceptions of musical notes, spoken words, melodies, and rhythms are acts of cognition, by-products of a listening mind; they are not direct sensory inputs as they might naively appear. Since they all unfold over time, they must involve some sort of memory. Since they are greatly simplified from the raw sensory data, they must be filtered and categorized with some kind of grouping or clustering mechanism. In addition, they influence and are influenced by our attention. Expectation projects current sensory impressions into the future. This anticipation is an act of (perceptual) imagination. Untangling the complex web of dependencies in our perceptual mechanisms is one of the great scientific adventures of our time.

4.1 How the Ear Works

The auditory system is a massively complex structure connected to an even more sophisticated processing unit. This section gives a simplified view of the operation of the ear that stresses some of the attributes that will be most useful in subsequent chapters when parsing musical signals for rhythmic patterns.

To the physicist, sound is a pressure wave that propagates through an elastic medium (i.e., the air). Molecules of air are alternately bunched together and then spread apart in a rapid oscillation that ultimately bumps up against the eardrum. When the eardrum vibrates, signals are sent to the brain, causing the perception of sound. Thus the word "sound" means one thing to the physicist (a waveform propagating through air) and quite another to the psychologist (a perception in the mind of a listener). As a practical matter, it is important to distinguish the physical attributes of a signal from the perceptions that the signal evokes:

(i) *frequency* is the physical correlate of the perception of *pitch*
(ii) *sound pressure level* (signal power) is the physical correlate of the perception of *loudness*
(iii) *waveshape* is the physical correlate of the perception of *timbre*
(iv) a *time interval* is the physical correlate of the perception of *duration*

The physical attributes are measurable properties of the signal whereas the perceptions are responses in the mind of the listener.

Sound waves can be pictured as graphs such as the sinuous oscillation in the upper left part of Fig. 4.2 where high-pressure regions are depicted lying above the nominal, and low-pressure regions are shown below. This particular waveshape, the *sine wave*, is characterized by three mathematical quantities: frequency, amplitude, and phase. The frequency of the wave is the number of complete oscillations that occur in one second. For example, a sine wave with a frequency of 100 Hz (short for Hertz, after the German physicist Heinrich Rudolph Hertz) oscillates 100 times each second. In the corresponding sound wave, the air molecules bounce back and forth 100 times each second.

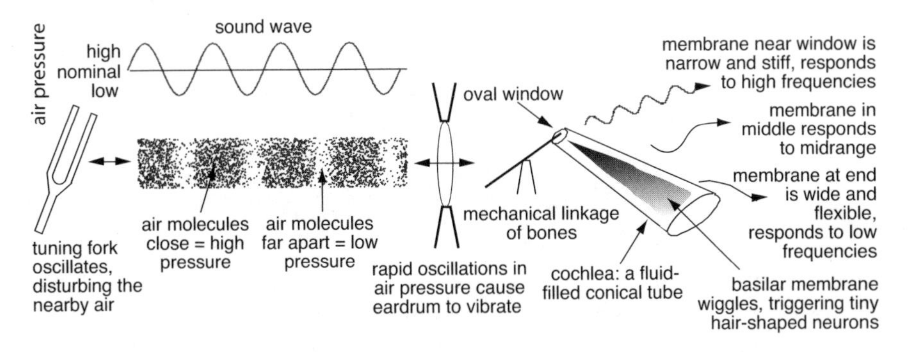

Fig. 4.2. The auditory system transforms a pressure wave into a frequency-selective spatial array. Peaks in the sound wave represent the clustering of air molecules and times when the pressure is high. Valleys represent lower than nominal pressure. The wave pushes against the eardrum during times of high pressure and pulls (like a slight vacuum) during times of low pressure, causing the drum to vibrate. A series of mechanical linkages connect to the oval window, which sits on the surface of the snail-shaped cochlea. As the window oscillates, fluid in the cochlea causes the basilar membrane to vibrate and wiggle like a flag flapping in the wind. This in turn causes tiny hair cells mounted on the membrane to vibrate and the waving of the hair cells is translated into tiny electrical impulses that stimulate neurons in the brain. Hair cells near the base of the membrane are perceived as having high pitch while those near the wider end are perceived as having low frequency.

Figure 4.2 shows a drastically simplified view of the auditory system. Sound or pressure waves, when in close proximity to the eardrum, cause it to vibrate. These oscillations are translated to the *oval window* through a mechanical linkage consisting of three small bones. The oval window is mounted at one end of the cochlea, which is a conical tube that is curled up like a snail shell (although it is straightened out in the illustration). The cochlea is filled with fluid, and it is divided into two chambers lengthwise by a thin layer of pliable tissue called the basilar membrane. The motion of the fluid rocks the membrane. Tiny hair-shaped neurons sit on the basilar membrane, sending messages toward the brain when they are jostled. The region nearest the oval window responds primarily to high frequencies, and the far end responds mostly to low frequencies. The high frequencies are perceived as having high pitch and the low frequencies are perceived as having low pitch. The phase of the sine wave essentially specifies when the wave starts with respect to some arbitrarily given starting time. For single sine waves, the phase has little effect on the sound.

4.1.1 Perception of Loudness

The ear reacts to variations about the nominal pressure (one atmosphere of pressure at sea level). Waves with larger fluctuations are generally perceived as louder, while waves with small deviations from nominal are heard as softer. To a first approximation, the ear responds to the sound pressure level (SPL), which is proportional to the average of the square of the pressure amplitude. Thus, a crude measure of the "loudness" of a sound is to calculate the power

$$p = \frac{1}{T}\int_{t}^{t+T} x^2(t)dt \quad \approx \quad \frac{1}{N}\sum_{i=1}^{N} x^2[i] \tag{4.1}$$

where $x(t)$ represents the amplitude of the waveform from time t to $t+T$ and the N values $x[i]$ represent a sampling of this wave.

For example, Fig. 4.3 shows a waveform constructed by alternating between one second of random numbers drawn from a Gaussian distribution and one second of random numbers drawn from a uniform distribution. The two processes are scaled so that each has power p equal to one. Clearly, the two *look* very different. But listen to sound example [S: 32]. It appears to contain a single continuous noise. The amplitude differences that are so prominent to the eye are inaudible to the ear.

Unfortunately, a complete characterization of loudness is not as simple as [S: 32] suggests. First, loudness depends on frequency: a sinusoid at 50 Hz must be much larger (be more powerful) than a sine wave at 500 Hz to be perceived as equally loud. Similarly, at very high frequencies, greater power is required. The "equal-loudness" contours of Fletcher and Munson [B: 183] quantify these observations.

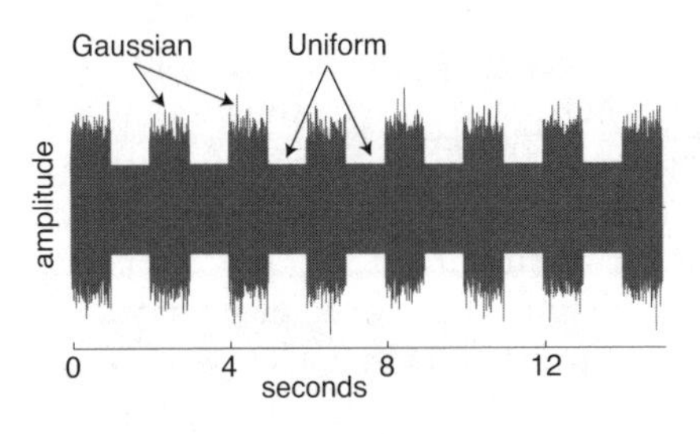

Fig. 4.3. Each second, the distribution of the noise changes, alternating between Gaussian and Uniform. The amplitudes are very different. This can be heard in example [S: 32], which appears to the ear as a steady undifferentiated noise.

4.1.2 Critical Band and JND

As shown in Fig. 4.2, sine waves of different frequencies excite different portions of the basilar membrane, high frequencies near the oval window and low frequencies near the apex of the conical cochlea. Early researchers such as Helmholtz [B: 94] believed that there is a direct relationship between the place of maximum excitation on the basilar membrane and the perceived pitch of the sound. This is called the "place" theory of pitch perception. When two tones are close enough in frequency so that their responses on the basilar membrane overlap, then the two tones are said to occupy the same *critical band*. The place theory suggests that the critical band should be closely related to the ability to discriminate different pitches. The critical band has been measured directly in cats and indirectly in humans in a variety of ways as described in [B: 168, B: 252]. The "width" of the critical band is roughly constant at low frequencies and increases linearly at higher frequencies, as shown in Fig. 4.4.

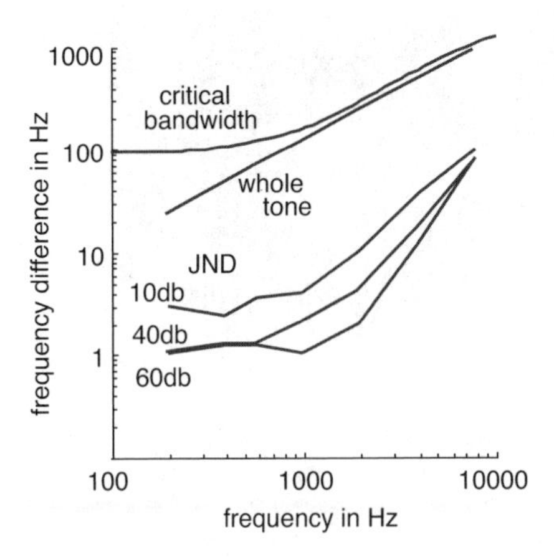

Fig. 4.4. Critical bandwidth is plotted as a function of its center frequency. Just Noticeable Differences at each frequency are roughly a constant percentage of the critical bandwidth, and they vary somewhat depending on the amplitude of the sounds. The frequency difference corresponding to a musical whole tone (the straight line) is shown for comparison. Data for critical bandwidth is from [B: 185] and for JND is from [B: 240].

The Just Noticeable Difference (JND) for frequency is the smallest change in frequency that a listener can detect. Careful testing such as [B: 251] has shown that the JND can be as small as two or three cents, although actual abilities vary with frequency, duration and intensity of the tones, training of the listener, and the way in which JND is measured. For instance, Fig. 4.4 shows the JND for tones with frequencies that are slowly modulated up and down. If the changes are made more suddenly, the JND decreases and even smaller differences are perceptible. As the JND is much smaller than the critical band at all frequencies, the critical band cannot be responsible for all pitch-detection abilities. On the other hand, the plot shows that JND is roughly a constant percentage of the critical band over a large range of frequencies.

4.1.3 Models of the Auditory System

Computational models of the auditory system such as those of [B: 125, B: 146] often begin with a bank of filters that simulate the action of the basilar membrane as it divides the incoming sound into a collection of signals in different frequency regions. Figure 4.5 schematizes a filter bank consisting of a collection of n bandpass filters with center frequencies $f_1, f_2, \ldots, f_n$. Typical models use between 20 and 40 filters, and the widths of the filters follow the critical bandwidth as in Fig. 4.4. Thus the lower filters have a bandwidth of about 100 Hz and grow wider as the center frequencies increase. A convenient way of picturing the output of such a filter bank is in a spectrogram-style plot which orients frequency on the vertical axis and the passage of time on the horizontal axis. The energy in each critical band is shown by the darkness of the shading.

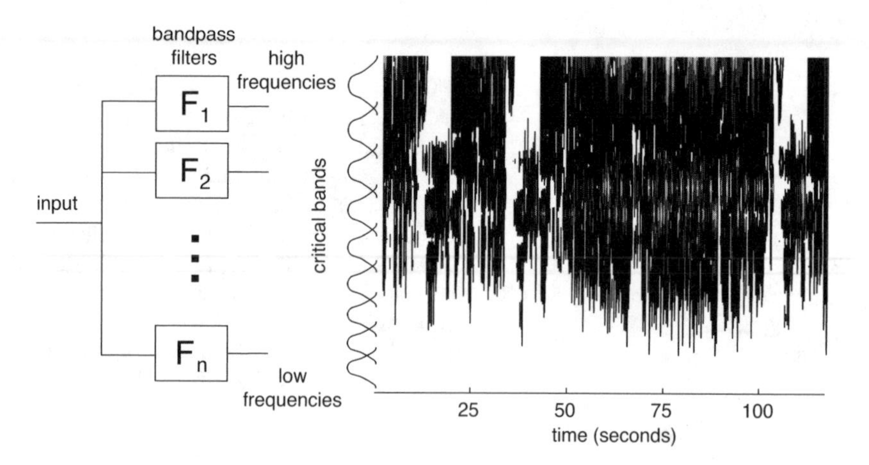

Fig. 4.5. The n filters separate the input sound into narrowband signals with bandwidths that approximate the critical bands of the basilar membrane. The output of the filterbank can be pictured in spectrogram-style where the energy in each critical band is indicated by the depth of the shading. Time evolves from left to right. This example shows about two minutes of the *Maple Leaf Rag*. [S: 5].

There are several reasons why the place theory cannot provide a complete model for the perception of pitch. There are no more than about forty critical bands on the basilar membrane. Yet even a piano has 88 keys, and it is obvious that people can distinguish far more pitches than this. The place theory also provides no possible explanation of phenomena like the missing fundamental (recall this "illusion" was illustrated in Fig. 1.7 on p. 12).

An alternative hypothesis, called the "periodicity" theory of pitch perception suggests that information is extracted directly from the time behavior of the sound. Neurons send electrical signals to each other in the form of sequences of pulses called spike trains. Detailed investigation into the neurons of the auditory system (these experiments are often conducted in the brains of anaesthetized cats, which have auditory systems similar to our own) show that the spike trains can synchronize to the waveform. For example, Fig. 4.6 shows a spike train that is phase-locked to an intrinsic feature of the waveform, in this case, the instantaneous peak values. This synchronization can occur reliably for frequencies up to a few hundred Hertz, but drops off at higher frequencies. Thus, the time interval over which a signal repeats is a feature that the auditory system may use to determine frequency.

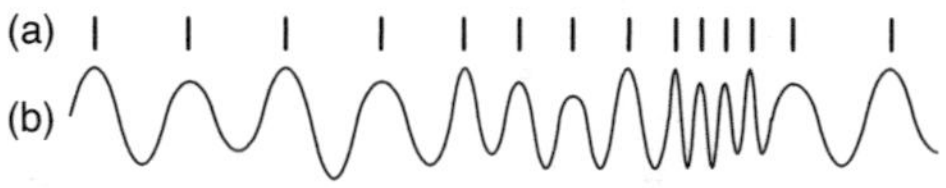

Fig. 4.6. The spike train (a), representing the firing of neurons, is synchronized to the maxima of the sound waveform (b)

There are several kinds of models that try to exploit temporal relationships between spike trains [B: 146, B: 159]. In reality, the pulses from individual neurons are erratic, but there are many neurons, and so it is common to build models that rely on the distribution of populations of neurons. Among the models that attempt to deduce pitch (and other) information from the intervals between collections of spike trains are connectionist networks and time-delay networks. Cariani [B: 26] has recently proposed *neural timing networks* such as in Fig. 4.7 as a possible structure.

Two spike trains pass in opposite directions. The nonzero time required for the signal to pass between adjacent neural elements causes the signals to be delayed. The coincidence detectors fire whenever spikes from both signals are present simultaneously, and the output of each detector is another spike train that moves downwards. Counting the spikes within any horizontal strip gives the convolution of the two signals. Counting the spikes in any vertical strip gives the cross-correlation at that delay time. Cariani observes that the array acts as a "temporal sieve" that passes temporal patterns common to both signals. The output is phase insensitive (like the ear) and can reproduce pitches associated with the missing fundamental. In certain simple situations it can recognize common pitches irrespective of the timbre, and recognize common timbres (in this case, certain vowel sounds) irrespective of pitch. In

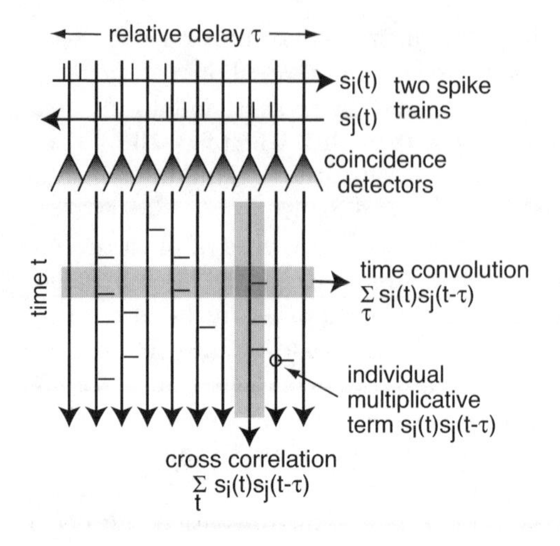

Fig. 4.7. This neural timing network (figure is used with permission from [B: 26]) can calculate the cross-correlation and the convolution between two signals (labeled $s_i(t)$ and $s_j(t)$) by summing up either vertically or horizontally. The resulting temporal patterns can mimic the perception of pitches, reproduce the missing fundamental, and recognize a common pitch independent of timbre (waveshape) and a common timbre independent of pitch.

addition, such period detection models can duplicate features of the auditory system like the pitch associated with time delayed pulse trains [B: 216].

There is now, and has been for the past 100 years or so, considerable controversy between advocates of the place and periodicity theories. It is probably safe to say that there are some aspects of pitch perception better explained by the place theory, and other factors better explained by the periodicity theory. Indeed, Pierce [B: 164] suggests that both mechanisms may operate in tandem, and a growing body of recent neurophysiological research reinforces this. Whatever the ultimate resolution of this argument, both kinds of models provide useful insights into the operation of the perceptual apparatus. The next sections turn to two central questions. What are the basic kinds of "events" that perception decodes, and what are the properties of these events?

4.2 Auditory Boundaries

People can perceive a wide range of auditory phenomenon and can distinguish many kinds of auditory stimuli. Moore [B: 150] points out that "the auditory system seems particularly well-suited to the analysis of changes in the sensory input." The idea of a boundary between two sounds attempts to pinpoint times at which changes are perceived in the audio stream.

> An *auditory boundary* occurs at a time t when the sound stimulus in the interval $[t - \epsilon, t]$ is perceptibly different from the sound stimulus in the interval $[t, t + \epsilon]$.

Boundaries are quite general, and they may occur on different time scales depending on the size of ϵ. For example, long scale auditory boundaries (with ϵ on the order of tens of seconds) occur when a piece of music on the radio is interrupted by an announcer, when a car engine starts, and when a carillon begins playing. Short scale auditory boundaries (with ϵ on the order of tenths of a second) occur when instruments change notes, when a speaker changes from one syllable to the next in connected speech, and each time a hammer strikes a nail. At yet smaller values of ϵ (on the order of milliseconds) the "grains of sound" [B: 181] occur too rapidly to be perceived individually. They merge into a perception of continuity without boundaries.

Perhaps the most common examples of short time scale auditory boundaries involve changes in amplitude (or power) such as occur when striking a drum. Before the strike, there is silence. At the time of the strike (and for a short period afterwards) the amplitude rises sharply, causing a qualitative change in the perception (from silence to sound). Shortly afterwards, the sound decays, and a second, weaker boundary is perceived (from sound into silence). Of course, other aspects of sound may also cause boundaries. For example, pitch (or frequency) changes are readily perceptible. An auditory boundary might occur when a sound changes pitch, as for example when a violin moves from an A note to an $A\sharp$. Before the boundary, the perception is dominated by the violin at fundamental $A = 440$ Hz while after the boundary the perception is dominated by the sound of the violin playing the $A\sharp = 466$ Hz. On the other hand, boundaries do not necessarily occur at all pitch changes. Consider the example of a pitch glide (say an oscillator sweeping from 100 Hz to 1000 Hz over a span of thirty seconds). While the pitch changes continuously, the primary perception is of the "glide" and so boundaries are not perceived except for the longer scale boundaries at the start and stop of the glide.

The perception of a change that occurs at an auditory boundary is called an *auditory event*, or, when the context is clear, an *event*. The preceding examples highlight several aspects of auditory boundaries and events. First, the boundaries and their related events may be of different strengths. Second, they may be caused by different kinds of physical phenomena which may correspond to different aspects of perception. Finally, the key phrase "perceptibly different" is not always transparent since exactly *which* aspect of perception dominates at any given time is a complex issue that depends on the training of the listener, on the focus of the listener's attention, and on a myriad of physical factors.

Isolated auditory boundaries divide time into before and after. They are typically perceived as the starting or stopping times of large scale events. For example, a refrigerator motor starts, creating a clearly perceptible boundary. Over time, the ear acclimates and the sound recedes into the background. When the motor stops, another boundary is formed and it is immediately clear that there is no longer any sound. The two boundaries delineate regions before the motor started, during the motors operation, and after the motor has ceased. There is no direct perceptual connection between the events, nor of the duration between the events, though of course it may be possible to estimate the time interval based on other intervening factors.

If a pair of events is separated by a shorter time interval (between about 100 ms and 3 s), then the interval is typically perceived as a duration; the two events become "connected" and the time between is perceived as the content of the duration. Durations may be empty (without sound) or filled (with sound). Either way, the events are perceived as bound together. For example, a drum hit twice in rapid succession; it is heard as a single entity *boom-boom* rather than one boom followed by another. Similarly, a flute plays a C note. The initiating event is the puff of air that begins the sound. The ending event is the return to silence. The content of the duration is the tonal, steady sound of the resonance of the flute body.

If the pair of events is separated by somewhere between 20 ms and 100 ms, it is typically perceived as a single (more complex) event; though the events are still distinguishable, there is no perception of duration, and it is often impossible to tell reliably which event occurred first. Finally, if the separation is less than 2 ms, then it becomes impossible to tell that two separate events have occurred; they have merged into a single perception.

With a single event there is a partitioning of time into before and after. With two events the perception of duration arises. Something even more amazing happens when there is a whole sequence of events!

4.3 Regular Successions

A new phenomenon emerges when auditory boundaries occur periodically or in a regular sequence: the perception of succession. Fraisse comments [B: 65]:

> The organization of successive elements into units of perception is such a fundamental part of our experience that we no longer notice it. It is the basis of our perception of rhythm, of melody, and even the sounds of speech.

A *regular succession* is a sequence of auditory boundaries or events that recur at (approximately) equal intervals of time. For example, suppose that a series of audio boundaries occur at times

$$\tau, \ \tau + T, \ \tau + 2T, \ \tau + 3T, \ \tau + 4T, \ldots \tag{4.2}$$

where T is the time between adjacent audio boundaries and τ specifies the starting time of the first event. Actual sounds may not be so precise: T might change (slightly) between repetitions, some of the $\tau + nT$ terms might be missing, and there may be some extra auditory boundaries interspersed among the T-width lattice. Thus, (4.2) is idealized and the term *regular succession* is used to emphasize that a sequence of auditory boundaries need not be strictly periodic.

4.3.1 Perceptions of Rate

Musicians and listeners often tap their foot to a piece of music. The word *beat* refers to times when the foot taps, and *tempo* refers to the rate at which the beats recur. In a simple sound sequence like *Regular Interval 750* [S: 3], each click typically corresponds to a tap and hence to a beat. Since the interval between beats is 750 ms, the tempo is 1.33 bps (beats per second) or 80 bpm (beats per minute). In the faster *Regular Interval 333* [S: 33], where the time interval between clicks is 333 ms, listeners tend to tap their foot to every other click, to cluster the clicks into pairs, and to count each pair as a single beat. Perhaps the most common way to count this example is

1 & 2 & 3 & 4 &,

and to tap the foot on 1-2-3-4 (but not on the &s). Thus the click rate is different from the tempo. A more extreme version is *Regular Interval 100*, where the interval between clicks is 100 ms, corresponding to a rate of 10 clicks per second. Listeners may cluster the clicks together into groups of four (or more) and so each beat lasts 400 ms (or more).

Even though the stimulus consists of exactly identical pulses following each other at identical intervals, the sounds are perceived as if they are grouped, usually in twos, threes, or fours. In the absence of any clearly articulated meter, the ear creates one, highlighting the spontaneous and idiosyncratic character of rhythmic groupings. Bolton [B: 17] showed that such *subjective rhythmization* occurs when the pulses are separated by (approximately) 0.1 to 2 s. Outside of this range (more rapidly than 0.1 s and more slowly than 2 s) the pulses are heard as identical.

Thus the perception of even a simple periodic series of events can change as the rate of presentation changes. What happens at even more extreme speeds? This question is explored in sound examples *Regular Intervals* T [S: 33], where T specifies the time interval between adjacent clicks for T between 2 ms and 5 s. Figure 4.8 shows that there are several perceptual regimes depending on the rate at which the clicks recur.

(a) A pitch is heard when the distance between successive pulses is less than about 50 ms.
(b) The clicks appear to be contiguous (no gap is heard between them). The overall effect of the sound is like the fast chirping of a cricket or the rapid ringing of an orchestral triangle.
(c) A rhythm is perceived between (about) 100 ms and 3 s. A gap is heard between successive clicks. The primary perception is of the regularity of a tatum, beat, or measure.
(d) There is no perceptual connection between the successive clicks when the interval between sounds is greater than about 3 s. In this region there is no direct perception of duration, though there is a sense of the distance between past and present events.

Sound example [S: 34] performs a slowly changing succession of clicks that covers most of these perceptual regions. Depending on the scale of the time parameter T, there are four different kinds of perceptions associated with regular successions: a sequence of disconnected events, perception of a rhythmic beat (at its corresponding tempo), perception of fluttering (at its corresponding rate), and the perception of a tone (at its corresponding pitch).

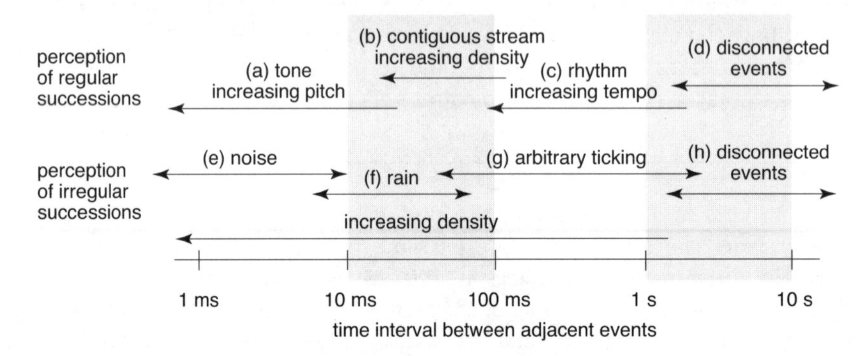

Fig. 4.8. Perception of a regular succession depends on the rate of repetition of the events; sensations range from (a) pitches, (b) contiguous clicks ("fluttering"), (c) beats, and (d) disconnected events. Perceptions of irregular successions also depend on the average rate of repetition of the events; sensations range from (e) noise, (f) "rain," (g) arbitrary clicks, and (h) disconnected events at the slowest rates. Arrows point in the direction of increasing pitch, speed, and density.

Regular successions are directly perceptible only within the perceptual present. Nonetheless, once the perception of a succession is established, it can readily extend beyond the present, both forwards and backwards in time. Thus the perception of a regular succession implies both an anticipation of future events (in particular, that the succession will continue) and it may project back into the past by reinterpreting prior sensory impressions. Said another way, once the "internal clock" of the listener has synchronized with the succession, there is an expectation that it will continue into the future, and previously occurring events may be retrospectively interpreted as leading towards the succession.

There is an inherent ambiguity in any periodic signal. If the sequence can be expressed as T periodic as in (4.2), then it is also $2T$ periodic (and nT periodic for any integer n). At the time scale of the beat, this ambiguity is the difference between the tatum, the beat and the measure (recall Fig. 1.9 on p. 15). At the time scale at which the perception of pitch occurs, an ambiguity of $2T$ corresponds to an octave jump.

What happens when the succession is not regular? *Irregular successions* are also perceived differently at different speeds as shown in Fig. 4.8.

(e) A noise is heard when the average distance between successive pulses is less than about 50 ms.
(f) The clicks appear to be contiguous (no gap is heard between them). The overall effect of the sound is like the sound of rain pounding on the roof.
(g) A collection of arbitrarily spaced clicks is perceived when the average interval between clicks is 200 ms to 2 s. A gap is heard between successive clicks.
(h) There is no perceptual connection between the successive clicks when the interval between sounds is greater than about 3 s.

The various perceptual regions for irregular successions are demonstrated in sound examples [S: 35] and [S: 36]. Observe that there is no perception of either tempo or of pitch as the rate of the events increases and decreases. Rather, the *density* of the sounds appears to increase and decrease. Thus irregular successions evoke qualitatively different kinds of perceptions than regular successions.

4.3.2 Regular Successions as a Single Perception

One of the most important forms of clustering is the ear's penchant for organizing the sound of a periodic waveform into "single sound" with a single pitch. Mathematically, a periodic sound is one which has all its overtones lying in a harmonic series, that is, a series in which there is a fundamental f and harmonics at $2f$, $3f$, $4f$, $5f$, etc. The perception when a sound changes from nonperiodic to periodic can be quite dramatic.

This is demonstrated in sound example [S: 37] which plays three sine waves, each sweeping linearly towards a harmonic series. The instantaneous frequencies of the sinusoids are given in Fig. 4.9. During the sweep, there are many complicated interactions between the sounds including difference tones and beating. But when the frequencies come into alignment with the harmonic series (with fundamental 200 Hz), the sound simplifies and merges into a single coherent object of perception. Observe how complicated the sound appears when the three sinusoids do not coalesce into a regular succession and how simple the sound appears once they have merged into a single entity.

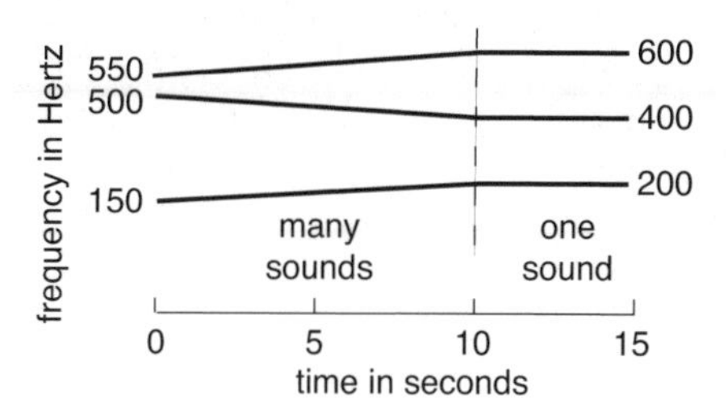

Fig. 4.9. The frequencies of three sine waves change linearly over time, moving towards the harmonic series at 200, 400, and 600 Hz. Many complex interactions occur during the sweep, but when they arrive at the harmonic series, they blend into a single note-like entity. Listen to [S: 37].

Such "single objects" are often called "notes" even though, strictly speaking, a "note" is a form of notation (recall Sect. 2.1.2) and not an auditory event. Thus periodicity (with its perceptual correlate pitch) is a perceptual cue that helps to glue auditory objects together. Even though a complex sound may contain several different fundamentals each with its own harmonics, the ear is often able to distinguish the number and pitches of the fundamentals. For example, trained listeners can readily discern the number of instruments playing at any moment in a string quartet.

Similar effects occur at different time scales, though they have different names. Figure 4.10 shows three click trains proceeding at three different rates. (a) begins at about 0.38 s per click, and, over the course of a minute smoothly changes its rate to 0.33 s per click. Similarly, (b) begins at 0.44 s per click and converges to 0.5 s, while (c) begins at 0.98 s per click and ends at 1 s.

Sound example [S: 38] performs four different versions of this. Each is 78 seconds long, and the synchronization (when all three rates have achieved their final values) occurs two-thirds of the way through. In the first, the synchronization occurs in both rate and phase, so that all three are aligned exactly as shown in Fig. 4.10. In the second and third, the relative phases of the three are arbitrary. In the fourth, the three sequences are performed on three different instruments (stick, clave, and tube).

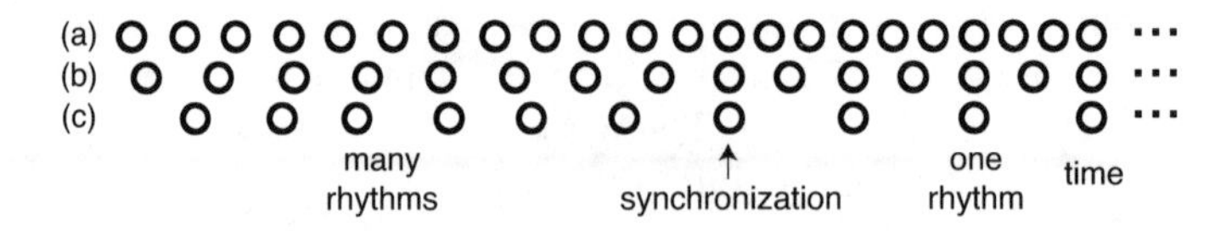

Fig. 4.10. The rates of three sequences of clicks change slowly over time, moving towards a single periodic repetition. Many complex interactions occur during the sweep, but when they synchronize at the periodic rate, they blend into a single rhythmic entity. Listen to [S: 38].

In all four cases, there is a qualitative change once synchronization occurs. Before, the hits seem random and unpredictable (though occasional short segments may achieve interesting rhythmic properties). After synchronization, there is a clear rhythmic pattern that repeats regularly. Again, the irregular successions correspond to complex perceptions where it is often not even clear how many simultaneous patterns are playing. There are always exactly three! When the periods align, the three patterns coalesce into a single readily perceivable rhythm. The first (phase aligned) case is the same as the three-against-two polyrhythm of Sect. 3.9 on p. 67 which can be heard in [S: 27]. The others correspond to various "rotations" of the timecycles (recall Fig. 3.11).

4.3.3 Perceptual Cues for Clustering of Notes

While periodicity is a strong factor, almost any perceptible feature of a sound can act as a clue for perceptual grouping. Intensity, and the change of intensity over time (often called the envelope of a sound) is also common. For example, the strike of a single note on the piano (or the strike of a bell) followed by the characteristic decay of the sound helps to fuse the partials of the string (or of the bell) into a single auditory object. Over time and with practice, such objects become familiar and easy to separate out from even a complex auditory stimulus. For example, people can readily distinguish the sound of a piano even when it is much quieter than an orchestra or band playing simultaneously.

Timbre is another factor that helps bind sounds together, and considerable research has been conducted to determine objective, measurable properties of signals that relate to the subjective notion of timbre. In a series of studies[1] investigating timbre, researchers generated sounds with various kinds of modifications, and asked subjects to rate their perceived similarity. A "multidimensional scaling algorithm" was then used to transform the raw judgments into a picture in which each sound is represented by a point so that closer points correspond to more similar sounds. The axes of the space can be interpreted as defining the salient features that distinguish the sounds. Attributes include:

(i) Degree of synchrony in the attack and decay of the partials
(ii) Amount of spectral fluctuation (coherent changes in the spectrum over time)
(iii) Presence (or absence) of high-frequency, inharmonic energy, especially in the attack
(iv) Bandwidth of the signal[2]
(v) Balance of energy in low versus high partials
(vi) Existence of formants[3]

One way to experiment with the perception of clustering is to generate collections of partials and ask listeners "how many notes" they hear.[4] Various features of the presentation reliably encourage tonal fusion. For instance, if the partials:

(vii) Begin at the same time (attack synchrony)
(viii) Have similar envelopes (amplitudes change similarly over time)
(ix) Are harmonically related (have the same fundamental period)
(x) Have the same vibrato rate (are modulated similarly)

---

[1] See [B: 57, B: 84, B: 85, B: 167].

[2] Roughly, the frequency range in which most of the energy lies.

[3] Resonances, which may be thought of as fixed filters through which a variable excitation is passed.

[4] This is an oversimplification of the testing procedures actually used by Bregman [B: 18] and his colleagues.

then they are more likely to fuse into a single perceptual entity. Almost any common feature of a subgroup of partials helps them to be perceived together. Perhaps the viola attacks an instant early, the vibrato on the cello is a tad faster, or an aggressive bowing technique sharpens the tone of the first violin. Any such quirks are clues that can help distinguish one instrument from another. Familiarity with the timbral quality of an instrument is also important when trying to segregate it from the surrounding sound mass, and there may be instrumental "templates" acquired with repeated listening.

The fusion and fissioning of sounds is easy to hear using a set of wind chimes with long sustain. I have a very beautiful set made of hollow metal tubes. When the clapper first strikes a tube, there is a "ding" that initiates the sound. After several strikes and a few seconds, the individuality of the tube's vibrations are lost. The whole set begins to "hum" as a single complex tone. The vibrations have fused. When a new ding occurs, it is initially heard as separate, but soon merges into the hum. This can be heard in [S: 39]. The same kind of separation can occur even for a single chime: [S: 40] repeats the sound of one chime at random intervals.

This section has discussed the features of sounds that allow individual auditory objects such as the notes produced by musical instruments or the phonemes of speech to be perceived as single entities distinct from the surrounding sound environment. The next section looks at the same kinds of questions at a longer time scale.

4.3.4 Perceptual Cues for Clustering of Rhythms

Suppose that the repetitions in a succession are not identical. Almost any kind of perceivable difference between the repetitions can induce a grouping effect. For example, suppose every third pulse is louder. Then the listener perceives the sequence as 1-2-3 1-2-3 1-2-3. If every fourth pulse is louder, the listener perceives 1-2-3-4 1-2-3-4. Of course, volume is not the only kind of difference that will effect grouping. Durations also play a role. For example, if every third pulse is followed by a short pause then groups of three are formed. If the pause occurs after every fourth pulse, groups of four are perceived. The way a sequence appears to be organized can be influenced by changes in

(i) Intensity: an increase in the volume of a pulse tends to define the start of a group
(ii) Duration (of the pulse): a lengthening of the pulse tends to define the start of a group
(iii) Duration (between pulses): a lengthening of the interval between two sounds tends to define the end of a group
(iv) Timbre: changes in tone quality (for example, changes in instrumentation) may signify key positions in the sequence
(v) Pitch: jumps in pitch tend to be heard as starting or ending points
(vi) Density: the number of events per second can signal a boundary

At first glance, (ii) and (iii) may appear to be contradictory, but in (ii) the increase in pulse length is perceived as analogous to an increase in intensity, while in (iii) the increased duration is perceived as analogous to a pause.

Figure 4.11 shows an example of grouping induced by timbre: □ represents the tone of a synthetic trumpet while ▲ is a synthesized flute. The example can be heard in [S: 41]. When played slowly (one note per second), this sounds like a three note ascending phrase that repeats regularly. But as it speeds up the two timbres break apart, each forms its own perceptual stream. Each of the melodies appears to be a descending pattern that occurs at half speed. There is also an interesting tempo in between where the pattern can be heard either way, depending on the focus of the listener. Similar examples of sound patterns reconfiguring at different tempos can be constructed from any of the kinds of features (i)–(vi) above. Important recent work in this area can be found in [B: 18].

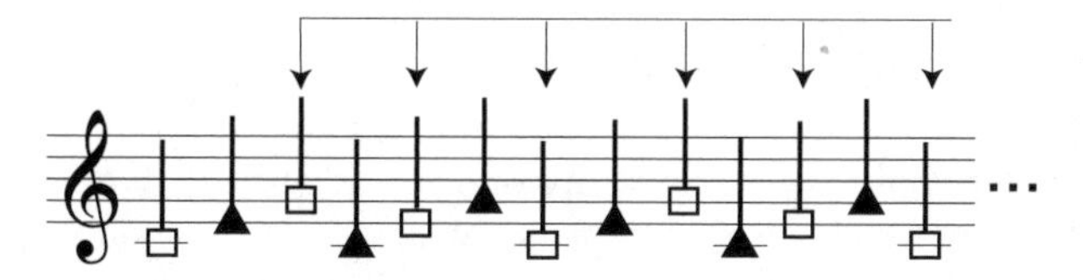

Fig. 4.11. The hollow squares and filled triangles represent two different timbres. In sound example [S: 41] they are a synthetic trumpet and a synthetic flute. When played slowly, the perception is just as it would appear: a repetition of an ascending major chord. When played rapidly, the sound breaks into two perceptual streams where each instrument plays it "own" descending melodic line. The line sounded by the □s is emphasized by the arrows.

4.3.5 Filled vs. Empty Durations

There are two basic ways to interpret a rhythmic passage; as a sequence of individual auditory boundaries (events) or as a sequence of durations. Though logically equivalent (since a duration is defined between any two adjacent events), it is important to ask if they are perceptually the same. This is illustrated in Fig. 4.12.

In a regular succession of short clicks, the two interpretations are identical since the time between the boundaries is empty, that is, the duration consists solely of the silence between clicks. A regular succession of filled durations, on the other hand, can be easily constructed using two different sounds A and B (in the notation of Fig. 4.12). Are successions of durations different from successions of events?

This question is explored in the sound examples *Regular Durations T* [S: 42] where T specifies the lengths of the durations for a variety of T between 2 ms and 5 s. In the examples, sound A is the pluck of an electric

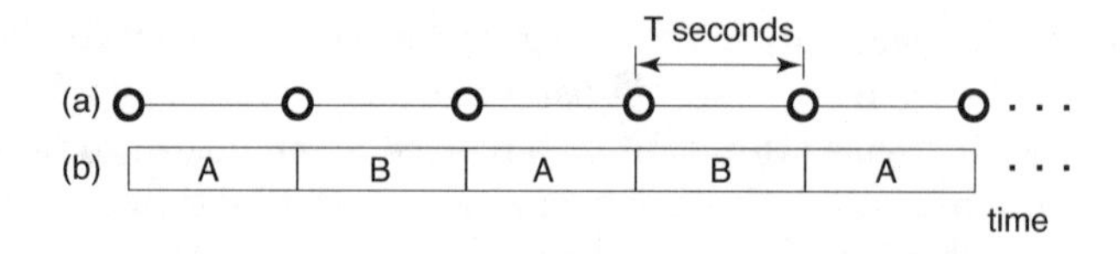

Fig. 4.12. In (a), the dark circles represent clicks or pulses separated by T seconds of empty (silent) time. In (b), A and B represent durations that are filled with two different sounds. The horizontal axis represents the passage of time. Rhythmic perceptions can be evoked by both filled and empty sound sequences.

guitar string and B is a note from a synthesizer. There are several perceptual regimes depending on T. These are the same as the regimes (a)–(d) of Fig. 4.8 for the regular succession of clicks: pitch, fluttering, beats, and isolated (disconnected) sequences. [S: 43] performs a slowly changing succession of durations that cover most of the above perceptual regions in a single sound example.

Thus, while the moment by moment perceptions are quite different when comparing regular successions of events and regular successions of durations, the kinds of rhythmic perceptions invoked are identical. It does not matter what the time is filled with, it matters when changes (boundaries) occur. Whether the duration consists of silence, a guitar pluck, a voice, or a synth, the primary perception is of the succession itself, not the constituent elements.

The same is also true of irregular successions of durations; the perceptual regimes mimic the regimes of irregular successions of events. As in (e)–(h) of Fig. 4.8, it is the density of the sound that most obviously increases as the lengths of the durations decrease. The various perceptual regions for irregular successions are demonstrated in sound example [S: 44]. The normal probability distribution is used to specify the durations of the events. The durations begin with a large (average) time interval, speed up until they have achieved an average rate of T durations per second, and then slow down again. T assumes the values 40 (for an average of 25 ms between events), 400, and 4000 durations in each second.

Observe that there is no perception of either tempo or of pitch as the rate of the durations increases and decreases. Rather, the *density* of the sounds appears to increase and decrease. As expected, irregular successions evoke qualitatively different kinds of perceptions from regular successions. But irregular successions of durations evoke qualitatively the same rhythmic perceptions as irregular successions of events. Again, it does not matter what the time is filled with, it matters when changes (boundaries) occur and the primary perception is of the succession itself, not the constituent elements.

4.3.6 Framework for Rhythm Perception

Regular successions provide the basis of Povel's "framework for rhythm perception" [B: 173] by assuming that listeners use an internal clock as a basis on which to build their perceptions. As expected from earlier studies (recall the identical clicks of example [S: 3] on p. 6), the completely regular pattern in Fig. 4.13(a) is heard (ambiguously) as grouped in twos, threes, fours, or more, depending on the listener, the speed of presentation, and a variety of other factors. When some of the hits are removed as in (b) and (c), the sequence disambiguates: (b) is heard as irrevocably grouped in threes while (c) is unalterably perceived in groups of four. In contrast, the irregularities in (d) cause it to be difficult for subjects to remember and reliably reproduce. Listen for yourself in [S: 45].

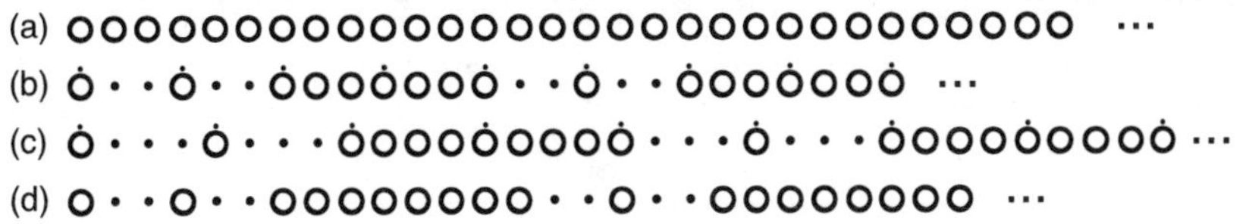

(a) o o o o o o o o o o o o o o o o o o o o o o o o o o o o o o ⋯
(b) ȯ • • ȯ • • ȯ o o ȯ o o ȯ • • ȯ • • ȯ o o ȯ o o ȯ ⋯
(c) ȯ • • • ȯ • • • ȯ o o o ȯ o o o ȯ • • • ȯ • • • ȯ o o o ȯ o o o ȯ ⋯
(d) o • • o • • o o o o o o o o • • o • • o o o o o o o o ⋯

Fig. 4.13. The completely regular succession (a) is grouped ambiguously. Removing certain events as in (b) and (c) removes the ambiguity. Removing events as in (d) leaves the sequence irregular and hard to accurately reproduce. These can be heard in [S: 45].

Povel models this effect using regular successions as an organizing principle. Of all possible temporal grids, the one with the "most economical description" is chosen as the best descriptor of the sequence. A grid is more economical if it covers more taps, is less economical if it predicts taps where none occur, and there is also a penalty for missed taps. For example, a grid that hits every third element of (b) and one that hits every fourth element of (c) do very well. These are indicated by the tiny black dots above the circles in Fig. 4.13. On the other hand, there is no economical grid that captures the essence of (d). When all goes well, the best grid corresponds to the regular tapping of a person listening to the sequence. Unfortunately, a formal definition of "most economical" that captures all the desired features is trickier than it might seem.

Povel and Essens [B: 174] hypothesize the presence of an internal clock that measures the passage of time. When hearing a rhythmic passage, the listener compares the sound to the beating of the internal clock. In this view, the essence of the perception lies in the relationship between the sound and the clock, between the regular succession within the sound and the regular succession of the internal clock.

Similarly, Parncutt [B: 158] suggests a measure of pulse saliency that models how listeners tap when confronted with simple sound patterns that vary in both rhythm and tempo. The salience of a single beat is proportional to its perceptual significance, and the regular succession with the highest salience is defined to be the beat (the rate at which a listener will tap) of the rhythmic sequence. Parncutt [B: 158] provides one of the clearest and most concise definitions of musical rhythm in terms of regular successions:[5]

> A *musical rhythm* is a sound that evokes the sensation of a regular succession of beats.

In this definition, rhythm is clearly a psychological phenomenon (a "sensation") evoked by the perception of a regular succession at a time scale where beats are perceived. Similarly, the American Standards Association [B: 7] defines a *tone* as a sound that evokes the sensation of pitch. Since pitch is the perception of a regular succession at small time intervals, the two definitions are parallel, differing mainly in time scale. Moreover, both definitions agree, roughly, with common usage. Like Povel and Essens' model of clock induction, this is an attempt to model rhythmic perception using a regular succession (the clock) as a part of the cognitive (and/or perceptual) process. See [B: 202] and [S: 60] for further discussion of the Povel and Essens model.

4.3.7 A Rhythmic Theory of Perception

Regular successions provide a concrete way to talk about the kinds of rhythmic activities that Mari Riess Jones [B: 103] views as fundamental to the way people interact with the world. In Jones' view, people

> possess their own temporal structures which are manifest psychologically in a series of tunable perceptual rhythms... by virtue of the principle of synchronization, [these rhythms] can become locked into corresponding periods in the time structure of world patterns.

Jones builds an axiomatic theory of perception which presumes that the mind can synchronize internal regular successions with those that exist in the world: fast successions synchronize with micropatterns (such as pitch percepts) while slow successions synchronize with macropatterns (such as rhythms). The rapid response of a listener (such as the ability to understand dozens of phonemes per second) arises from a priming of the appropriate rhythmic patterns. Moreover, since time intervals and durations cannot be judged apart from the sounds used to delineate them, Jones couples pitch and loudness to the temporal dimension. The resulting theory couples rhythmic activity to the most basic processes of the mind: perception, expectation, and attention.

Thus Jones views perception as a dynamic process characterized by an entrainment between the mind and the environment. This process defines (and is defined by) sensations of regular motion, creates expectations of future events, and allows continuity of perception through time spans larger than the perceptual present. Rhythm imposes temporal structure on cognition.

---

[5] Parncutt uses the word "pulse" instead of regular succession.

4.3.8 Rhythm Without Notes

It is tempting to view musical notes as individual entities each with a start, a steady state sustain, and a final decay into silence, since this is precisely how a musician creates a musical phrase on an instrument or with the voice. Everyday common sense suggests that loudness changes are a necessary part of the structure of musical rhythm. The following examples show, however, that rhythms can exist independently of loudness contours and of individually identifiable notes. As we have seen, there are a great number of factors that can cause sounds to cluster into perceptible entities.

> Any factors that can create auditory boundaries can be used to create patterns in time that can be perceived as rhythmic.

The examples demonstrate auditory boundaries without loudness changes using a variety of perceptual clues such as pitch, noise bandwidth, and modulations.

(i) *Changing only Pitch:* In [S: 46], the loudness is equal at all times. Pitches change every 0.25 s.
(ii) *Changing only Bandwidth:* In [S: 47], white noise is passed through a linear filter. The bandwidth of the filter changes every 0.25 s and the loudness is equalized throughout. The three different versions use different kinds of filters.
(iii) *Amplitude Modulation:* In the first part of [S: 48], a "rhythmic pattern" uses a single pitch that is amplitude modulated at different rates. The loudness of each 0.25 s interval is equalized, though amplitude modulations inherently involve changes in loudness within the sound. The second example applies the amplitude modulated contours to the pitches from [S: 46] and the noises from [S: 47]. These are called "smooth rhythms" in [B: 74].
(iv) *Frequency Modulations:* in [S: 49] the modulation on the carrier is changed every 0.25 s, causing the perception of a rhythmic pattern. The three examples use the same set of parameters but with different carriers: sine, square, and sawtooth waves. All the carriers are fixed at 300 Hz.

An extended exploration of these techniques is presented in *Pulsing Silences* [S: 50], which is taken from *Exomusicology* [D: 43]. The piece begins with a rich harmonic tone that is unvarying for three and a half minutes. A variety of time varying filters and modulations are applied. Some fairly complex rhythmic motifs appear. The CD booklet states: "A single note changes without moving, grows while remaining still. Even if there were only one note, there would still be music."

From a common sense perspective, it is tempting to think of a regular succession as a sequence of equidistant notes. This suggests, for example, that when searching a signal for the presence of a regular beat, it would be necessary to first identify the locations (in time) of the notes, and to then parse the note list to decipher the beat. This approach has figured prominently in the literature (see Chap. 8), and tends to work very well in situations where the notes can be accurately identified. Unfortunately, these methods tend to fail when the notes are not easily identified. This may limit the applicability of such systems to monophonic performances (where it is easy to identify note events) and to situations where musical scores and/or MIDI data is available.

Somewhat counterintuitively, it may be easier to identify the presence of a regular succession than to identify individual note events that define the succession. The beat tracking problem requires finding the best regular lattice of points (for instance, the τ and T in (4.2)) that fit the sequence of auditory boundaries evoked by the performance of a particular piece of music. It does not require the location of the individual note components.

To motivate this, Fig. 4.14 shows several seconds of a piano rendition of the *Maple Leaf Rag*. The waveform looks much like any musical waveform; it jiggles up and down in complex and mystifying patterns. Clearly, the eye is impressed by different kinds of features than the ear. Nonetheless, some aspects of the signal are easy to interpret. For example, some of the changes correspond to notes, which are transcribed from the musical score. At each of the three times labeled (a) there is a new note and there is a noticeable change in the waveform: it gets larger, it becomes more (or less) dense, it changes shape. But not all the notes are as clearly displayed. For example, the notes at positions (b) and (d) are nearly indistinguishable from the waveform alone.

Figure 4.14 also shows a feature vector (in this case, based on the change in the spectrum) that is calculated from the waveform. It has prominent values at many of the most significant events in the passage. Each of the notes labeled (a) has a large spike in the feature vector. Similarly, the note (d) appears prominently in the feature vector even though there is little visual change in the waveform. Perhaps even more importantly, points such as (f) and (c) stand out significantly from the surrounding data. These occur at a beat and a tatum point where there are no notes! Feature vectors can show aspects of a performance that even the note list does not.

Of course, no feature vector is perfect. This one shows extraneous pulses (at least from the point of view of notes, tatum and beats) at points such as (e).[6] While the feature vector does pinpoint the note (d) handily, the note (b) remains invisible. This occurs in this case because (b) is exactly one octave above the previous note (which is still sounding). (b) remains invisible among all the simultaneously sounding harmonics. Feature vectors are discussed further in Sect. 4.4.

Many auditory boundaries (and their related events) are apparent in this feature vector, and it is not hard to imagine that it is possible to process

[6] Careful listening reveals that the leftmost point (e) is caused by a soft click, probably from the pedal of the piano.

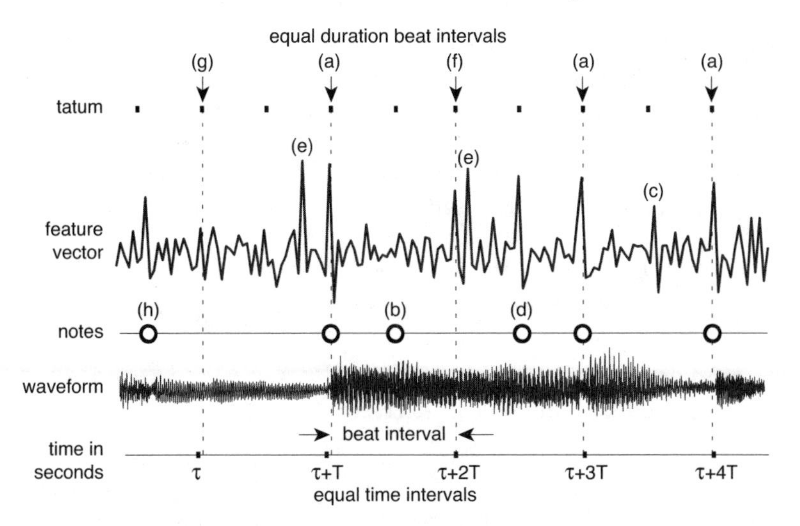

Fig. 4.14. The waveform shows about 3 s of the *Maple Leaf Rag*. The notes, as marked in the score, are placed at the appropriate times and the foot-tapping beat (marked with the down arrows) occurs at every other timepoint of the tatum. The feature vector (in this case, a measure of the change between successive spectra) succeeds in delineating many of the most important events; notes, times of beats, and tatum timepoints. The feature vector also misses some important events such as the note marked (b), and it reports other changes (e) that do not correspond to the score or the metric pulse. Observe that the word *duration* refers to the perception of a time interval, not to the time interval itself. Thus the beats may have equal duration even though the elapsed times may vary.

the feature vector in order to locate the tatum and the beats; several such approaches will be examined in Chaps. 5–7. In contrast, this feature vector is inadequate for the purpose of reliable note identification.

4.3.9 Changes to Regular Successions

When two rays of colored light overlap, they appear to meld together into a single beam with a new color. In contrast, when two pitched sounds occur simultaneously, they typically appear to interpenetrate, each retaining its own characteristic pitch and timbre. This difference is caused by the human perceptual apparatus, and not by any intrinsic properties of light or sound waves and is why there are harmonies in sound (where two or more notes can be heard simultaneously), but not in light. Almost any piece of music provides an example of the interpenetration of multiple layers of sound.

The most familiar example of the merging of regular successions into their common period is when a collection of harmonically related sine waves merge together to form the perception of a single tone with a complex timbre, as was illustrated in sound example [S: 37]. At the level of notes (instead of partials) this same kind of reasoning leads to Rameau's "fundamental bass" [B: 178], the implied "root" or fundamental of all simultaneously occurring pitches.

Just as sounds may merge together when the periods align so that the common period lies in the auditory region, a single sound may split into two when the periods become misaligned. For example, by changing its frequency slightly, a harmonic may stand out from (be separately perceptible from) a complex tone.

There are three kinds of change that may occur to a regular succession:

(i) the instantaneous amplitude values may change from repetition to repetition
(ii) the phase may shift
(iii) the period may increase or decrease

At a slow time scale, a pulse train that divides the measure into two equal parts may be played simultaneously with a pulse train that divides the measure into three equal parts. The result is a "new" more complex rhythm, the three against two polyrhythm, as shown in [S: 38] and as discussed in Sect. 3.9. On the other hand, if the two rhythms can be segregated into separate perceptual streams, they remain perceptually distinct. Thus there are two perceptual possibilities at each time scale: the sounds may merge or they may superimpose. When tones merge together, they create a new tone with a different (generally more complex) timbre.

Similarly, when two "rhythms" merge, they create a new (more complex) rhythm. The simplest case occurs when both have the same period. Each row of Fig. 4.15 shows two regular successions with period T. The parameter Δ

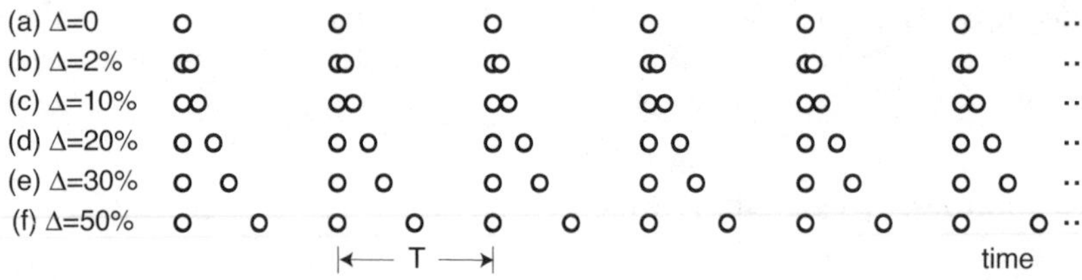

Fig. 4.15. Several perceptual regimes occur when two periodic sequences, each of period T, are sounded simultaneously. When $\Delta = 0$, there is effectively only one sequence of period T. At $\Delta = 50\%$, the sequences are evenly spaced and the net effect is of a single sequence with period $\frac{T}{2}$. In between are a flam effect ($\Delta = 2\%$), rapid doublets ($\Delta = 10\%$), doublets ($\Delta = 20\%$), and a galloping rhythm at $\Delta = 30\%$.

specifies the phase offset, the percentage of T that the second sequence is shifted relative to the first. At $\Delta = 0$ the two are synchronous and the result is indistinguishable from a single succession of period T. At $\Delta = 50\%$, the two sequences are $T/2$ apart and the result is indistinguishable from a single succession with period $\frac{T}{2}$. For very small Δ, the two clicks overlap. In this "flamming" effect, only one sound is heard per period, but it has a greatly changed character compared to the original clicks. Once the time interval between the pulses is more than about 50 ms, the two clicks are perceived separately, though they still appear connected: this might be called a rapid doublet. As Δ increases further, the doublet effect eventually changes into a new rhythmic effect: the gallop. The galloping rhythm persists through approximately $\Delta = 45\%$, where it takes on the appearance of a more rapid single succession. These can be heard in sound examples [S: 51] which plays each of the rows in Fig. 4.15 for about ten seconds.

Suppose next that the two regular successions have almost the same period. Even if they begin together, they will eventually drift apart, as shown in schematic form in Fig. 4.16. Perhaps the simplest way to think of this is in terms of an instantaneous phase shift Δ between two successions with equal periods. Effectively, this is the same as beginning with (a) of Fig. 4.15, moving to (b), through (c), (d), (e), and (f), and then returning (in reverse alphabetical order) back to (a). Thus the effective Δ between the two sequences increases at a linear rate and the sound passes through all the possible perceptual regimes encountered above. This is performed in sound example [S: 52], which clearly demonstrates the various rhythmic perceptions (a)–(f) possible with two regular successions.

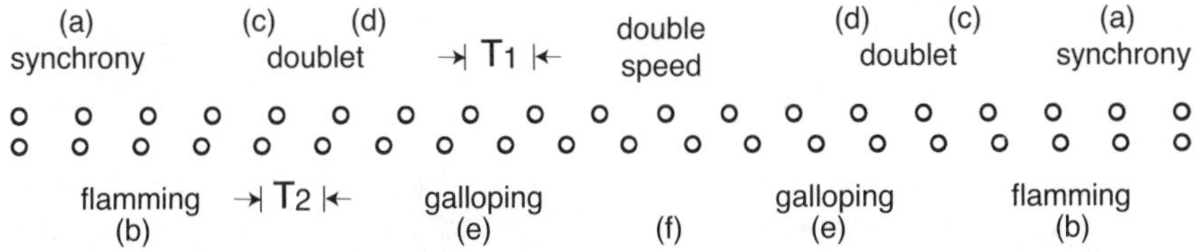

Fig. 4.16. Two periodic sequences with periods $T_1 \approx T_2$ are sounded simultaneously. Over time, the sound shifts through all the perceptual regimes of Fig. 4.15: flamming, doublets, galloping, and double speed. For a more musical discussion of this effect, see Sect. 3.13.

The idea of exploiting all the rhythmic complexities of a pair of regular successions has been pursued by a number of composers in recent years. For example, Steve Reich's 1967 *Piano Phase* [D: 33] repeats a pair of simple piano lines that retrogress against each other to create a fascinating panorama of melodic and rhythmic patterns despite the simplicity of the source material. The most obvious feature of *Piano Phase* is its repetitiveness since it contains little or no variation in pitch, tempo, dynamics or timbre. This is a classic example of what Kramer [B: 117] calls "nonlinear" music where the listener is faced with a slowly evolving sonic structure that does not progress towards a goal; rather, it simply *is*.

4.3.10 Multiple Regular Successions

When there are three or more regular successions, the perceptions can become very complex. Sound example [S: 53] explores the situation with three simultaneous sequences. In (a), the three rates are $T_1 = 0.5$ s, and $T_{2,3} = 0.5 \pm 0.003$ s. There are a number of sensible rhythms perceived over the course of a single repetition. In (b) and (c), however, when the T_i differ by larger amounts, the appearance rapidly becomes chaotic, losing the compelling rhythmic feel of (a) and the earlier two-period examples.

Using computers allows careful control over the exact timing of rhythmic parts. The idea of multiple near-equal periods is explored in *Nothing Broken in Seven* [S: 54]. The melody in 7-tone equal temperament[7] repeats the same six notes throughout. *Phase Seven* [S: 55], also in 7-tone equal temperament, uses an eight note melody. In both cases, the melody line is played against itself at five different tempos, two of which are speeded up (by 1% and 2%) and two of which are slowed down (also by 1% and 2%). This creates raw material that repeats fully only after several days. In order to create pieces of manageable size, selected bits are culled, orchestrated using various bell-like sounds, and then rejoined. In both cases, although the original pattern is monotonously simple, the result increases and decreases in complexity as the melodies phase against themselves. When there are five phasing lines, a very large number of "different" rhythms are perceptible.

4.3.11 One-hundred Metronomes

A striking experiment in multiple regular successions is György Ligeti's *Poeme Symphonique* [D: 31] which is performed by ten players each operating ten metronomes. At the start of the piece, the players wind their metronomes, and then set them in motion, each at a different tempo. The sound is rapid, chaotic, and random. This is simulated in [S: 56], which plays 100 simultaneous regular successions. Unlike Ligeti's version, the example begins with a single metronome, adds a second, then a third, until all 100 are sounding. The digital metronomes do not wind down over time: instead, the piece ends once all 100 have entered.

The overall impression of 100 metronomes is not dissimilar from randomly generated irregular successions. Compare, for example, [S: 56] with [S: 36]. The complexity of the intervals and durations along with the rapid arrival rate becomes overwhelming. Too many regular successions become indistinguishable from randomly generated irregular successions with a similar density, that is, a similar number of ticks per second.

[7] A tuning where the octave is divided into seven equal parts instead of the standard twelve.

4.4 Feature Vectors: Perceptually Motivated Preprocessing

A good feature vector highlights relevant properties of a signal and de-emphasizes irrelevant aspects. For instance, the feature vector in Fig. 4.14 shows how note onsets, tatum timepoints, and beat locations can be emphasized while disregarding less interesting details of the signal. In perceptual terms, this feature vector spotlights auditory boundaries and downplays the perceptually static signal between.

This section presents two strategies for the creation of feature vectors. The first exploits auditory models by simulating (part of) the behavior of the auditory apparatus. For example, Fig. 4.5 filters the signal into a number of channels corresponding to critical bands and this process can be mimicked to construct feature vectors. The second strategy is based on the observation from Sect. 4.3.8 that almost any factors that can create auditory boundaries can be used to create rhythmic patterns. Similarly, almost any mathematical operation that can extract auditory boundaries from a signal can be useful as a feature vector. Section 4.4.3 suggests several feature vectors aimed at extracting particular kinds of auditory boundaries.

The kinds of regularities associated with musical pulse, meter, and rhythm occur on time scales between tenths of a second and tens of seconds. The standard audio sampling rate of 44.1 kHz with its 22 kHz bandwidth is not needed to capture such slow fluctuations, and some kind of data reduction can be used to reduce the amount of data that must be processed. Accordingly, all of the feature vectors are downsampled from the audio rate. The goal is to drastically reduce the amount of data in a perceptually relevant way.

4.4.1 Critical Band Feature Vectors

In computational models of the auditory system such as those of Patterson [B: 160] and Leman [B: 125], the operation of the basilar membrane is commonly modeled as a collection of bandpass filters that divide the sound into (roughly) $\frac{1}{3}$-octave regions over the audio range. These filters simulate the action of the critical bands (recall Fig. 4.4) and Fig. 4.17 shows one way to customize the model to generate a collection of feature vectors that describe the variation of energy in each critical band.

The signal $s(k)$ is passed through a window and then transformed by an FFT.[8] The succeeding block calculates the RMS energy in each frequency band by summing the appropriate terms in the magnitude spectrum.[9] Both the size of the window (which must match the size of the FFT) and the amount

[8] The windowed short-time Fourier transform is discussed further in Sect. 5.3.3.

[9] This FFT-based energy measure is functionally analogous to the energy accumulation of the lowpass filtering of the rectification nonlinearity associated with inner hair cell models [B: 160].

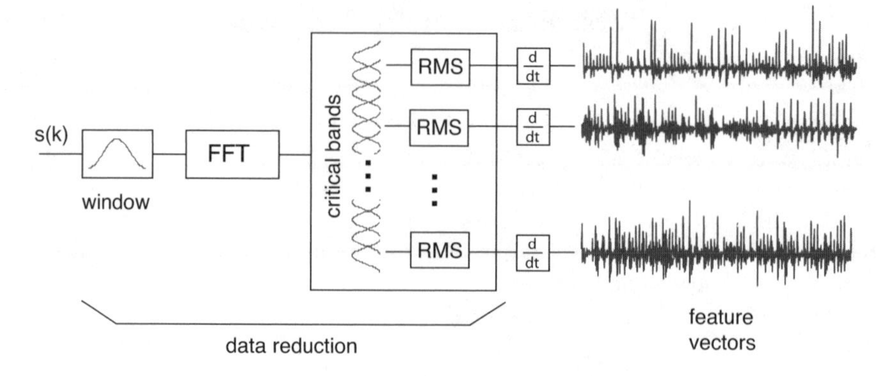

Fig. 4.17. A set of feature vectors based on the idea that rhythmically important events are correlated with repetitive variations of the energy in various frequency bands. The sampled audio data is partitioned into frequency bands using the FFT. The output is a collection of feature vectors (one for each critical band) representing the energy fluctuations in each band. In a typical implementation, each feature vector has an effective sampling rate between 50 and 200 Hz. Each feature vector contains significantly fewer data points than the audio.

of overlap between successive blocks (which defines the "effective" resampling rate) are free parameters. The derivative is used because the auditory system is more finely attuned to changes in the signal than to the signal itself, though in some applications it may be more effective to use the signal before the derivative. Clearly, it is possible to replace the FFTs with a wavelet transform or with a bank of bandpass filters.

4.4.2 Listening to Feature Vectors I

All feature vectors are not created equal; some reflect the underlying rhythm of a piece more accurately than others. For instance, in the *Maple Leaf Rag* there is only sporadic energy in the bottom two critical bands; these two feature vectors are unlikely to be useful in subsequent processing. On the other hand, many pop songs contain a repetitive bass drum that sits in the bottom two bands. In this case. the two lowest feature vectors are ideal candidates for further rhythmic analysis.

Scheirer [B: 188] creates a modulated noise signal from the amplitude envelopes of the outputs of a collection of filterbanks and observes that this noise signal often elicits "the same rhythmic percept" as the original audio signal. In essence, Scheirer suggests listening to the feature vectors and letting the ear decide if the feature vector retains the desired rhythmic information. Of course, it is not reasonable to listen to the feature vector directly as if it were an audio signal. Rather, the feature vector can be used to define an amplitude envelope that modulates a noise. The simplest procedure is to "hold" each value of the feature vector for the duration of the effective sampling interval.[10] Fig. 4.18 illustrates.

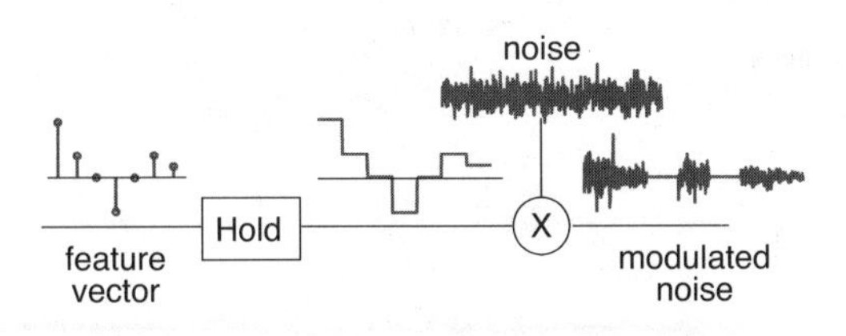

Fig. 4.18. A feature vector is upsampled and then modulated with noise. If the sound evokes the same rhythmic feel as the original, the feature vector has not lost the "essence" of the rhythm. Listen to a number of feature vectors in [S: 57, S: 58].

For example, [S: 57] presents several individual feature vectors derived from the *Maple Leaf Rag*. Examples (i) and (ii) demonstrate that individual feature vectors can indeed retain the rhythmic feel of the original. Examples (iii) and (iv) play the feature vector from the second critical band; as expected, these do not evoke the original rhythm. Sound example [S: 58] plays all the feature vectors together, that is, the method of Fig. 4.18 is applied to each feature separately and the results are summed. The rhythmic feel is clear. Of course, the tone quality of these examples has changed radically from the original: all melody and harmony have been lost. Basic rhythmic information (including the beat) remains intact.

4.4.3 Extracting Auditory Boundaries from a Signal

Another approach to generating feature vectors is to choose aspects of a signal that directly reflect auditory boundaries. Typically, this involves segmenting the signal and extracting a single number for each segment, a procedure shown graphically in Fig. 4.19. A study of a wide variety of feature vectors along with a technique for ascertaining their appropriateness in the beat tracking task can be found in [B: 205]. This section details four simple methods.

Energy Measure: The simplest of the feature vectors is particularly appropriate for audio in which the envelope of the sound clearly displays the beat. Let $x(t)$ represent the audio waveform, which is sampled at a constant interval to give the sequence $x[k]$. Group the sampled data into M overlapping segments each containing N consecutive terms. Let $x_n[k]$ represent the kth element (out of N) in the nth segment. The energy in the nth segment, which is effectively a measure of the instantaneous loudness (recall (4.1)), is

$$e[n] = \sum_{k=0}^{N-1} x_n^2[k], \quad n = 1, 2, \ldots, M. \tag{4.3}$$

[10] i.e., the number of samples between adjacent windows divided by the audio sampling rate.

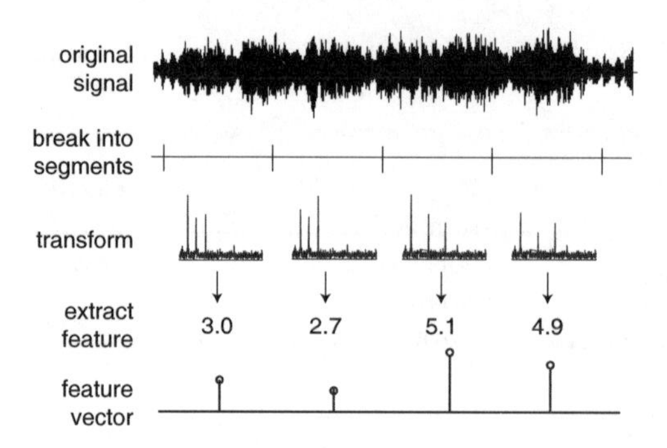

Fig. 4.19. The signal is partitioned into (possibly overlapping) segments which are processed independently. The output is a sequence (the feature vector) extracted from the segments.

Then the terms of the "energy" feature vector are defined to be the change in (the derivative of) the $e[n]$. Though numerical derivatives can be poorly conditioned, the action of the summing, combined with sensible overlapping, ensures that the numerical problems do not overwhelm the data. An example is provided in Fig. 4.20(b).

Group Delay: The remaining methods operate in the frequency domain and share a common notation. With $x(t)$, $x_n[k]$, N, M as above, let $X_n[j]$ be the FFT of x_n. Each of the frequency domain methods processes X_n in a different way to form a scalar value for each segment. The sequence of such values defines the feature vector.

Structurally, the transform X_n consists of N complex numbers that are most commonly represented as magnitude and phase pairs, with the phase unwrapped (meaning that factors of 2π are added or subtracted so as to make the phase angle continuous across boundaries at integer multiples of $\pm\pi$). For many musical waveforms, the unwrapped phase lies close to a straight line. The slope of this line defines the "group delay" method of creating feature vectors. An example is provided in Fig. 4.20(c). Appendix A of [B: 206] shows how the slope is proportional to a time shifted version of the energy. It is not dependent on the total energy in the window, but rather on the distribution of the energy within the window.

Spectral Center: With notation inherited from the previous sections, the "spectral center" method of creating feature vectors locates the frequency f_c where half of the energy in the spectrum lies below f_c and half lies above. This is

$$\int_{f=0}^{f_c} X_n^2(f)\, df = \int_{f=f_c}^{\infty} X_n^2(f)\, df. \tag{4.4}$$

The feature vector value is then defined as the change in (i.e., the derivative of) f_c from segment to segment. The spectral center is sensitive to pitch changes and to changes in the distribution of energy such as might occur when different instruments enter or leave, or when one instrument changes registers. Like the group delay, it is insensitive to amplitude changes in the audio. A numerical example is provided in Fig. 4.20(d).

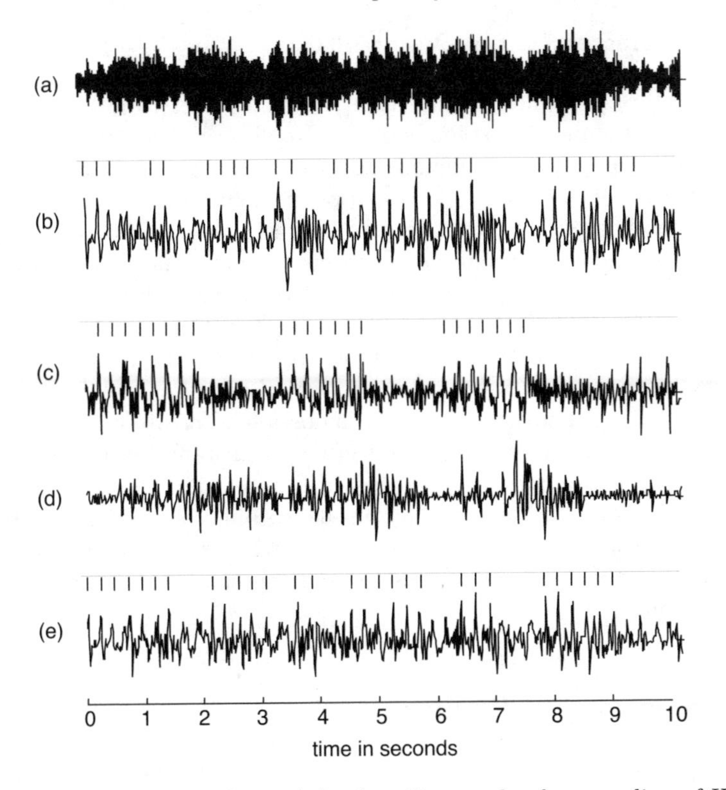

Fig. 4.20. The audio waveform of the first 10 seconds of a recording of Handel's *Water Music* is shown in (a). The various feature vectors are: (b) the energy method, (c) the group delay, (d) the change in the center of the spectrum, and (e) the dispersion of spectral energy. Tick marks emphasize beat locations that are visually prominent.

Spectral Dispersion: The spectral dispersion gives a measure of the spread of the spectrum about its center. Let

$$sd[n] = \sum_{j=0}^{N-1} X_n^2[j]|j - f_c| \tag{4.5}$$

define the spectral dispersion $sd[n]$ of the nth segment about the spectral center f_c. It weights energy at remote frequencies more than those close to the spectral center. The feature vector is then defined as the change in (the derivative of) $sd[n]$. This provides a crude measure of how the spectral energy is distributed: small values mean that the energy is primarily concentrated near the center while large values mean that the energy is widely dispersed. For example, near the percussive attack of a violin the spectral dispersion is large, while it is small in the (relative) steady state between attacks. An example is provided in Fig. 4.20(e).

4.4.4 Listening to Feature Vectors II

Feature vectors are much smaller than the audio files from which they are derived. For example, 110 seconds of the *Maple Leaf Rag* contains 4,894,208 samples of audio; the corresponding feature vectors contain 9560 samples.[11] After removing 99.8% of the information in a file, does anything remain? Using the modulated noise technique of Fig. 4.18 shows that each of the four feature vectors (energy, group delay, spectral center, and dispersion) clearly evokes the rhythm of the rag. Listen to sound example [S: 59].

These are just four of the many possible low level audio features that could be used in the creation of feature vectors. While these four do not enjoy statistical independence, it is easy to see that they measure different features of the underlying audio stream since it is possible to create a sound for which any three of the feature vectors are (essentially) constant, but the fourth varies significantly. For example, an idealized trill on a violin has constant energy, constant dispersion, and constant group delay, but varying center. Similarly, if a short sine wave burst alternates with a white noise burst, they can be chosen so that the energy, group delay, and center remain the same but the dispersion varies widely.

4.5 Perception vs. Reality

The auditory (and visual) illusions of the first chapter reveal that perception is not a direct reflection of the external world. When trying to understand the operation of the perceptual mechanism, scientists must simplify and isolate particular aspects of sound in order to make progress. Real music is very different from the "music-like" stimuli that scientists studying the human auditory system typically use. Fraisse [B: 66] says that the fundamental laws of perception... "do not explain music any more than gravity explains the art of architecture. But there is not an architect who ignores gravity any more than there is a musical rhythm that does not respect these perceptual laws." Meyer (as quoted in [B: 108]) says:

> It is an inexcusable error to equate acoustical phenomenon with qualitative experiences. The former are abstract scientific concepts, the latter are psychological perceptions. There is no one-to-one relationship between an acoustical event and its concomitant perceptual experience.

Indeed, it would be foolish to believe that feature vectors directly reflect perceptions. But it would be equally foolish to ignore what is known about the mechanisms of perception in a machine intended to make music for beings limited to perceiving sounds through a sensory apparatus. One of the primary

[11] The overlap of successive FFTs is 512 and the effective sampling rate is 86 Hz.

goals of *Rhythm and Transforms* is to exploit the results of research into the perception of rhythm in order to create machines that more closely reflect our own abilities. The feature vectors of this chapter will play a key role in the automatic location of auditory boundaries and in the subsequent task of locating regular successions (beats) from a musical performance.

5

Transforms

Transforms model a signal as a collection of waveforms of a particular form: sinusoids for the Fourier transform, mother wavelets for the wavelet transforms, periodic basis functions for the periodicity transforms. All of these methods are united in their use of inner products as a basic measure of the similarity and dissimilarity between signals, and all may be applied (with suitable care) to problems of rhythmic identification.

Suppose there are two signals or sequences. Are they the same or are they different? Do they have the same orientation or do they point in different directions? Are they periodic? Do they have the same periods? The inner product is one way of quantifying the similarity of (and the dissimilarity between) two signals. It can be used to find properties of an unknown signal by comparing it to one or more known signals, a technique that lies at the heart of many common transform methods. The inner product is closely related to (cross) correlation, which is a simple form of pattern matching useful for aligning signals in time. A special case is autocorrelation which is a standard way of searching for repetitions or periodicities. The inner product provides the basic definitions of a variety of transform techniques such as the Fourier and wavelet transforms as well as the nonorthogonal periodicity transforms.

The first section reviews the basic ideas of the angle between two signals or sequences in terms of the inner product and sets the mathematical notations that will be used throughout the chapter. Section 5.2 defines the cross-correlation between two signals or sequences in terms of the inner product and interprets the correlation as a measure of the fit or alignment between the signals. Section 5.3 shows how the Fourier transform of a signal is a collection of inner products between the signal and various sinusoids. Some cautionary remarks are made regarding the applicability of transforms to the rhythm-finding problem. Two signal processing technologies, the short time Fourier transform and the phase vocoder are then described in Sects. 5.3.3 and 5.3.4. Wavelet transforms are discussed in Sect. 5.4 in terms of their operation as an inner product between a "mother wavelet" and the signal of interest. The final section describes the Periodicity transforms, which are again introduced in terms of an inner product, and some advantages are noted in terms of rhythmic processing.

5.1 Inner Product: The Angle Between Two Signals

The angle between two vectors gives a good indication of how closely aligned they are: if the angle is small then they point in nearly the same direction; if the angle is near 90 degrees, then they point in completely different directions (they are at right angles). The generalization of these ideas to sequences and signals uses the *inner product* to define the "angle." When the inner product is large, the sequences are approximately the same ("point in the same direction") while if the inner product is zero (if the two are *orthogonal*) then they are like two vectors at right angles.

A common definition of the inner product between two vectors x and y is

$$\langle x, y\rangle = \sum_k x[k]y[k]. \tag{5.1}$$

The length (or *norm*) of a vector is the square root of the sum of the squares of its elements, and can also be written in terms of the inner product as

$$||x|| = \left(\sum_k x^2[k]\right)^{\frac{1}{2}} = \sqrt{\langle x, x\rangle}. \tag{5.2}$$

For example, consider the two vectors $x = (2,1)$ and $y = (1,0)$ shown in Fig. 5.1. The lengths of these vectors are $||x|| = \sqrt{2^2+1^2} = \sqrt{5}$ and $||y|| = \sqrt{1^2+0^2} = 1$ and the inner product is $\langle x, y\rangle = 2\cdot 1 + 1\cdot 0 = 2$. The angle between x and y is the θ such that

$$\cos(\theta) = \frac{\langle x, y\rangle}{||x||\ ||y||}. \tag{5.3}$$

For the vectors in Fig. 5.1, $\cos(\theta) = \frac{2}{\sqrt{5}}$ and so $\theta = 0.46$ radians or about 26 degrees.

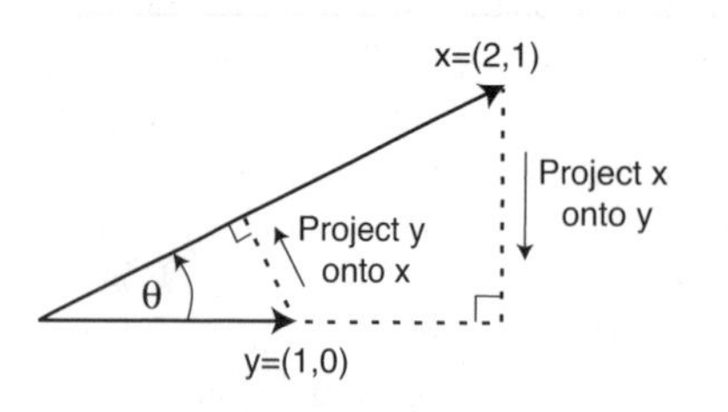

Fig. 5.1. The angle θ between two vectors x and y can be calculated from the inner product using (5.3). The projection of x in the direction y is $\cos(\theta)||x||$, which is the same as $\langle x, y\rangle/||y||$. This is the dotted line forming a right angle with y. The projection of y onto x, given by $\langle x, y\rangle/||x||$, is also shown. If a projection is zero, then x and y are already at right angles (orthogonal).

The inner product is important because it extends the idea of angle (and especially the notion of a right angle) to a wide variety of signals. The definition (5.1) applies directly to sequences (where the sum is over all possible k) while

$$\langle x(t), y(t) \rangle = \int_{-\infty}^{\infty} x(t)y(t)dt \tag{5.4}$$

defines the inner product between two functions $x(t)$ and $y(t)$ by replacing the sum with an integral.[1] As before, if the inner product is zero, the two signals are said to be orthogonal. For instance, the two sequences

$$\begin{aligned} x &= \quad \ldots 1, 1, -1, -1, 1, 1, -1, -1, \ldots \\ y &= \quad \ldots 1, -1, 1, -1, 1, -1, 1, -1, \ldots \end{aligned}$$

are orthogonal, and

$$z = \quad \ldots 1, 1, 1, 1, 1, 1, 1, 1, \ldots$$

is orthogonal to both x and y. Taking all linear combinations of x, y, and z (i.e., the set of all $a_1 x + a_2 y + a_3 z$ for all real numbers a_i) defines a subspace with three dimensions. Similarly, two sinusoids

$$x(t) = \sin(2\pi f_1 t) \text{ and } y(t) = \sin(2\pi f_2 t)$$

with frequencies f_1 and f_2 are orthogonal whenever $f_1 \neq f_2$. The set of all linear combinations of sinusoids for all possible frequencies f_i is at the heart of the Fourier transform of Sect. 5.3 and orthogonality plays an important role because it simplifies many of the calculations. If the signals are complex-valued, then $y(t)$ in (5.4) (and $y[k]$ in (5.1)) should be replaced with their complex conjugates.

Suppose there is a set of signals x_i that all have the same norm, so that $||x_i||^2 = ||x_k||^2$ for all i and k. Given any signal y, the inner product can be used to determine which of the x_is is closest to y where "closeness" is defined by the norm of the difference. Since

$$||y - x_i||^2 = ||y||^2 - 2\langle y, x_i \rangle + ||x_i||^2 \tag{5.5}$$

and since $||y||$ and $||x_i||$ are fixed, the i that minimizes the norm on the left hand side is the same as the i that maximizes the inner product $\langle y, x_i \rangle$.

5.2 Correlation and Autocorrelation

The (cross) correlation between two signals $x(t)$ and $y(t)$ with shift τ can be defined directly or in terms of the inner product:

$$\begin{aligned} R_{xy}(\tau) &= \int_{-\infty}^{\infty} x(t)y(t+\tau)dt \\ &= \langle x(t), y(t+\tau) \rangle. \end{aligned} \tag{5.6}$$

[1] Observe that writing the t inside the inner product is an abuse of notation; nonetheless, it is useful because there are many situations (such as (5.6)) where the arguments of the x and y need to be distinguished.

When the correlation $R_{xy}(\tau)$ is large, x and y point in (nearly) the same direction. If $R_{xy}(\tau)$ is small (near zero), $x(t)$ and $y(t+\tau)$ are nearly orthogonal. The correlation can also be interpreted in terms of similarity or closeness: large $R_{xy}(\tau)$ mean that $x(t)$ and $y(t+\tau)$ are similar (close to each other) while small $R_{xy}(\tau)$ mean they are different (far from each other). These follow directly from (5.5).

In discrete time, the (cross) correlation between two sequences $x[k]$ and $y[k+j]$ with time shift j is

$$\begin{aligned} R_{xy}(j) &= \sum_{k=-\infty}^{\infty} x[k]y[k+j] \\ &= \langle x[k], y[k+j] \rangle. \end{aligned} \tag{5.7}$$

Correlation shifts one of the sequences in time and calculates how well they match (by multiplying point by point and summing) at each shift. When the sum is small then they are not much alike; when the sum is large, many terms are similar. Equations (5.6) and (5.7) are recipes for the calculation of the correlation. First, choose a τ (or a j). Shift the function $y(t)$ by τ (or $y[k]$ by j) and then multiply point by point times $x(t)$ (or times $x[k]$). The area under the resulting product (or the sum of the elements) is the cross-correlation. Repeat for all possible τ (or j).

Seven pairs of functions $x(t)$ and $y(t)$ are shown in Fig. 5.2 along with their correlations. In (a), a train of spikes is correlated with a Gaussian pulse. The correlation reproduces the pulse, once for each spike. In (b), the spike train is replaced by a sinusoid. The correlation smears the pulse and inverts it with each undulation of the sine. In (f), two random signals are generated: their correlation is small (and random) because the two random sequences are independent.

One useful situation is when x and y are two copies of the same signal but displaced in time. The variable τ shifts y and at some shift τ^* they become aligned. At this τ^*, $x(t)$ is the same as $y(t+\tau^*)$ and the product is positive everywhere: hence, when integrated, $R_{xy}(\tau^*)$ achieves its largest value. This situation is depicted in Fig. 5.2(g) which shows a y that is a shifted version of x. The maximum value of the correlation occurs at the τ^* where $x(t) = y(t+\tau)$, where the signals are closest. Correlation is an ideal tool for aligning signals in time.

A special case that can be very useful is when the two signals x and y happen to be the same. In this case, $R_{xx}(\tau) = \langle x(t), x(t+\tau) \rangle$ is called the *autocorrelation* of x. For any x, the largest value of the autocorrelation always occurs at $\tau = 0$, that is, when there is no shift. This is particularly useful when x is periodic since then $R_{xx}(\tau)$ has peaks at values of τ that correspond precisely to the period. For example, Fig. 5.2(c) shows a periodic spike train with one second between spikes. The autocorrelation has a series of peaks that are precisely one second apart. Similarly, in (d) the input is a sinusoid with

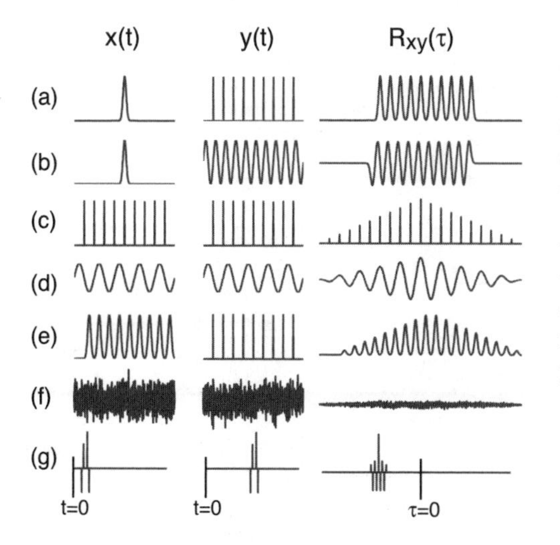

Fig. 5.2. Seven examples of the crosscorrrelation between two signals x and y. The examples consider spike trains, Gaussian pulses, sinusoids, pulse trains, and random signals. When $x = y$ (as in (c) and (d)), the largest value of the correlation occurs at a shift $\tau = 0$. The distance between successive peaks of $R_{xx}(\tau)$ is directly related to the periodicity in the input.

frequency 0.5 Hz. The peaks of the autocorrelation occur 2 seconds apart, exactly the periodicity of the sine wave.

5.3 The Fourier Transform

Computer techniques allow us to look inside a sound; to dissect it into its constituent elements. But what are the fundamental elements of a sound? Are they sine waves, sound grains, wavelets, notes, beats, or something else? Each of these kinds of elements requires a different kind of processing to detect the regularities, the frequencies, scales, or periods.

As sound (in the physical sense) is a wave, it has many properties that are analogous to the wave properties of light. Think of a prism, which bends each color through a different angle and so decomposes sunlight into a family of colored beams. Each beam contains a "pure color," a wave of a single frequency, amplitude, and phase.[2] Similarly, complex sound waves can be decomposed into a family of simple sine waves, each of which is characterized by its frequency, amplitude, and phase. These are called the *partials*, or the *overtones* of the sound, and the collection of all the partials is called the *spectrum*. Figure 5.3 depicts the *Fourier transform* in its role as a "sound prism."

This prism effect for sound waves is achieved using the Fourier transform. Mathematically, the Fourier transform of a function $x(t)$ is defined as

[2] For light, frequency corresponds to color, and amplitude to intensity. Like the ear, the eye is blind to the phase of a single sinusoid.

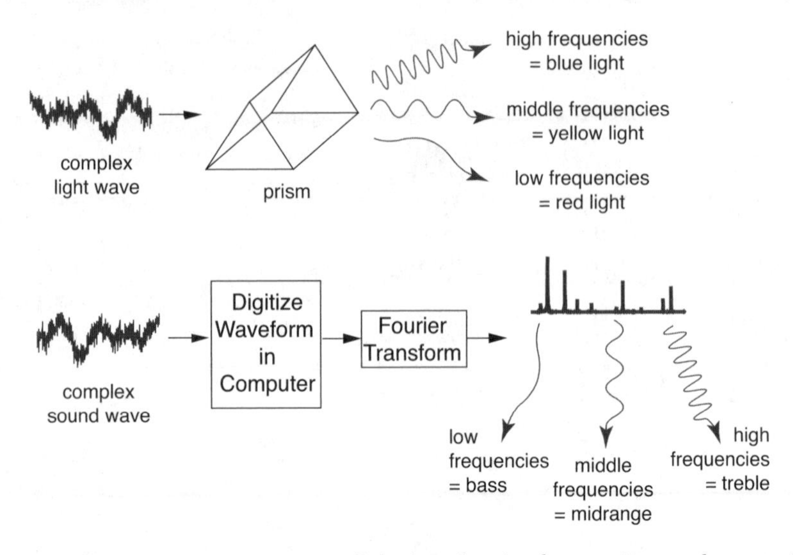

Fig. 5.3. Just as a prism separates light into its simple constituent elements (the colors of the rainbow), the Fourier Transform separates sound waves into simpler sine waves in the low (bass), middle (midrange), and high (treble) frequencies. Similarly, the auditory system transforms a pressure wave into a spatial array that corresponds to the various frequencies contained in the wave, as shown in Fig. 4.2 on p. 79.

$$\begin{aligned} X(f) &= \int_{-\infty}^{\infty} x(t) e^{-j2\pi f t} dt \\ &= \langle x(t), e^{j2\pi f t} \rangle \end{aligned} \tag{5.8}$$

for all real f. It is the inner product of the signal $x(t)$ and the complex-valued sinusoid[3] $e^{-j2\pi f t}$.

Consider the meaning of the Fourier transform (5.8). First, $X(f)$ is a function of frequency: for each f the integral defined by the inner product is evaluated to give a complex-valued number with magnitude m and angle θ. Since $X(f)$ is the correlation (inner product) between the signal $x(t)$ and a sinusoid of frequency f, m is the magnitude (and θ the phase) of the sine wave that is closest[4] to $x(t)$. Since sine waves of different frequencies are orthogonal[5] there is no interaction between different frequencies and m is the amount of the frequency f present in the signal $x(t)$. The Fourier transform shows how $x(t)$ can be uniquely decomposed into (and rebuilt from) sums of sinusoids.

Second, the Fourier transform is invertible. The inversion formula $x(t) = \int_{-\infty}^{\infty} X(f) e^{j2\pi f t} df = \langle X(f), e^{-j2\pi f t} \rangle$ reverses the role of the time and fre-

[3] Euler's formula specifies the relationship between real and complex sinusoids: $e^{\pm j\theta} = \cos(\theta) \pm j \sin(\theta)$.

[4] Recall (5.5).

[5] That is, the inner product of two sinusoids is $\langle e^{-j2\pi f_1 t}, e^{j2\pi f_2 t} \rangle = \delta(f_1 - f_2)$ where $\delta(z)$ is the "delta function" that has unit area and is zero except when $z = 0$.

quency variables and ensures that the transform neither creates nor destroys information.

5.3.1 Frequency via the DFT/FFT

The spectrum gives important information about the makeup of a sound and is most commonly implemented in a computer by the Discrete Fourier Transform (DFT) or the more efficient Fast Fourier Transform (FFT). Standard versions of the DFT and/or the FFT are available in audio processing software and in numerical packages (such as MATLAB® and Mathematica) that can manipulate sound data files.

Like the Fourier transform, the DFT decomposes a signal into its constituent sinusoidal elements. Like the Fourier transform, the DFT is an invertible, information preserving transformation. But the DFT differs from the Fourier transform in three useful ways. First, it applies to discrete-time sequences which can be stored and manipulated directly in computers (rather than to functions or analog waveforms). Second, it is a sum rather than an integral, and so is easy to implement in either hardware or software. Third, it operates on a finite data record (rather than operating on a function that must be defined over all time). Given a sequence $x[k]$ of length N, the DFT is defined by

$$\begin{aligned} X[n] &= \sum_{k=0}^{N-1} x[k] e^{-j2\pi nk/N} \qquad n = 0, 1, 2, ..., N-1 \\ &= \langle x[k], e^{j2\pi nk/N} \rangle. \end{aligned} \tag{5.9}$$

For each value n, (5.9) multiplies each term of the data by a complex exponential and then sums. Compare this to the Fourier transform; for each frequency f, (5.8) multiplies each point of the waveform by a complex exponential and then integrates. Thus $X[n]$ is a function of frequency in the same way that $X(f)$ is a function of frequency. Indeed, the term $e^{-j2\pi nk/N}$ is a discrete-time sinusoid with frequency proportional to n.

A good example of the use of the DFT/FFT for spectral analysis appears in Fig. 2.19 on p. 44 which shows the waveform and corresponding spectrum of the pluck of a guitar string. While the time evolution of the signal is clear from the waveform, the underlying nature of the sound as a sum of a number of harmonically related sinusoids is clear from the spectrum. The two plots are complementary and display different aspects of the same sound.

One potential source of confusion is that the frequency f in the Fourier transform can take on any value while the frequencies present in (5.9) are all integer multiples n of $2\pi/N$. This "fundamental frequency" is precisely the frequency of the sine wave with period equal to the length N of the window over which the DFT is taken. Thus the frequencies in (5.9) are constrained to a discrete set and the frequencies are separated by a constant difference.

This resolution is equal to the sampling rate divided by the window size (the number of samples used in the calculation), that is,

$$\text{resolution in Hz} = \frac{\text{sampling rate}}{\text{window size}}. \tag{5.10}$$

For example, the window used for the guitar pluck in Fig. 2.19 contains 32,000 samples and the sampling rate is 44.1 kHz. Thus the resolution is 1.38 Hz. The peak of the spectrum occurs at entry $n = 142$ in the output of the FFT, which corresponds to a frequency of $142 \cdot 1.38$ which is approximately (but not exactly) 196 Hz, as annotated in the figure. Observe that the units of (5.10) are inverse seconds (the units of the numerator are samples per second while the denominator has units of samples). Thus an accuracy of 10 Hz requires a duration of only 0.01 s and an accuracy of 1.38 Hz requires a time window of 0.72 s, as used with the guitar pluck. To achieve an accuracy of $\frac{1}{10}$ Hz would require a time window of at least 10 s, irrespective of the sampling rate.

Why not simply use long windows for increased resolution? Because long windows do not show when (in time) events occur. For example, Fig. 5.4 shows a signal that consists of two sinusoids: a sine wave with frequency 150 Hz is followed by a somewhat larger wave with frequency 100 Hz. The magnitude spectrum shows peaks near the expected values of 100 and 150 Hz. But it does not show the order of the sine waves. Indeed, the magnitude spectrum is the same if the sine waves are reversed in order, and even if they both sound for the entire time interval.[6] Thus, use of the FFT requires a compromise: long windows are desired in order to have good frequency resolution while short windows are desired in order to locate events accurately in time. This is a kind of uncertainty principle in which it is necessary to trade off measurements of the frequency and the temporal location.

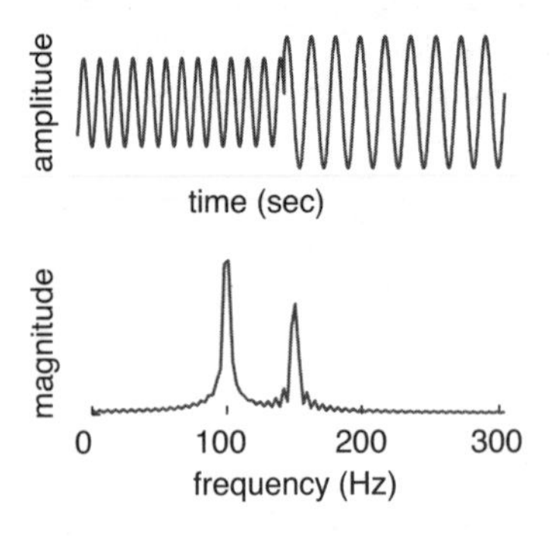

Fig. 5.4. A signal consists of two sine waves. The first, at 150 Hz, lasts for 0.5 s and the second, at 100 Hz, begins when the first ends. The magnitude spectrum shows peaks corresponding to both sine waves but does not show their temporal relationship. The magnitude spectrum would look the same if the order of the sine waves were reversed or if they occurred simultaneously (rather than successively).

Windowing also influences the accuracy of frequency estimation through the effect called "smearing." Figure 5.5 shows two different analyses of the

[6] The phase spectrum of the three cases differs, but the relationship between the phase and the temporal order is notoriously difficult to decipher.

same 200 Hz sine wave. In the top case, the window size is 0.5 seconds and so exactly 100 repetitions of the wave fit into the window. Accordingly, all of the inner products in (5.9) are zero except for one that has frequency exactly equal to 200 Hz. The algorithms in MATLAB® report these values as less than 10^{-14}, which is numerically indistinguishable from zero. In contrast, the bottom analysis uses a window of 0.503 s and so an integer number of waves does not fit exactly within the window. This implies that none of the terms in the inner product have frequency exactly 200 Hz. A large number of terms become nonzero in order to compensate, to represent a frequency that falls between the cracks of its resolution.[7]

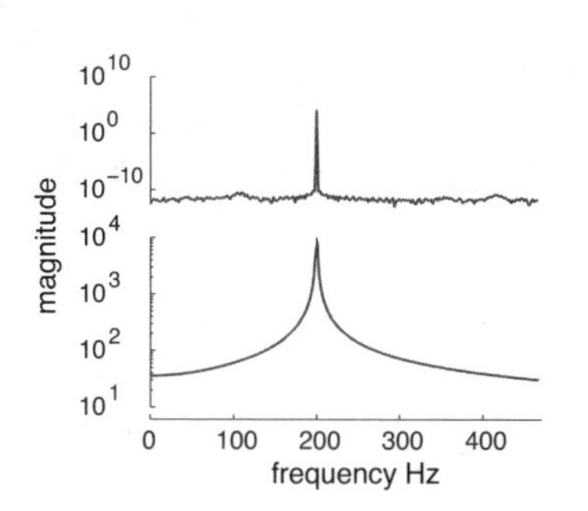

Fig. 5.5. A sine wave of frequency 200 Hz is analyzed twice, resulting in two spectra. The window used in top spectrum is 0.5 s, and so an integer number of periods of the signal fits exactly. This means that one of the terms in the inner product (5.9) has frequency exactly equal to 200 Hz: this one is large and all others are (numerically) zero. In the bottom spectrum, the window width 0.503 does not support an integer number of periods. No single term in the inner product has frequency 200 Hz and the representation is "smeared."

5.3.2 Three Mistakes

Over the years, the Fourier transform has found many uses throughout science and engineering and it is easy to develop a naive overconfidence in its use. In terms of the rhythm finding goals of *Rhythm and Transforms*, the naive argument goes something like this:

> The Fourier transform is an ideal tool for finding frequencies and/or periodicities in complex data sets. The beat of a piece of music and the larger rhythmic structures are, at heart, different frequencies and/or periodicities that exist in the sound. Accordingly, it should be straightforward to apply the Fourier transform to find the beat and higher metrical structures within a musical passage.

[7] The effect of smearing can be studied by observing that the windowed signal is equal to the product of the signal and the window. Consequently, the spectrum of the windowed signal is equal to the convolution of the spectrum of the signal (the spike as in the top part of Fig. 5.5) with the spectrum of the window (in this case, the rectangular window has a spectrum that is a sinc function). Thus the smearing can be controlled, but never eliminated, by careful choice of window function. See [B: 170] for details.

This section shows that this argument is fundamentally flawed in three separate ways.

The first flaw is the easiest to see since it has been repeatedly emphasized throughout the earlier chapters: rhythmic phenomenon are only partially defined by the sound itself, they are heavily influenced by the perceptual apparatus of the listener. Accordingly, it is only sensible to expect to be able to locate the part of the rhythm that is in the sound itself using a technique such as the Fourier transform.

The second flaw arises from a misunderstanding of the nature of rhythmic phenomena. Consider naively applying the FFT to the first 100 s of an audio CD in the hopes of finding "the beat" of a performance that occurs at (say) two times per second. As shown in Fig. 1.5 on p. 7, the phenomenon of musical beats occur at rates between about 0.2 Hz and 2 Hz. Formula (5.10) shows that 100 s corresponds to a frequency resolution of 1/100 Hz which should allow detection within the needed range with a fair degree of accuracy. But surprise! The output of this FFT contains no measurable energy below 20 Hz. How can this be? We clearly *hear* the beat at 2 Hz, how can the FFT show nothing near 2 Hz?

The FFT says that there is no match between the signal (in this case the sound) and sinusoids with frequencies near 2 Hz. This should come as no surprise, since human hearing extends from a high of 20 kHz down to a low of about 20 Hz and we cannot directly perceive a 2 Hz sinusoid.[8] Yet we clearly perceive *something* occurring two times each second. In other words, the perception of rhythm is not a perception of sinusoids at very low frequencies. Rather, it is a perception of auditory boundaries (such as changes in energy) at the specified rate. Thus the "hearing" of a pitch at 200 Hz is a very different phenomenon from the "hearing" of a rhythm at 2 Hz. While the Fourier transform is adept at displaying the physical characteristics of the sine waves associated with the perception of pitch, it does not straightforwardly display the physical characteristics of patterns of auditory boundaries associated with rhythmic perception.

Using this insight, it is easy to modify the sound wave so that the transform does reveal something. The simplest approach is to take the FFT of the energy of the sound wave (rather than of the sound wave itself). This is a primitive kind of perceptually-motivated data preprocessing that might lead to better replication of the ear's abilities. But it is a slippery slope: what kind of criteria will specify the best kind of preprocessing to use? Maybe it would be better to take the absolute value of the sound wave? Or to take the percentage change in the absolute value of the energy? There are many possibilities, and it is hard to know what criteria for success look like.

The third flaw in the argument arises from the nature of the FFT itself. Consider the simplest situation where a drum beats at a regular rhythm. Some

[8] It is common practice to filter out all frequencies below about 20 Hz on recordings. Even in live situations, music contains no purposeful energy at these frequencies.

kind of simple preprocessing (such as taking the energy of the sound wave) is applied. The input to the transform looks like a train of somewhat noisy pulses. The output of the FFT is: a train of somewhat noisy-looking pulses. Figure 5.6 shows three cases. In each case the signal is a set of regularly spaced noises with period T_i seconds. The transform is a set of regularly spaced pulses separated by $\frac{1}{T_i}$. As the time-pulses grow further apart, the frequency-pulses grow closer together.

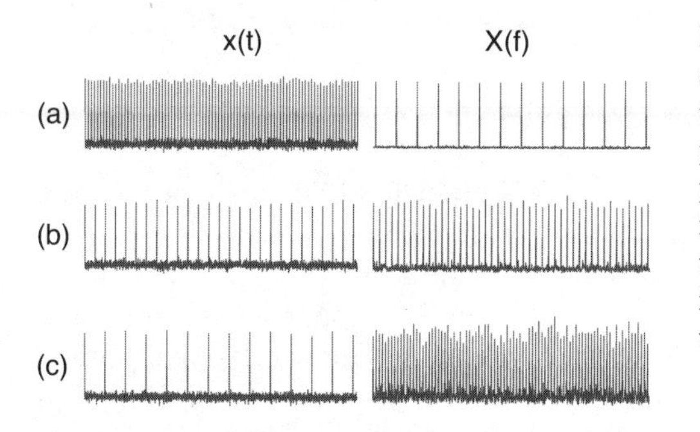

Fig. 5.6. The FFT is applied to a train of noisy pulses. The spectrum is again a train of noisy pulses. Close pulses in time imply widely separated pulses in frequency and distant pulses in time imply small separation in frequency. Three cases are shown with progressively longer period.

Students of the Fourier transform will recognize Fig. 5.6 as somewhat noisy versions of a fundamental result from Fourier series. Let $\delta(t)$ represent a single spike at time t. Then a train of such spikes with T s between spikes is the sum $s(t) = \sum_k \delta(t - kT)$. The Fourier transform of $s(t)$ is[9]

$$S(f) = \frac{1}{T} \sum_{n=-\infty}^{\infty} \delta(f - \frac{n}{T})$$

which is itself a spike train in frequency. Thus the behavior in Fig. 5.6 is not a pathology of deviously chosen numerical parameters: it is simply how the transform works.

The goal of the analysis is to locate a regular succession within the input. In the case of a pulse train this requires locating the distance between successive pulses. As Fig. 5.6 suggests, it is no easier to locate the distance between pulses in the transformed data than in the original data itself. Thus, at least in the situation of simple regular inputs like the pulse train, there is no compelling reason to believe that the transform provides any insight: it simply returns another problem with the same character as the original.

To summarize: application of the Fourier transform to the problem of describing rhythmic phenomena is neither straightforward nor obvious. First is the problem that only part of the perception of rhythm is located in the sound wave (this critique applies to all such signal-based approaches). Second is the problem that some kind of preprocessing of the audio signal is required

[9] This result can be found in most tables of Fourier transforms since it is the key to the sampling theorem. See [B: 102].

in order for the transform to show anything. Finally, even in the idealized case where a rhythm consists of exact repetitions of a brief sound, the Fourier transform provides little insight.

These critiques do not, however, mean that the Fourier transform is incapable of playing a role in the interpretation of rhythmic phenomena. Rather, they show that it is necessary to carefully consider proper uses of the FFT within a larger system. For example, it can be used as part of a method of preprocessing the audio so as to emphasize key features of a sound and to locate auditory boundaries where the character of a sound changes.

5.3.3 Short-time Fourier Transform

The short-time Fourier transform (STFT) is often used when a signal is too long to be analyzed with a single transform or when it is desirable to have better time-localization. The idea is to use a window function $w(t)$ that zeroes all but a short time interval. All events in the FFT are then localized to that interval. The windows are shaped so that when they are overlapped (shifted by S samples and summed) their sum $\sum_n w(t-nS)$ is constant for all t. This is shown schematically in Fig. 5.7 where the windows are overlapped by half their support.[10]

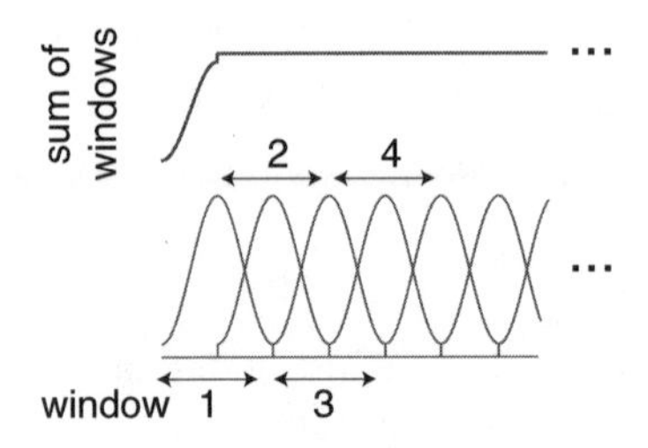

Fig. 5.7. A set of overlapping windows is used to zero all but a short segment of the signal. The FFT can then be applied to that segment in order to localize events in time. An overlap factor of 2 is shown; 4 might be more common in applications.

Using the window functions, the STFT can be described mathematically in much the same way as the Fourier transform itself

$$X_{STFT}(f, \tau) = \langle x(t), w(t-\tau)e^{j2\pi ft} \rangle.$$

Observe that X_{STFT} is a function of both time (τ specifies where the window is nonzero) and frequency (f has the same meaning as in (5.8)). Similarly, the discrete-time version parallels the definition of the DFT in (5.9)

$$X_{STFT}[n, i] = \langle x[k], w[k-i]e^{j2\pi nk/N} \rangle$$

[10] Let $W(f)$ be the Fourier transform of the window $w(t)$ and $X(f)$ be the transform of the data within the time span of the window. Then the convolution of $W(f)$ and $X(f)$ describes the effect of the windowing on the data analysis. See [B: 170] or [B: 218] for a detailed comparison of various window functions.

where n is the frequency variable and i specifies (via the window $w[k-i]$) where in time the FFT is taken. Thus the STFT provides a series of spectral snapshots that move through time. Plotting the snapshots sequentially is like looking at a multi-banded graphic equalizer. A common plotting technique is to change the magnitude of the spectra into colors (or into grayscale). Placing the frequency on the vertical axis and time on the horizontal axis leads to a spectrogram such as that depicting the *Maple Leaf Rag* in Fig. 2.20 on p. 45.

The operation of an STFT-based signal processor is diagrammed in Fig. 5.8. The signal is partitioned into segments by the windows. The FFT is applied to each segment separately (only one processing path is shown). The resulting spectrum may then be manipulated in any desired way; later chapters demonstrate some of the possibilities. In the figure, no changes are made to the spectrum and so the inverse transform (the IFFT block) rebuilds each segment without modification. When summed together, the segments reconstruct the original signal. Thus the STFT is invertible: it is possible to break the signal into spectral snapshots and then reconstruct the original signal from the snapshots.

In typical use, the support of the window (the region over which it is nonzero) is between 512 and 4096 samples. Using a medium window of size 2048 and a sampling rate of 44.1 kHz, the resolution in frequency is, from (5.10), about 21.5 Hz. The resolution in time is $\frac{2048}{44100} \approx 46$ ms (about 1/20 of a second). This may be adequate to specify high frequencies (where 21.5 Hz is a small percentage of the frequency in question) but it is far too coarse at the low end. A low note on the piano may have a fundamental near 80 Hz. The resolution of this FFT is only good to within 25%! For comparison, the distance between consecutive notes on the piano is a constant 6%. Musical keyboards and scales are designed so that all equidistant musical intervals are a constant percentage apart in frequency, mimicking the constant percentage pitch perception of the auditory system.

This discussion raises two questions. First, is there a way to improve the frequency resolution of the STFT without overly harming the time resolution? The phase vocoder makes improved frequency estimates by using phase information that the STFT ignores; this is explored in Sect. 5.3.4. Second, is there a way to create a transform that operates at constant percentages (like the ear) rather than at constant differences? Brown's "constant-Q spectral transform" [B: 20] uses variable length windows (small ones to analyze high frequencies and large ones to capture low frequencies) that are tuned logarithmically like the steps of the 12-tone equal tempered scale (the chromatic scale). But it has not become popular, probably due to its noninvertibility (hence it cannot be used in a signal processing system like the STFT of Fig. 5.8). Perhaps the most successful method that can operate with constant percentages is the wavelet transform, which is discussed in Sect. 5.4.

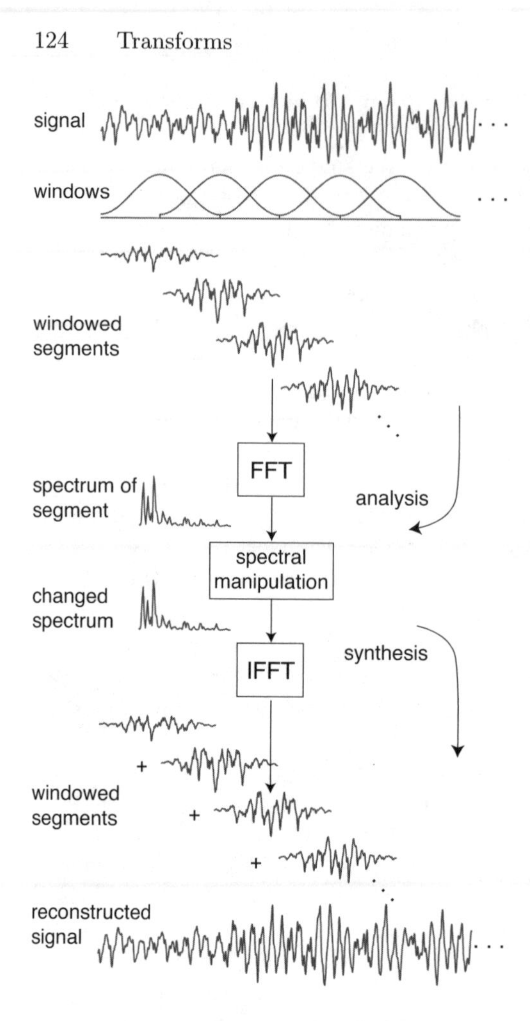

Fig. 5.8. A short-time Fourier transform (STFT) signal processor is an analysis/synthesis method that begins by windowing a signal into short segments. The FFT is applied to each segment separately and the resulting spectral snapshot can be manipulated in a variety of ways. After the desired spectral changes, the resynthesis is handled by the inverse FFT to return each segment to the time domain. The modified segments are then summed. For the special case where no spectral manipulations are made (as shown), the output of the STFT is identical to the input.

5.3.4 The Phase Vocoder

Like the short-time Fourier transform, the phase vocoder (PV) is an analysis-resynthesis technique based on the FFT. The analysis portion of the PV begins by slicing the signal into windowed segments that are analyzed using the FFT. If the PV used only the magnitude spectrum, the frequency resolution of each segment would be dictated by (5.10). Instead, the PV compares the phases of corresponding partials in successive segments and uses the comparison to improve the frequency estimates. The gains can be considerable. The resynthesis of the PV calculates a vector that can be inverted using the IFFT. The resulting signal has the same frequency content as the original but can be stretched or compressed in time.

Phase vocoders based on banks of (analog) filters were introduced by Flanagan [B: 62] for the compression of speech signals. Portnoff [B: 171] showed how the same idea can be implemented digitally using the FFT, and

Dolson's tutorial [B: 49] helped bring the method to the attention of the computer music community. Recent work such as Laroche [B: 124] focuses on fine-tuning the resynthesis portion of the algorithm for various applications such as pitch shifting and time-stretching. Several MATLAB® implementations are currently available on the internet: see Brandorff and Møller-Nielsen's `pVoc` [W: 7] and Ellis' `pvoc.m` [W: 14]. Also notable is Klingbeil's graphical interface called SPEAR (Sinusoidal Partial Editing Analysis and Resynthesis) [B: 113] [W: 22]. There is also a version on the CD in the software folder.

Analysis Using the Phase Vocoder

To see how the analysis portion of the PV can use phase information to make improved frequency estimates, suppose there is a sinusoid of unknown frequency but with known phases: at time t_1 the sinusoid has phase θ_1 and at time t_2 it has phase θ_2. The situation is depicted in Fig. 5.9. The sinusoid may have a frequency that moves it directly from θ_1 to θ_2 in time $\Delta t = t_2 - t_1$. Or it may begin at θ_1, move completely around the circle, and end at θ_2 after one full revolution. Or it may revolve twice, or n times.[11] Thus the frequency must be

$$f_n = \frac{(\theta_2 - \theta_1) + 2\pi n}{2\pi \Delta t} \tag{5.11}$$

for some integer n. Without more information, it is not possible to uniquely determine f, though it is constrained to one of the above values.

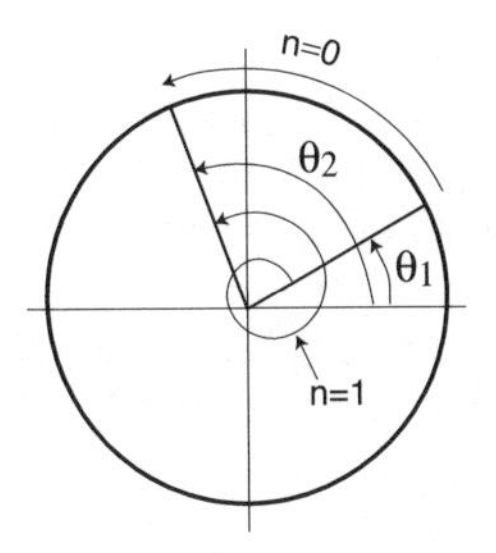

Fig. 5.9. The phases θ_1 and θ_2 of a sinusoid are known at two different times t_1 and t_2. The frequency must then fulfill f_n of (5.11), where the integer n specifies the number of revolutions around the circle (the number of periods of the sinusoid that occur within the specified time Δt). The two cases $n = 0$ (the lowest possible frequency) and $n = 1$ (one complete revolution) are shown.

The phase vocoder exploits (5.11) by locating a common peak in the magnitude spectrum of two different segments. It then chooses the f_n that is closest to the frequency of that peak. This is shown diagrammatically in Fig. 5.10 where the signal is assumed to be a single sinusoid that spans the time interval over which the calculations are made. The output of the windowing is a collection of short sinusoidal bursts. The FFT is applied to each burst,

[11] In other words, the frequency multiplied by the change in time must equal the change in angle, that is, $2\pi f(t_2 - t_1) = \theta_2 - \theta_1$ or some 2π multiple. Solving for f gives (5.11).

resulting in magnitude and phase spectra. For the case of a pure sinusoidal input, the magnitude spectra of the successive spectra are the same (as shown). But the phase spectra differ, and these provide the needed values of θ_1 (the phase corresponding to the peak of the first magnitude spectrum) and θ_2 (the phase corresponding to the peak of the second magnitude spectrum). The time difference Δt can be determined directly from the window length, the overlap factor, and the sampling rate. These values are then substituted into (5.11) and the f_n that is closest in frequency to the peak is the PVs frequency estimate.

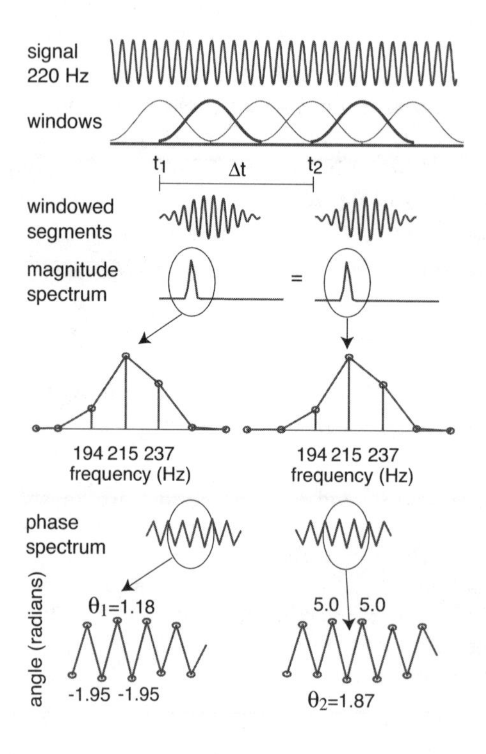

Fig. 5.10. The analysis portion of the phase vocoder rests on the assumption that a sinusoid remains fixed in frequency for the duration of the calculation. The input (shown here as a 220 Hz sinusoid) is windowed and the FFT is taken of the resulting bursts. The common peaks in the magnitude spectra are located (in this case at 215 Hz) and their phases recorded (in this case, the phase corresponding to the first and second bursts are $\theta_1 = 1.18$ and $\theta_2 = 1.87$). Information about sampling rate, window size, and overlap factor specify the time interval between the bursts (in this case, $\Delta t = 0.07$). These parameters are entered into (5.11) and the f_n closest to the frequency of the peak is chosen as the frequency estimate. In more interesting signals, when there are many sinusoids, the method is repeated for each peak in the magnitude spectrum.

To see the PV in action, and to give an idea of its accuracy, consider the problem of estimating the frequency of a 220 Hz sinusoid using a 2K FFT (assuming a sampling rate of 44.1 kHz). According to (5.10), the resolution of the FFT is 21.5 Hz, that is, it is only possible to find the frequency of the sinusoid to within about 10 Hz. Indeed, the nearby frequencies that are exactly representable are 193.8, 215.3, and 236.9, as shown in the enlargement of the magnitude spectrum in Fig. 5.10. Since the peak at 215.3 is the largest, an actual error of 4.7 Hz occurs when using only the FFT magnitude. The PV improves this by exploiting phase information. The phases corresponding to the peaks at 215.3 are $\theta_1 = 1.18$ and $\theta_2 = 1.87$ and so

$$f_n = \frac{(\theta_2 - \theta_1) + 2\pi n}{2\pi \Delta t} = \frac{(1.18 - 1.87) + 2\pi n}{2\pi \cdot 0.07}$$

since[12] $\Delta t = \frac{2048}{2} \cdot \frac{1}{44100} = 0.023$. With these values, the first six f_n are

$$47.7472,\ 90.8136,\ 133.8800,\ 176.94649,\ 220.01290,\ \text{and}\ 263.0793.$$

Clearly, the fifth term is closest to 215.3, and the error in the frequency estimate is 0.0129 Hz, a vast improvement over 4.7 Hz. This kind of accuracy is typical and is not just a numerical fluke. In fact, [B: 177] shows that, under certain conditions (for a signal consisting of a single sinusoid and with a Δt corresponding to a single sample) the phase vocoder estimate of the frequency is closely related to the maximum likelihood estimate.

In more complex situations, when the input signal consists of many sine waves, the phase manipulations are repeated for each peak individually, which is justified as long as the peaks are adequately separated in frequency. Once the analysis portion is complete, it is possible to change the signal in a variety of ways: by modifying the rate at which time passes (spacing the output bursts differently from the input bursts), by changing the frequencies in the signal (so that the output will contain different frequencies than the input), by adding or by subtracting partials.

Resynthesis Using the Phase Vocoder

Once the modifications are complete, it is possible to synthesize the output waveform. One possibility is to use a straightforward additive-synthesis where the partials (each with its desired frequency and amplitude) are generated individually and then summed together. This is computationally intensive[13] when there are a large number of partials. Fortunately, there is a better way: the PV creates a complex-valued (frequency) vector. This is inverted using the IFFT and the resulting output bursts are time-shifted and summed as in the STFT.

Specification of the magnitude spectrum of the output is straightforward since it can be inherited directly from the input. The phase values are chosen to ensure continuity of the most prominent partials through successive bursts, as shown in Fig. 5.11 for a single sinusoid. For each peak j in the magnitude spectrum, the required phase can be calculated directly from the frequency f_j and the time interval Δt between frame k and frame $k-1$. This is

$$\theta_k^j = \theta_{k-1}^j + 2\pi f^j \Delta t.$$

It is also necessary to choose the nearby phases (those under the same peak in the magnitude spectrum). If these are chosen to be $\theta_k^j + \mathrm{mod}(n, 2)\pi$ (where

[12] The window width is 2048 with an overlap of 2, the sampling rate is 44.1 kHz, and the second burst is one step ahead of the first.

[13] There is still the problem of assigning appropriate phase values to the generated sine waves, a problem that the phase vocoder handles elegantly.

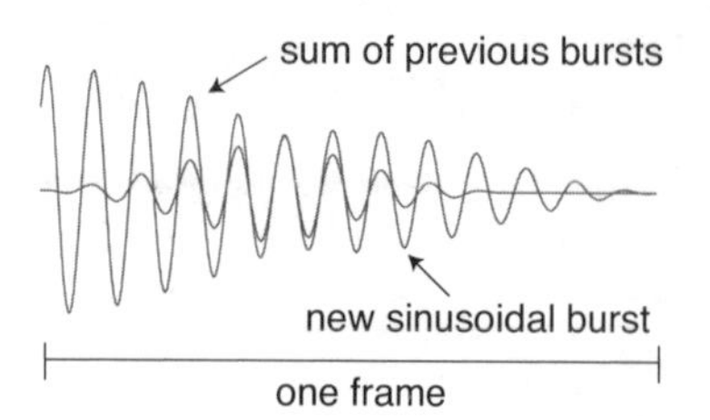

Fig. 5.11. Each new windowed sinusoidal element (burst) is added in phase with the already existing signal. The result is a continuous sinusoid of the specified frequency.

n is the number of bins away from the peak value), the burst generated by the IFFT will be windowed with tapered ends, as in Fig. 5.11. For example, in the phase spectrum plots of Fig. 5.10, the values to the left and right of θ_1 and θ_2 are (approximately) either π or 0 away.[14] An implementation of the phase vocoder called `PV.m` can be found in the MATLAB® folder on the CD.

5.4 Wavelet Transforms

In the STFT and the phase vocoder, sinusoids are used as basis functions and windows are used to localize the signal to a particular time interval. In the wavelet transforms, the windows are incorporated directly into the basis functions and a variety of nonsinusoidal shapes are common. Several different "mother wavelets" are shown in Fig. 5.12.

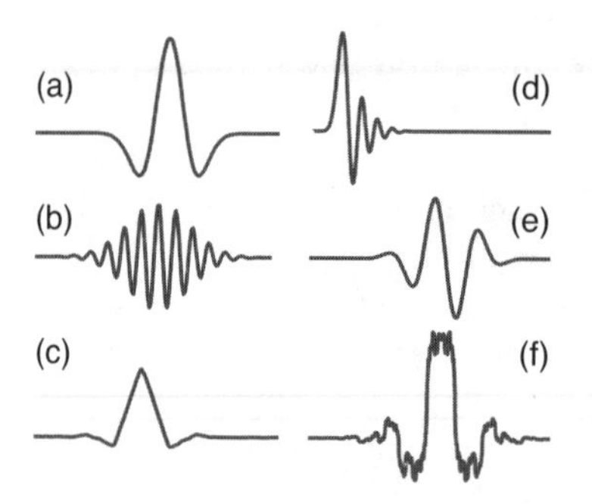

Fig. 5.12. There are many kinds of wavelet basis functions, including the (a) Mexican Hat wavelet, (b) complex Morlet wavelet, (c) Coiflets wavelet, (d) Daubechies wavelet, (e) complex Gaussian wavelet, and the (f) biorthogonal spline wavelet. The wavelet transform operates by correlating the signal with scaled and shifted versions of a basis function.

The wavelet transform uses the mother wavelet much as the STFT uses a windowed sinusoid: one parameter specifies where in time the wavelet is centered (analogous to the windowing) and another parameter stretches or compresses the wavelet. This latter is called the *scale* of the wavelet and is analogous to frequency in the STFT.

[14] A formal justification of this choice requires observing that the phase values of a Bartlett (and a Parzen window) have exactly this pattern of values. Other patterns of 0 and π, such as that in [W: 33], correspond to different choices of output windowing functions.

Let $\psi(t)$ be the mother wavelet (for example, any of the signals[15] in Fig. 5.12), and define

$$\psi_{a,b}(t) = \frac{1}{\sqrt{|a|}} \psi\left(\frac{t-b}{a}\right).$$

The parameter b shifts the wavelet in time while the parameter a scales the wavelet by stretching or compressing it (and also by adjusting the amplitude). Figure 5.13 illustrates the effects of the two parameters for several different values.

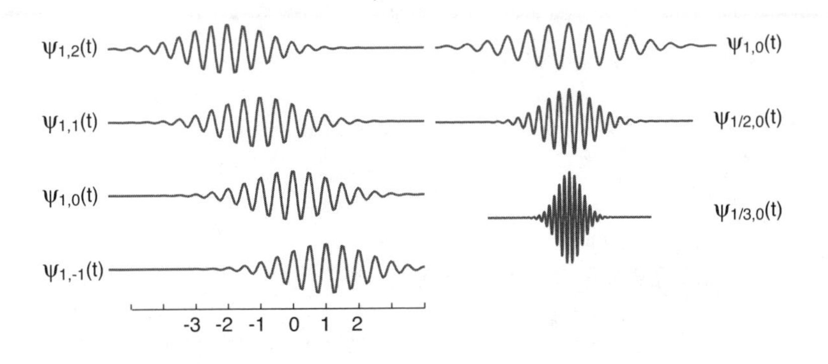

Fig. 5.13. The complex Morlet wavelet is a complex-valued sinusoid windowed by a Gaussian envelope $\psi_{a,b}(t) = \frac{1}{\sqrt{3\pi|a|}} e^{j2\pi 2t} e^{-\frac{(t-b)^2}{3a}}$. These plots show the real part of the Morlet wavelet for a variety of shifts b and scales a. As b decreases, the wavelet moves to the right; as a decreases, the wavelet compresses and grows.

The continuous wavelet transform uses the shifted and scaled functions as a basis for representing a signal $x(t)$ via the inner product

$$W(a,b) = \langle x(t), \psi^*_{a,b}(t) \rangle. \tag{5.12}$$

For every (a,b) pair, the coefficient $W(a,b)$ is the inner product of the signal with the appropriately scaled and shifted basis function $\psi_{a,b}(t)$. Where the signal is aligned with the basis function (when $x(t)$ locally looks like the basis function), the coefficient is large. Where the signal is very different from the basis function (the extreme being orthogonal) then the coefficient is small. As a and b change, the inner product scans through the signal looking for places (in time) and values (in scale) where the signal correlates well with the wavelet. This suggests that prior information about the shape or general character of the signal can be usefully exploited by the wavelet transform by

[15] To be considered a wavelet function, $\psi(t)$ must be orthogonal to the function $x(t) = 1$ and must have finite energy. Thus $\langle 1, \psi^*(t) \rangle = 0$ and $\langle \psi(t), \psi^*(t) \rangle < \infty$.

tailoring the wavelet to the signal. When the information is correct, the set of parameters $W(a,b)$ can provide a concise and informative representation of the signal. When the information is incorrect, the wavelet representation may be less useful.

When the wavelet is real-valued, $W(a,b)$ is real; when the wavelet is complex-valued (like the Morlet wavelet used in Fig. 5.13) then $W(a,b)$ is complex. It is common to plot the magnitude of $W(a,b)$ (using a color or grayscale mapping) with axes defined by the scale a and time b, much as the STFT and the spectrogram are plotted with axes defined by frequency and time. When $W(a,b)$ is complex, it is also common to plot a similar contour with the phase, though sometimes plots of the real and/or imaginary parts are useful. For example, Fig. 5.14 shows separate plots of the magnitude and phase when the complex Gaussian wavelet (from Fig. 5.12(e)) is applied to an input that is a train of spikes separated by one second. The temporal locations of the spikes are readily visible in both the magnitude and the phase plots (the vertical stripes).

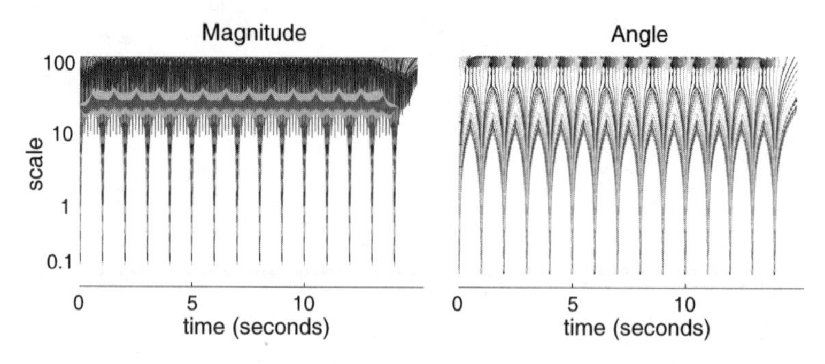

Fig. 5.14. The complex Gaussian wavelet is applied to an input spike train with period one second. The left plot shows values of the magnitude of the wavelet coefficients $W(a,b)$ of (5.12) while the right hand plot shows the phase. The location in time of the spikes is easy to see.

There is an interesting parallel between the wavelet transform of the spike train in Fig. 5.14 and the corresponding Fourier transform of a spike train in Fig. 5.6. In both cases, the transform returns a display (plot) that contains data of the same general character as the input. The output of the FT is a spike train; the output of the wavelet transform is a collection of regularly spaced ridges in a two-dimensional field. This suggests that the wavelet transform is not going to be able to magically solve the rhythm finding problem. In many cases (such as the spike train) it is no simpler to determine regularity from the output of the wavelet transform than it is to determine regularity directly from the input itself. Again, as with the FT, this does imply that wavelet transforms cannot play a role in rhythm analysis. Rather, it means that they must be used thoughtfully and in proper contexts.

This discussion has stressed the similarities between the STFT and the wavelet transforms. There are also important differences. In the windowed FFT and the granular techniques (such as the "Gabor grains" of Fig. 2.23 on p. 47), the frequency of the waveform is independent of the grain duration. In wavelets, there is an inverse relation maintained between the frequency of the waveforms and the duration of the wavelet. Unlike a typical grain, most wavelets contain the same number of cycles irrespective of the scale (roughly, frequency) of the wavelet. Thus the duration of the wavelet window grows or shrinks as a function of the scale; wavelets that capture low frequency information are dilated (wide in time) while those that represent high frequencies are contracted. This allows more precise localization of the high frequency components. This can be seen in Fig. 5.13 where the Morlet wavelet maintains the same number of cycles at all scale values.

5.5 Periodicity Transforms

The Periodicity Transforms (PT) decompose data into a sum of periodic sequences by projecting onto a set of "periodic subspaces" $\mathcal{P}_p$, leaving residuals whose periodicities have been removed. As the name suggests, this decomposition is accomplished directly in terms of periodic sequences and not in terms of frequency or scale, as do the Fourier and Wavelet Transforms. In consequence, the representation is linear-in-period, rather than linear-in-frequency or linear-in-scale. Unlike most transforms, the set of basis vectors is not specified a priori, rather, the Periodicity Transforms find their own "best" set of basis elements. In this way, it is analogous to the approach of Karhunen-Loeve [B: 23], which transforms a signal by projecting onto an orthogonal basis that is determined by the eigenvectors of the covariance matrix. In contrast, the periodic subspaces $\mathcal{P}_p$ lack orthogonality, which underlies much of the power of (and difficulties with) the Periodicity Transforms. Technically, the collection of all periodic subspaces forms a *frame* [B: 24], a more-than-complete spanning set. The PT specifies ways of sensibly handling the redundancy by exploiting some of the general properties of the periodic subspaces.

This section describes the PT and compares its output to other transforms in a number of examples. Later chapters will detail how the PT may be applied to the problem of detecting rhythmic patterns in a musical setting. Much of this is based on the work with Tom Staley documented in [B: 207] and [B: 208], which may be found on the accompanying CD.

5.5.1 Periodic Subspaces

A sequence of real numbers $x[k]$ is called *p-periodic* if there is an integer p with $x[k+p] = x[k]$ for all integers k. Let

$\mathcal{P}_p$ be the set of all p-periodic sequences, and
$\mathcal{P}$ be the set of all periodic sequences.

In practice, a data vector x contains N elements. This can be considered to be a single period of an element $x_N \in \mathcal{P}_N \subset \mathcal{P}$, and the goal is to locate smaller periodicities within x_N, should they exist. The strategy is to "project" x_N onto the subspaces $\mathcal{P}_p$ for $p < N$. When x_N is "close to" some periodic subspace $\mathcal{P}_p$, then there is a p-periodic element $x_p \in \mathcal{P}_p$ that is close to the original x. This x_p is an ideal choice to use when decomposing x. To make these ideas concrete, it is necessary to understand the structure of the various spaces, and to investigate how the needed calculations can be realized.

Observe that $\mathcal{P}_p$ is closed under addition since the sum of two sequences with period p is itself p-periodic. Similarly, $\mathcal{P}$ is closed under addition since the sum of x_1 with period p_1 and x_2 with period p_2 has period (at most) p_1p_2. Thus, with scalar multiplication defined in the usual way, both $\mathcal{P}_p$ and $\mathcal{P}$ form linear vector spaces, and $\mathcal{P}$ is equal to the union of the $\mathcal{P}_p$.

For every period p and every "time shift" s, define the sequence $\delta_p^s[j]$ for all integers j by

$$\delta_p^s[j] = \begin{cases} 1, & \text{if } (j-s) \bmod p = 0 \\ 0, & \text{otherwise} \end{cases}. \tag{5.13}$$

The sequences δ_p^s for $s = 0, 1, 2, ..., p-1$ are called the p-periodic basis vectors since they form a basis for $\mathcal{P}_p$.

Example 5.1. For $p = 4$, the 4-periodic basis vectors

j	$\cdots$	-4	-3	-2	-1	0	1	2	3	4	5	6	7	$\cdots$
$\delta_4^0[j]$	$\cdots$	1	0	0	0	1	0	0	0	1	0	0	0	$\cdots$
$\delta_4^1[j]$	$\cdots$	0	1	0	0	0	1	0	0	0	1	0	0	$\cdots$
$\delta_4^2[j]$	$\cdots$	0	0	1	0	0	0	1	0	0	0	1	0	$\cdots$
$\delta_4^3[j]$	$\cdots$	0	0	0	1	0	0	0	1	0	0	0	1	$\cdots$

span the 4-periodic subspace $\mathcal{P}_4$.

An inner product can be imposed on the periodic subspaces by considering the function from $\mathcal{P} \times \mathcal{P}$ into $\Re$ defined by

$$\langle x, y \rangle = \lim_{k \to \infty} \frac{1}{2k+1} \sum_{i=-k}^{k} x[i]y[i], \tag{5.14}$$

for arbitrary elements x and y in $\mathcal{P}$. For the purposes of calculation, observe that if $x \in \mathcal{P}_{p_1}$ and $y \in \mathcal{P}_{p_2}$, the product sequence $x[i]y[i] \in \mathcal{P}_{p_1p_2}$ is p_1p_2-periodic, and (5.14) is equal to the average over a single period, that is,

$$\langle x, y \rangle = \frac{1}{p_1p_2} \sum_{i=0}^{p_1p_2-1} x[i]y[i]. \tag{5.15}$$

The corresponding norm on $\mathcal{P}$ is called the *Periodicity Norm*

$$||x|| = \sqrt{\langle x, x \rangle}. \tag{5.16}$$

These definitions of inner product and norm are slightly different from (5.1) and (5.2). The extra term ($\frac{1}{p_1 p_2}$ in the example above) ensures that the norm gives the same value whether x is considered to be an element of $\mathcal{P}_p$, of $\mathcal{P}_{kp}$ (for positive integers k), or of $\mathcal{P}$.

Example 5.2. Let $x \in \mathcal{P}_3$ be the 3-periodic sequence $\{\cdots, 1, 2, 3, \cdots\}$ and let $y \in \mathcal{P}_6$ be the 6-periodic sequence $\{\cdots, 1, 2, 3, 1, 2, 3, \cdots\}$. Using (5.16), $||x|| = ||y||$.

As usual, the signals x and y in $\mathcal{P}$ are said to be orthogonal if $\langle x, y \rangle = 0$.

Example 5.3. The periodic basis elements δ_p^s for $s = 0, 1, ..., p-1$ are orthogonal, and $||\delta_p^s|| = \sqrt{1/p}$.

The idea of orthogonality can also be applied to subspaces. A signal x is orthogonal to the subspace $\mathcal{P}_p$ if $\langle x, x_p \rangle = 0$ for all $x_p \in \mathcal{P}_p$, and two subspaces are orthogonal if every vector in one is orthogonal to every vector in the other. Unfortunately, the periodic subspaces $\mathcal{P}_p$ are not orthogonal to each other.

Example 5.4. If p_1 and p_2 are mutually prime, then

$$\langle \delta_{p_1}^s, \delta_{p_2}^s \rangle = \langle \delta_{p_1 p_2}^s, \delta_{p_1 p_2}^s \rangle = \frac{1}{p_1 p_2} \neq 0.$$

Suppose that $p_1 p_2 = p_3$. Then $\mathcal{P}_{p_1} \subset \mathcal{P}_{p_3}$ and $\mathcal{P}_{p_2} \subset \mathcal{P}_{p_3}$, which restates the fact that any sequence that is p-periodic is also np-periodic for any integer n. But $\mathcal{P}_{p_3}$ can be strictly larger than $\mathcal{P}_{p_1} \cup \mathcal{P}_{p_2}$.

Example 5.5. Let $x = \{\cdots, 1, 2, 1, -1, -2, -1, \cdots\} \in \mathcal{P}_6$. Then x is orthogonal to both $\mathcal{P}_2$ and $\mathcal{P}_3$, since direct calculation shows that x is orthogonal to δ_2^s and to δ_3^s for all s.

In fact, no two subspaces $\mathcal{P}_p$ are linearly independent, since $\mathcal{P}_1 \subset \mathcal{P}_p$ for every p. This is because the vector $\mathbf{1}$ (the 1-periodic vector of all ones) can be expressed as the sum of the p periodic basis vectors

$$\mathbf{1} = \sum_{s=0}^{p-1} \delta_p^s$$

for every p. In fact, $\mathcal{P}_1$ is the only commonality between $\mathcal{P}_{p_1}$ and $\mathcal{P}_{p_2}$ when p_1 and p_2 are mutually prime. More generally, $\mathcal{P}_{np} \cap \mathcal{P}_{mp} = \mathcal{P}_p$ when n and m are mutually prime. The structure of the periodic subspaces reflects the structure of the integers.

5.5.2 Projection onto Periodic Subspaces

The primary reason for formulating this problem in an inner product space is to exploit the projection theorem. Let $x \in \mathcal{P}$ be arbitrary. Then a minimizing vector in $\mathcal{P}_p$ is an $x_p^* \in \mathcal{P}_p$ such that

$$||x - x_p^*|| \leq ||x - x_p||, \quad \text{for all } x_p \in \mathcal{P}_p.$$

Thus x_p^* is the p-periodic vector "closest to" the original x. The projection theorem from Luenberger [B: 135], stated here in slightly modified form, shows how x_p^* can be characterized as an orthogonal projection of x onto $\mathcal{P}_p$.

Theorem 5.6 (The Projection Theorem). *Let $x \in \mathcal{P}$ be arbitrary. A necessary and sufficient condition that x_p^* be a minimizing vector in $\mathcal{P}_p$ is that the error $x - x_p^*$ be orthogonal to $\mathcal{P}_p$.*

Since $\mathcal{P}_p$ is a finite (p-dimensional) subspace, x_p^* will in fact exist, and the projection theorem provides, after some simplification, a simple way to calculate it. The optimal $x_p^* \in \mathcal{P}_p$ can be expressed as a linear combination of the periodic basis elements δ_p^s as

$$x_p^* = \alpha_0 \delta_p^0 + \alpha_1 \delta_p^1 + \cdots + \alpha_{p-1} \delta_p^{p-1}.$$

According to the projection theorem, the unique minimizing vector is the orthogonal projection of x on $\mathcal{P}_p$, that is, $x - x_p^*$ is orthogonal to each of the δ_p^s for $s = 0, 1, ..., p-1$. Thus

$$0 = \langle x - x_p^*, \delta_p^s \rangle = \langle x - \alpha_0 \delta_p^0 - \alpha_1 \delta_p^1 - ... - \alpha_{p-1} \delta_p^{p-1}, \delta_p^s \rangle.$$

Since the δ_p^s are orthogonal to each other, this can be rewritten using the additivity of the inner product as

$$\begin{aligned} &= \langle x - \alpha_s \delta_p^s, \delta_p^s \rangle \\ &= \langle x, \delta_p^s \rangle - \alpha_s \langle \delta_p^s, \delta_p^s \rangle \\ &= \langle x, \delta_p^s \rangle - \frac{\alpha_s}{p}. \end{aligned}$$

Hence α_s can be written as

$$\alpha_s = p \, \langle x, \delta_p^s \rangle.$$

Since $x \in \mathcal{P}$, it is periodic with some period N. From (5.15), the above inner product can be calculated

$$\alpha_s = p \, \frac{1}{pN} \sum_{i=0}^{pN-1} x[i] \delta_p^s[i].$$

But δ_p^s is zero except when $(s - i) \bmod p = 0$, and this simplifies to

$$\alpha_s = \frac{1}{N} \sum_{n=0}^{N-1} x[s+np]. \tag{5.17}$$

If, in addition, N/p is an integer, then this reduces to

$$\alpha_s = \frac{1}{N/p} \sum_{n=0}^{N/p-1} x[s+np]. \tag{5.18}$$

Example 5.7. With $N = 14$ and $p = 2$, let $x \in \mathcal{P}_{14}$ be the 14-periodic sequence

$$x = \{\cdots, 2, -1.1, -1.1, 2, -1.2, -1.2, 2, -1.1, -1.1, 2, -1.2, -1.1, 2, -1.1, \cdots\}.$$

Then the projection of x onto $\mathcal{P}_2$ is $x_2 = \{\cdots, 0.2, -0.228, \cdots\}$.

This sequence x_2 is the 2-periodic sequence that best "fits" this 14-periodic x. But looking at this x closely suggests that it has more of the character of a 3-periodic sequence, albeit somewhat truncated in the final "repeat" of the $2, -1, -1$. Accordingly, it is reasonable to project x onto $\mathcal{P}_3$.

Example 5.8. With $N = 14$ and $p = 3$, let $x \in \mathcal{P}_{14}$ be as defined in example 5.7. Then the projection of x onto $\mathcal{P}_3$ (using (5.17)) is $x_3 = -0.2\{\cdots, 1, 1, 1, \cdots\}$.

Clearly, this does not accord with the intuition that this x is "almost" 3-periodic. In fact, this is an example of a rather generic effect. Whenever N and p are mutually prime, the sum in (5.17) cycles through all the elements of x, and so $\alpha_s = \frac{1}{N}\sum_{i=0}^{N-1} x[i]$ for all s. Hence the projection onto $\mathcal{P}_p$ is the vector of all ones (times the mean value of the x). The problem here is the incommensurability of the N and p.

What does it mean to say that x (with length N) is p-periodic when N/p is not an integer? Intuitively, it should mean that there are $\lfloor N/p \rfloor$ complete repeats of the p-periodic sequence (where $\lfloor z \rfloor$ is the largest integer less than or equal to z) plus a "partial repeat" within the remaining $\bar{N} = N - p\lfloor N/p \rfloor$ elements. For instance, the $N = 14$ sequence

$$x_1,\ x_2,\ x_3,\ x_1,\ x_2,\ x_3,\ x_1,\ x_2,\ x_3,\ x_1,\ x_2,\ x_3,\ x_1,\ x_2$$

can be considered a (truncated) 3-periodic sequence.

There are two ways to formalize this notion: to "shorten" x so that it is compatible with p, or to "lengthen" δ_p^s so that it is compatible with N. Though roughly equivalent (they differ only in the final $\bar{N}$ elements), the first approach is simpler since it is possible to replace x with $x_{\tilde{N}}$ (the $\tilde{N}$-periodic sequence constructed from the first $\tilde{N} = p\lfloor N/p \rfloor$ elements of x) whenever the projection operator is involved. With this understanding, (5.17) becomes

$$\alpha_s = \frac{1}{\lfloor N/p \rfloor} \sum_{n=0}^{\lfloor N/p \rfloor - 1} x_{\tilde{N}}[s+np]. \tag{5.19}$$

Example 5.9. With $N = 14$ and $p = 3$, let $x \in \mathcal{P}_{14}$ be as defined in example 5.7. Then the projection of x onto $\mathcal{P}_3$ (using (5.19)) is $x_3 = \{\cdots, 2, -1.14, -1.125, \cdots\}$.

Clearly, this captures the intuitive notion of periodicity far better than example 5.8, and the sum (5.19) forms the foundation of the Periodicity Transforms. The calculation of each α_s thus requires $\lfloor N/p \rfloor$ operations (additions). Since there are p different values of s, the calculation of the complete projection x_p requires $\tilde{N} \approx N$ additions. MATLAB® routines that carry out the needed calculations are available on the CD.

Let $\pi(x, \mathcal{P}_p)$ represent the projection of x onto $\mathcal{P}_p$. Then

$$\pi(x, \mathcal{P}_p) = \sum_{s=0}^{p-1} \alpha_s \, \delta_p^s \tag{5.20}$$

where the δ_p^s are the (orthogonal) p-periodic basis elements of $\mathcal{P}_p$. Clearly, when $x \in \mathcal{P}_p$, $x = \pi(x, \mathcal{P}_p)$. By construction, when x is projected onto $\mathcal{P}_{np}$ it finds the best np-periodic components within x, and hence the residual $r = x - \mathcal{P}_{np}$ has no np-periodic component. The content of the next result is that this residual also has no p-periodic component. In essence, the projection onto $\mathcal{P}_{np}$ "grabs" all the p-periodic information.

Theorem 5.10. *For any integer n, let $r = x - \pi(x, \mathcal{P}_{np})$ be the residual after projecting x onto $\mathcal{P}_{np}$. Then $\pi(r, \mathcal{P}_p) = 0$.*

All proofs are found in [B: 207] which can also be found on the CD. The next result relates the residual after projecting onto $\mathcal{P}_p$ to the residual after projection onto $\mathcal{P}_{np}$.

Theorem 5.11. *Let $r_p = x - \pi(x, \mathcal{P}_p)$ be the residual after projecting x onto $\mathcal{P}_p$. Similarly, let $r_{np} = x - \pi(x, \mathcal{P}_{np})$ denote the residual after projecting x onto $\mathcal{P}_{np}$. Then*

$$r_{np} = r_p - \pi(r_p, \mathcal{P}_{np}).$$

Combining the two previous results shows that the order of projections doesn't matter in some special cases, that is

$$\pi(x, \mathcal{P}_p) = \pi(\pi(x, \mathcal{P}_p), \mathcal{P}_{np}) = \pi(\pi(x, \mathcal{P}_{np}), \mathcal{P}_p),$$

which is used in the next section to help sensibly order the projections.

5.5.3 Algorithms for Periodic Decomposition

The Periodicity Transforms search for the best periodic characterization of the length N signal x. The underlying technique is to project x onto some periodic subspace giving $x_p = \pi(x, \mathcal{P}_p)$, the closest p-periodic vector to x.

This periodicity is then removed from x leaving the residual $r_p = x - x_p$ stripped of its p-periodicities. Both the projection x_p and the residual r_p may contain other periodicities, and so may be decomposed into other q-periodic components by projection onto $\mathcal{P}_q$. The trick in designing a useful algorithm is to provide a sensible criterion for choosing the order in which the successive ps and qs are chosen. The intended goal of the decomposition, the amount of computational resources available, and the measure of "goodness-of-fit" all influence the algorithm. The analysis of the previous sections can be used to guide the decomposition by exploiting the relationship between the structure of the various $\mathcal{P}_p$. For instance, it makes no sense to project x_p onto $\mathcal{P}_{np}$ because $x_p \in \mathcal{P}_{np}$ and no new information is obtained. This section presents several different algorithms, discusses their properties, and then compares these algorithms with some methods available in the literature.

One subtlety in the search for periodicities is related to the question of appropriate boundary (end) conditions. Given the signal x of length N, it is not particularly meaningful to look for periodicities longer than $p = N/2$, even though nothing in the mathematics forbids it. Indeed, a "periodic" signal with length $N-1$ has $N-1$ degrees of freedom, and surely can match x very closely, yet provides neither a convincing explanation nor a compact representation of x. Consequently, we restrict further attention to periods smaller than $N/2$.

Probably the simplest useful algorithm operates from small periods to large, as shown in Table 5.1. The Small-To-Large Algorithm is simple because there is no need to further decompose the basis elements x_p; if there were significant q-periodicities within x_p (where "significant" is determined by the threshold T), they would already have been removed by x_q at an earlier iteration. The algorithm works well because it tends to favor small periodicities, to concentrate the power in $\mathcal{P}_p$ for small p, and hence to provide a compact representation.

Table 5.1. Small-To-Large Algorithm

pick threshold $T \in (0,1)$
let $r = x$
for $p = 2, 3, ..., N/2$
    $x_p = \pi(r, \mathcal{P}_p)$
    if $\frac{||r - x_p||}{||x||} > T$
        $r = r - x_p$
        save x_p as basis element
    end
end

Thinking of the norm as a measure of power, the threshold is used to insure that each chosen basis element removes at least a fraction T of the power from the signal. Of course, choosing different thresholds leads to different decompositions. If T is chosen too small (say zero) then the decomposition will simply pick the first linear independent set from among the p-periodic basis vectors

$$\overbrace{\delta_2^1,\ \delta_2^2}^{\mathcal{P}_2},\ \overbrace{\delta_3^1,\ \delta_3^2,\ \delta_3^3}^{\mathcal{P}_3},\ \overbrace{\delta_4^1,\ \delta_4^2,\ \delta_4^3,\ \delta_4^4}^{\mathcal{P}_4},\ \delta_5^1,\ \delta_5^2, ...,$$

which defeats the purpose of searching for periodicities. If T is chosen too large, then too few basis elements may be chosen (none as $T \to 1$). In between "too small" and "too large" is where the algorithm provides interesting descriptions. For many problems, $0.01 < T < 0.1$ is appropriate, since this allows detection of periodicities containing only a few percent of the power, yet ignores those p which only incidentally contribute to x.

An equally simple "Large-To-Small" algorithm is not feasible, because projections onto x_p for composite p may mask periodicities of the factors of p. For instance, if $x_{100} = \pi(x, \mathcal{P}_{100})$ removes a large fraction of the power, this may in fact be due to a periodicity at $p = 20$, yet further projection of the residual onto $\mathcal{P}_{20}$ is futile since $\pi(x - x_{100}, \mathcal{P}_{20}) = 0$ by Theorem 5.10. Thus an algorithm that decomposes from large p to smaller p must further decompose both the candidate basis element x_p as well as the residual r_p, since either might contain smaller q-periodicities.

The M-Best Algorithm deals with these issues by maintaining lists of the M best periodicities and the corresponding basis elements. The first step is to build the initial list. This is described in Table 5.2. At this stage, the algorithm has compiled a list of the M periodicities q_i that remove the most "energy" (in the sense of the norm measure) from the sequence. But typically, the q_i will be large (since by Theorem 5.10, the projections onto larger subspaces np contain the projections onto smaller subspaces p). Thus the projections x_{q_i} can be further decomposed into their constituent periodic elements to determine whether these smaller (sub)periodicities remove more energy from the signal than another currently on the list. If so, the new one replaces the old.

Table 5.2. M-Best Algorithm (step 1)

```
pick size M
let r_0 = x
for i = 1, 2, ..., M
        find q_i with ||π(r_{i-1}, P_{q_i})|| ≥ ||π(r_{i-1}, P_q)|| ∀q ∈ {1, 2, ..., N/2}
        r_i = r_{i-1} − π(r_{i-1}, P_{q_i})
        concatenate q_i and x_{q_{i-1}} = π(r_i, P_{q_i}) onto respective lists
end
```

Fortunately, it is not necessary to search all possible periods $p < q_i$ when decomposing, but only the factors. Let $\rho_i = \{n; q_i/n \text{ is an integer}\}$ be the set of factors of q_i. The second step in the algorithm, shown in Table 5.3, begins by projecting the x_{q_i} onto each of its factors $Q \in \rho_i$. If the norm of the new projection x_Q is larger than the smallest norm in the list, and if the sum of all the norms will increase by replacing x_{q_i}, then the new Q is added to the list and the last element x_{q_M} is deleted. These steps rely heavily on Theorem 5.11. For example, suppose that the algorithm has found a strong periodicity in (say) $\mathcal{P}_{140}$, giving the projection $x_{140} = \pi(x, \mathcal{P}_{140})$. Since $140 = 2^2 \cdot 5 \cdot 7$, the factors are $\rho = \{2, 4, 5, 7, 10, 14, 20, 28, 35, 70\}$. Then the inner loop in step 2 searches over each of the $\pi(x_{140}, \mathcal{P}_Q)$ $\forall Q \in \rho$. If x_{140} is "really" composed of a significant periodicity at (say) 20, then this new periodicity is inserted in the list and will later be searched for yet smaller periodicities. The M-Best Algorithm is relatively complex, but it removes the need for a threshold parameter by maintaining the list. This is a sensible approach and it often succeeds in building a good decomposition of the signal. A variation called the M-Best algorithm with γ-modification (or M-Best$_\gamma$) is described in [B: 207] (which can be found on the CD), where the measure of energy removed is normalized by the (square root of) the length p.

Table 5.3. M-Best Algorithm (step 2)

repeat until no change in list
  for $i = 1, 2, ..., M$
    find Q^* with $||\pi(x_{q_i}, \mathcal{P}_{Q^*})|| \geq ||\pi(x_{q_i}, \mathcal{P}_Q)||$ $\forall$ $Q \in \rho_i$
    let $x_{Q^*} = \pi(x_{q_i}, \mathcal{P}_{Q^*})$ be the projection onto $\mathcal{P}_{Q^*}$
    let $x_{q^*} = x_{q_i} - x_{Q^*}$ be the residual
    if $(||x_{q^*}|| + ||x_{Q^*}|| > ||x_{q_M}|| + ||x_{q_i}||)$
     & $(||x_{q^*}|| > \min_k ||x_{q_k}||$ & $||x_{Q^*}|| > \min_k ||x_{q_k}||)$
      replace q_i with q^* and x_{q_i} with x_{q^*}
      insert Q^* and x_{Q^*} into lists at position $i - 1$
      remove q_M and x_{q_M} from end of lists
    end if
  end for
end repeat

Another approach is to project x onto all the periodic basis elements δ_p^s for all p and s, essentially measuring the correlation between x and the individual periodic basis elements. The p with the largest (in absolute value) correlation is then used for the projection. This idea leads to the Best-Correlation Algorithm of Table 5.4, which presumes that good p will tend to have good correlation with at least one of the p-periodic basis vectors. This method tends to pick out periodicities with large regular spikes over those that are more uniform.

A fourth approach is to determine the best periodicity p by Fourier methods, and then to project onto $\mathcal{P}_p$. Using frequency to find periodicity is cer-

Table 5.4. Best-Correlation Algorithm

$M =$ number of desired basis elements
let $r = x$
for $i = 1, 2, ..., M$
  $\rho = \underset{p}{\text{argmax}} \; | < r, \delta_p^s > |$
  save $x_\rho = \pi(r, \mathcal{P}_\rho)$ as basis element
  $r = r - x_\rho$
end

tainly not always the best idea, but it can work well, and has the advantage that it is a well understood process. The interaction between the frequency and periodicity domains can be a powerful tool, especially since the Fourier methods have good resolution at high frequencies (small periodicities) while the periodicity transforms have better resolution at large periodicities (low frequencies).

Table 5.5. Best-Frequency Algorithm

$M =$ number of desired basis elements
let $r = x$
for $i = 1, 2, ..., M$
  $y = ||DFT\{r\}||$
  $p =$ Round$(1/f)$, where $f =$ frequency at which y is max
  save $x_p = \pi(r, \mathcal{P}_p)$ as basis element
  $r = r - x_p$
end

At present, there is no simple way to guarantee that an optimal decomposition has been obtained. One foolproof method for finding the best M subspaces would be to search all of the possible $\binom{N}{M}$ different orderings of projections to find the one with the smallest residual. This is computationally prohibitive in all but the simplest settings, although an interesting special case is when $M = 1$, that is, when only the largest periodicity is of importance.

5.5.4 Signal Separation

When signals are added together, information is often lost. But if there is some characteristic that distinguishes the signals, then they may be recoverable from their sum. Perhaps the best known example is when the spectrum of x and the spectrum of y do not overlap. Then both signals can be recovered from $x + y$ with a linear filter. But if the spectra overlap significantly, the situation is more complicated. This example shows how, if the underlying signals are periodic in nature, then the Periodicity Transforms can be used to recover signals from their sum. This process can be thought of as a way to extract a "harmonic template" from a complicated spectrum.

Consider the signal z in Fig. 5.15, which is the sum of two zero mean sequences, x with period 13 and y with period 19. The spectrum of z is quite complex, and it is not obvious just by looking at the spectrum which parts of the spectrum arise from x and which from y. To help the eye, the two lattices marked A and B point to the spectral lines corresponding to the two periodic sequences. These are inextricably interwoven and there is no way to separate the two parts of the spectrum with linear filtering.

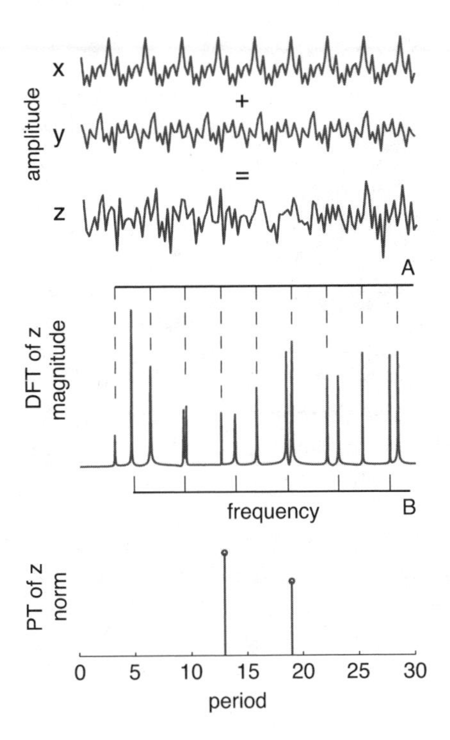

Fig. 5.15. The signal z is the sum of the 13-periodic x and the 19-periodic y. The DFT spectrum shows the overlapping of the two spectra (emphasized by the two lattices labeled A and B), which cannot be separated by linear filtering. The output of the M-Best$_\gamma$ Periodicity Transform, shown in the bottom plot, locates the two periodicities (which were a priori unknown) and reconstructs (up to a constant offset) both x and y given only z.

When the Periodicity Transform is applied to z, two periodicities are found, with periods of 13 and 19, with basis elements that are exactly $x_{13} = x + c_1$ and $y_{19} = y + c_2$, that is, both signals x and y are recovered, up to a constant. Thus the PT is able to locate the periodicities (which were assumed a priori unknown) and to reconstruct (up to a constant offset) both x and y given only their sum. Even when z is contaminated with 50% random noise, the PT still locates the two periodicities, though the reconstructions of x and y are noisy. To see the mechanism, let η be the noise signal, and let $\eta_{13} = \pi(\eta, \mathcal{P}_{13})$ be the projection of η onto the 13-periodic subspace. The algorithm then finds $x_{13} = x + c_1 + \eta_{13}$ as its 13-periodic basis element. If the

x and y were not zero mean, there would also be a component with period one.

For this particular example, all four of the PT variants behave essentially the same, but in general they do not give identical outputs. The Small-To-Large algorithm regularly finds such periodic sequences. The Best-Correlation algorithm works best when the periodic data is spiky. The M-Best algorithm is sometimes fooled into returning multiples of the basic periodicities (say 26 or 39 instead of 13) while the M-Best$_\gamma$ is overall the most reliable and noise resistant. The Best-Frequency algorithm often becomes 'stuck' when the frequency with the largest magnitude does not closely correspond to an integer periodicity. The behaviors of the algorithms are explored in detail in four demonstration files that accompany the periodicity software.[16]

Two aspects of this example deserve comment. First, the determination of a periodicity and its corresponding basis element is tantamount to locating a "harmonic template" in the frequency domain. For example, the 13-periodic component has a spectrum consisting of a fundamental (at a frequency f_1 proportional to $1/13$), and harmonics at $2f_1, 3f_1, 4f_1, \ldots$. Similarly, the 19-periodic component has a spectrum consisting of a fundamental (at a frequency f_2 proportional to $1/19$), and harmonics at $2f_2, 3f_2, 4f_2, \ldots$. These are indicated in Fig. 5.15 by the lattices A and B above and below the spectrum of z. Thus the PT provides a way of finding simple harmonic templates that may be obscured by the inherent complexity of the spectrum. The process of subtracting the projection from the original signal can be interpreted as a multi-notched filter that removes the relevant fundamental and its harmonics. For a single p, this is a kind of "gapped weight" filter familiar to those who work in time series analysis [B: 116].

The offsets c_1 and c_2 occur because $\mathcal{P}_1$ is contained in both $\mathcal{P}_{13}$ and in $\mathcal{P}_{19}$. In essence, both of these subspaces are capable of removing the constant offset (which is an element of $\mathcal{P}_1$) from z. When x and y are zero mean, both c_1 and c_2 are zero. If they have nonzero mean, the projection onto (say) $\mathcal{P}_{13}$ grabs all of the signal in $\mathcal{P}_1$ for itself (Thus $c_1 = \text{mean(x)} + \text{mean(y)}$, and further projection onto $\mathcal{P}_{19}$ gives $c_2 = -\text{mean(y)}$). This illustrates a general property of projections onto periodic subspaces. Suppose that the periodic signals to be separated were $x_{np} \in \mathcal{P}_{np}$ and $x_{mp} \in \mathcal{P}_{mp}$ for some mutually prime n and m. Since $\mathcal{P}_{np} \cap \mathcal{P}_{mp}$ is $\mathcal{P}_p$, both $\mathcal{P}_{np}$ and $\mathcal{P}_{mp}$ are capable of representing the common part of the signal, and x_{np} and x_{mp} can only be recovered up to their common component in $\mathcal{P}_p$. In terms of the harmonic templates, there is overlap between the set of harmonics of x_{np} and the harmonics of x_{mp}, and the algorithm does not know whether to assign the overlapping harmonics to x_{np} or to x_{mp}. The four different periodicity algorithms make different choices in this assignment.

[16] MATLAB® versions of the periodicity software can be found on the CD and online at [W: 53]. The demos are called `PTdemoS2L`, `PTdemoBC`, `PTdemoMB`, and `PTdemoBF`.

It is also possible to separate a deterministic periodic sequence $z \in \mathcal{P}_p$ from a random sequence y when only their sum $x = y + z$ can be observed. Suppose that y is a stationary (independent, identically distributed) process with mean m_y. Then $E\{\pi(y, \mathcal{P}_p)\} = m_y \cdot \mathbf{1}$ (where $\mathbf{1}$ is the vector of all ones), and so

$$E\{\pi(x, \mathcal{P}_p)\} = E\{\pi(y + z, \mathcal{P}_p)\} = E\{\pi(y, \mathcal{P}_p)\} + E\{\pi(z, \mathcal{P}_p)\} = m_y \cdot \mathbf{1} + z$$

since $E\{\pi(z, \mathcal{P}_p)\} = E\{z\} = z$. Hence the deterministic periodicity z can be identified (up to a constant) and removed from x. Such decomposition will likely be most valuable when there is a strong periodic "explanation" for z, and hence for x. In some situations such as economic and geophysical data sets, regular daily, monthly, or yearly cycles may obscure the underlying signal of interest. Projecting onto the subspaces $\mathcal{P}_p$ where p corresponds to these known periodicities is very sensible. But appropriate values for p need not be known a priori. By searching through an appropriate range of p (exploiting the various algorithms of Sect. 5.5.3), both the value of p and the best p-periodic basis element can be recovered from the data itself.

5.5.5 Choice of Effective Sampling Rate

While the Periodicity Transforms are, in general, robust to modest changes in the amplitude of the data [B: 207], they are less robust to changes in the period. To see why, consider a signal with a 1 s periodicity. If this were sampled at $T = 0.1$ s, the periodicity would be readily detected at $p = 10$. But suppose the signal were sampled at $T = 0.0952$ s, corresponding to a desired periodicity "at" $p = 10.5$. Since the algorithms are designed to search for integer periodicities, they detect the periodicity at two repetitions of $p = 10.5$ samples, that is, at $q = 21$ samples.

This "integer periodicity" limitation of the PT can be mitigated by proper choice of a highly factorable integer. Fortunately, it is not necessary to know the underlying periodicity in order to make this choice; the output of the PT can help specify a good effective sampling rate T. A simple procedure is:

(i) using a convenient T, apply the PT to the sampled data
(ii) locate a major peak (e.g., the largest) with periodicity q
(iii) pick a composite number c (an integer with many factors) near q
(iv) let the new sampling interval be $\hat{T} = \frac{q}{c} T$
(v) resample the data with sampling interval $\hat{T}$
(vi) use the PT to locate the periodicities in the resampled data

Continuing the example with $T = 0.0952$, the PT in (i) detects a peak at $q = 21$ in (ii). A composite number near 21 is 24 ($= 2 \cdot 2 \cdot 2 \cdot 3$), which gives a $\hat{T} = \frac{21}{24} T = 0.0833$ for (iv). The data is resampled in (v) and then (vi) detects the peak at $p = 12$, which corresponds to the desired periodicity at 1 s. This technique is useful when resampling audio feature vectors as in Sect. 8.3.1.

5.5.6 Discussion of PT

The Periodicity Transforms are designed to locate periodicities within a data set by projecting onto the (nonorthogonal) periodic subspaces. The methods decompose signals into their basic periodic components, creating their own "basis elements" as linear combinations of delta-like p-periodic basis vectors.

In some cases, the PTs can provide a clearer explanation of the underlying nature of the signals than standard techniques. For instance, the signal z of Fig. 5.15 is decomposed into (roughly) 14 complex sinusoids by the DFT, or into two periodic sequences by the PT. In a strict mathematical sense, they are equivalent, since the residuals are equal in norm. But the PT "explanation" is simpler and allows the recovery of the individual elements from their sum. When periodicity provides a better explanation of a signal or an event than does frequency, then the PT is likely to outperform the DFT. Conversely, when the signal incorporates clear frequency relationships, the DFT will likely provide a clearer result. In general, an analysis of truly unknown signals will benefit from the application of all available techniques.

Like the Hadamard transform [B: 248], the PT can be calculated using only additions (no multiplications are required). As shown in Sect. 5.5.2, each projection requires approximately N operations. But the calculations required to project onto (say) $\mathcal{P}_p$ overlap the calculations required to project onto $\mathcal{P}_{np}$ in a nontrivial way, and these redundancies can undoubtedly be exploited in a more efficient implementation.

Several methods for finding the "best" basis functions from among some (possibly large) set of potential basis elements have been explored in the literature [B: 24], many of which are related to variants of general "projection pursuit" algorithms [B: 98]. Usually these are set in the context of choosing a representation for a given signal from among a family of prespecified frame elements. For instance, a Fourier basis, a collection of Gabor functions, a wavelet basis, and a wavelet packet basis may form the elements of an overcomplete "dictionary." Coifman [B: 33] proposes an algorithm that chooses a basis to represent a given signal based on a measure of entropy. In [B: 139], a greedy algorithm called "matching-pursuit" is presented that successively decomposes a signal by picking the element that best correlates with the signal, subtracts off the residual, and decomposes again. This is analogous to (though somewhat more elaborate than) the Best-Correlation algorithm of Sect. 5.5.3. Nafie [B: 152] proposes an approach that maintains "active" and "inactive" dictionaries. Elements are swapped into the active dictionary when they better represent the signal than those currently active. This is analogous to the M-Best algorithm. The "best basis" approach of [B: 118] uses a thresholding method aimed at signal enhancement, and is somewhat analogous to the Small-To-Large algorithm. Using an l^1 norm, [B: 29] proposes a method that exploits Karmarkar's interior point linear programming method. The "method of frames" [B: 39] essentially calculates the pseudo-inverse of a (large rectangular) matrix composed of all the vectors in the dictionary.

While these provide analogous approaches to the problems of dealing with a redundant spanning set, there are two distinguishing features of the Periodicity Transforms. The first is that the p-periodic basis elements are inherently coupled together. For instance, it does not make any particular sense to choose (say) δ_3^1, δ_4^3, δ_7^3, and δ_9^2 as a basis for the representation of a periodic signal. The p-periodic basis elements are fundamentally coupled together, and none of the methods were designed to deal with such a coupling. More generally, none of the methods is able (at least directly) to exploit the kind of structure (for instance, the containment of certain subspaces and the equality of certain residuals) that is inherent when dealing with the periodic subspaces of the PT.

5.6 Summary

A transform must ultimately be judged by the insight it provides and not solely by the elegance of its mathematics. Transforms and the various algorithms encountered in this chapter are mathematical operations that have no understanding of psychoacoustics or of the human perceptual apparatus. Thus a triangle wave may be decomposed into its appropriate harmonics by the Fourier transform irrespective of the time axis. It makes no difference whether the time scale is milliseconds (in which case we would hear pitch) or on the order of seconds (in which case we would hear rhythm). It is, therefore, up to us to include such extra information in the interpretation of the transformed data.

6

Adaptive Oscillators

One way to model biological clocks is with oscillators that can adapt their period and phase to synchronize to external events. To be useful in the beat tracking problem, the oscillators must be able to synchronize to a large variety of possible input signals and they must be resilient to noises and disturbances. Clock models can be used to help understand how people process temporal information and the models are consistent with the importance of regular successions in cognition and perception. This chapter expands on the presentation in [B: 6].

We perceive light with the eyes, sound with the ears, smells with the nose. We perceive the passage of time: the eternity of a boring class or the rapid passage of an exciting ballgame. How is this possible? What organ senses time?

One possible answer is that we may have internal (biological) clocks.[1] Cycles of such clocks could be accumulated to explain our perception of time intervals. Successions of clock ticks might synchronize with phenomena in the world, heightening our expectations and perceptions at significant times and relaxing between. Not only can we sense the passage of time, we are also capable of marking periods of time. Beating a drum, walking rhythmically, and maintaining regular breaths and heartbeats are easy tasks. Internal clocks help to explain these abilities, since the clocks could regulate the muscular motions that allow the creation of such regular successions.

One way to model such internal clocks is via oscillators: systems which generate (roughly) periodic signals. To be useful, the clocks must do more than just mark time, they must also respond to external events. When we encounter a regular signal such as the beat of a dance, we can choose to synchronize to that beat. Thus the frequency and phase of internal clocks must be malleable, capable of synchronizing to a variety of stimuli. The models must be similarly flexible. They must:

(i) Generate a periodic waveform
(ii) Have an input
(iii) Be capable of adapting to the frequency (or period) and phase of the input when it changes

[1] Such internal clocks form the basis of Povel's framework for rhythm perception and Jones' rhythmic theory of perception, as discussed in Sects. 4.3.6 and 4.3.7.

(iv) Be robust to noisy periods, missing events, and unexpected events

This chapter begins with an overview of the basic ideas of synchronization and entrainment and Sect. 6.2 reviews the mathematical notation used to describe the models. Several oscillators are presented in Sect. 6.3. The various ways that inputs can drive the oscillators and the ways that the oscillators can adapt their settings to achieve synchronous actions are discussed in Sect. 6.4. The taxonomy of adaptive oscillators includes the pendulum, the Van der Pol and Fitzhugh–Nagumo oscillators, as well as a variety of structures that are closely related varieties of adaptive wavetable oscillators and phase-reset oscillators. Section 6.5 examines the behavior of the oscillators as the parameters adapt to follow various input signals. When the model is chosen properly for the class of input signal, the oscillators can adapt quickly and robustly. When the input signals deviate from the expected class, the oscillators may fail to achieve synchronization.

6.1 Entrainment and Synchronization

In 1665, the Dutch scientist Christian Huygens was working on the design of a pendulum clock. He noticed that when two clocks were mounted near each other on a wooden beam, the pendulums began to swing in unison. Whenever one pendulum swung left, the other swept right; when the one moved right, the other swayed left. When he unmounted one of the clocks from the support, they gradually fell out of step. Huygens concluded that tiny vibrations caused by the swinging of the pendulums were conducted through the beam, coupling the motions of the two clocks. This kind of process, where oscillators interact to achieve synchronous behavior, is called entrainment.

Figure 6.1 shows two oscillators with frequencies f_1 and f_2 with phases θ_1 and θ_2. At first, the two vibrate independently, but over time the coupling causes the frequencies to move closer together until eventually (at the time marked "synchronization achieved") the frequencies stabilize and the phases remain locked together. As shown, the oscillators after entrainment are 180 degrees out of phase like the motion in Huygens's entrained pendulum clocks.

Since Huygens's time, entrainment has been observed throughout nature. Individual heart muscle cells each pulse at their own rate; when two are placed close together, they begin pulsing in synchrony. The period of rotation of the planet Mercury is in a 3:2 relation to its period of revolution around the Sun; this can be explained in terms of tidal forces. Women who live in the same household often find that their menstrual cycles coincide [B: 145]. In Thailand, thousands of fireflies gather in the trees at nightfall. At first, their flashing is random and scattered, but over time they become increasingly entrained until eventually they are all flashing simultaneously on and off. This, and many other examples are documented in [B: 163].

Closely related to entrainment is the idea of one oscillator synchronizing to another (but without feedback from the second to the first). This occurs,

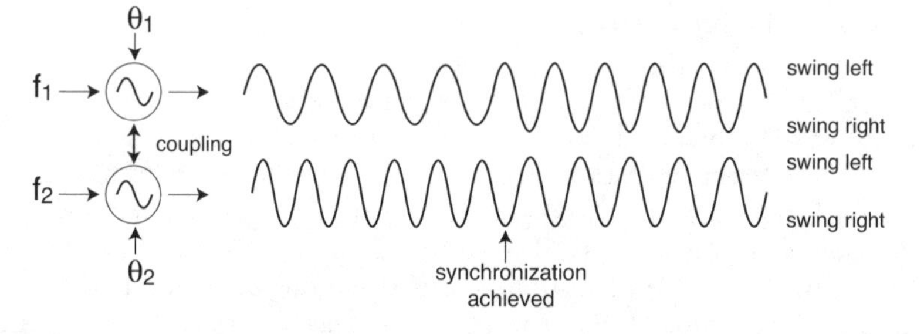

Fig. 6.1. When two oscillators are coupled together, their frequencies may influence each other. When the outputs synchronize in frequency and lock in phase, they are said to be entrained.

for instance, when the sleep cycle becomes synchronized with the daily cycle of light and dark. In mammals, circadian rhythms are generated by a pacemaker in the hypothalamus [B: 149] which contains about 16000 neurons. Researchers believe that each neuron acts as an oscillator with an average period of about 24.3 hours. The neurons are coupled together and driven by light, which causes them (in the normal healthy individual) to synchronize to the solar day. Certain species of clams synchronize their behavior to the tides; even after being removed from the ocean they continue to open and close for several days at their habitual rate.

Both entrainment and synchronization play important roles in music. It is easy for people to synchronize body motions to repetitive stimuli. Picture music as a (very complex) oscillator with a frequency defined by the period of the beat. Then the listener is a second (very complex) oscillator that must synchronize to the first in order to dance, to march in step, or to clap hands with the beat. Such tasks are easy for most people and synchronization occurs rapidly, typically within a few repetitions.

Entrainment occurs between the performers of an improvisational ensemble. Picture each member as a separate (very complex) oscillator. In order to play together, the players listen and respond to each other. Players influence and are in turn influenced by each other, coupled by actively listening and entrained to their common beat.

The processes of synchronization and entrainment are central to any kind of beat tracking machine. Indeed, picture the computer as (yet another) oscillator that must synchronize to the music in order to accomplish the beat tracking goal. If the computer can also control a drum machine (or has other auditory output) then the machine can become a true player in the ensemble only when it can achieve entrainment.

6.2 Systems Notation

A dynamical "system" takes an input signal and maps it into an output signal. The word "dynamical" indicates that the system is not memoryless, that is, that the current operation of the system depends not only on the input at the present moment but on the history of the input (and possibly also on the history of the output). Typically, this means that the system can be described by a set of ordinary differential equations or by a set of difference equations, depending on whether time is modeled as continuous or discrete. The oscillators discussed in this chapter are just one kind of dynamical system: the story of differential and difference equations is vast, and there are a number of very good books such as Luenberger's classic [B: 134] and the recent *Synchronization* [B: 163].

There are two different ways of modeling time in a system: as continuous or as discrete. The mathematics reflects this dichotomy: continuous-time systems are described by differential equations and discrete-time systems are described by difference equations. This section reviews the basic mathematical notations.

In continuous time, t is a real number that represents a point in time. A function (or signal) $x(t)$ assigns a numerical value to each timepoint. Often these functions are differentiable, and if so, then the symbols $\dot{x}$ and $\ddot{x}$ represent the first and second derivatives of $x(t)$, that is, $\dot{x} \equiv \frac{dx(t)}{dt}$ and $\ddot{x} \equiv \frac{d^2x(t)}{dt^2}$. A differential equation is an implicit relationship between a signal and its derivatives. For instance, consider the equation $\dot{x} = -ax$. This means that there is some function $x(t)$ which has the following property: its derivative is equal to $-a$ times itself. It is possible that there is an explicit representation for the function $x(t)$. For this particular case, a little calculus shows that $x(t) = e^{-at}$ since $\dot{x} = \frac{dx(t)}{dt} = \frac{de^{-at}}{dt} = -ae^{-at} = -ax(t)$.

One particularly useful differential equation is the second order linear equation

$$\ddot{x} + \omega^2 x = 0 \tag{6.1}$$

which defines a function whose second derivative is equal to the minus of itself times the constant ω^2. This also has an explicit solution, $x(t) = \cos(\omega t)$, which can be verified by taking the second derivative of $x(t)$ and checking that it fulfills (6.1). Thus this system describes an oscillator with frequency f where $\omega = 2\pi f$. Even when it is not possible to write down an explicit solution, it is almost always possible to simulate the differential equation, to write a computer program that can calculate an unknown $x(t)$.

In discrete time systems, integers k are used to represent points in time. A sequence $x[k]$ assigns a numerical value to each point of time. Discrete time systems can exhibit the same wide range of behaviors as their continuous time counterparts. For example, the system $x[k] = ax[k-1]$ has the closed form solution $x[k] = a^k x[0]$. For $|a| < 1$, the signal decays exponentially from the starting value $x[0]$ towards a resting value at 0.

A particularly useful discrete system is the linear oscillator

$$x[k] = \beta x[k-1] - x[k-2].$$

Depending on the values of β and the initial conditions $x[0]$ and $x[1]$, this generates a sampled sinusoidal signal. The resulting oscillators are studied in [B: 2, B: 64]. Unlike differential equations, which define the signal implicitly, such discrete systems are explicit recipes for the creation of the signal. Given any β and any two starting values $x[0]$ and $x[1]$, the system provides a recipe for $x[2]$ (as $\beta x[1] - x[0]$). Next, the recipe is applied again to find $x[3]$ (as $\beta x[2] - x[1]$), and so on. While only the simplest difference equations have explicit closed form solutions, it is straightforward to follow the recipe and write computer code that simulates the evolution of the system.

Speaking loosely, systems in the world often refer to variables such as position, momentum, and velocity, which can take on values at any time. Hence they are often best modeled continuously using differential equations. In contrast, computer-based systems most naturally represent quantities in discrete time as sequences of numbers. Fortunately, there is a close relationship between the two kinds of models. When taking samples of a signal, they must be taken fast enough so that important information is not lost. Suppose that a signal has no frequency content above B Hz. The *sampling theorem* [B: 102] states that if sampling occurs at a rate greater than $2B$ samples per second, it is possible to reconstruct the original signal from the samples alone. Thus, as long as the samples are taken rapidly enough, no information is lost. On the other hand, when samples are taken too slowly, an arbitrary signal cannot be reconstructed exactly from the samples; the resulting distortion is called *aliasing*.

To make this concrete, suppose that a continuous-time signal $x(t)$ has bandwidth less than B (its spectrum contains no energy above B Hz). If the time interval T_s between the samples of $x(t)$ is less than $\frac{1}{2B}$, all is well: the original $x(t)$ can be completely rebuilt from just its samples. For example, the sinuous curve in Fig. 6.2 is sampled every T_s seconds and the kth sample occurs at time $x(kT_s) = x[k]$.

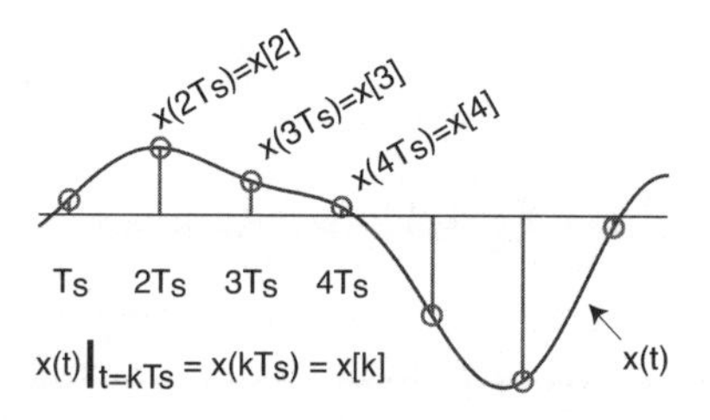

Fig. 6.2. The sampling of a continuous signal can be accomplished without loss of information as long as the signal is bandlimited and the sampling is rapid enough. A common notation uses square brackets to indicate the discretized version of the signal. Thus $x(t)$ evaluated at time $t = kT_s$ is equal to $x(kT_s)$, which is notated $x[k]$.

The significance of the sampling theorem is that only the samples $x[k]$ need to be saved and it is still possible to calculate the value of $x(t)$ at any point. Fortunately, this result for signals also transfers to systems, and any continuous system can be mapped into an equivalent discrete-time system which will have inputs and outputs that are the same at the sampling times. For example, the continuous system $\dot{x} = ax$ behaves the same as the discrete system $x[k+1] = e^{aT_s} x[k]$.

6.3 Oscillators

The Latin word *oscillare* means "to ride in a swing." It is the origin of *oscillate*, which means to move back and forth in steady unvarying rhythm. Thus, a device that creates a signal that moves repeatedly back and forth is called an *oscillator*.

The Pendulum: A mass suspended from a fixed point that swings freely under the action of gravity is called a pendulum. To build a mathematical model of the pendulum, let θ be the angle between the rod and an imaginary vertical (dotted) line and let ℓ be the length of the rod, as shown in Fig. 6.3. Using Newton's laws of motion, it is a fun exercise to write down the equations of motion that specify how the pendulum moves. The result is the differential equation

$$\ddot{\theta} + \omega^2 \sin(\theta) = 0 \tag{6.2}$$

that describes the evolution of the angle over time where $\omega^2 = \frac{g}{\ell}$ and where g is the gravitational constant. While (6.2) does not have a simple solution, it can be approximated quite closely when the angle is small since $\sin(\theta) \approx \theta$ for small θ. This reduces the nonlinear pendulum equation (6.2) to the linear oscillator (6.1) and the frequency of oscillation of the pendulum is approximately $f = \frac{1}{2\pi}\sqrt{\frac{g}{\ell}}$.

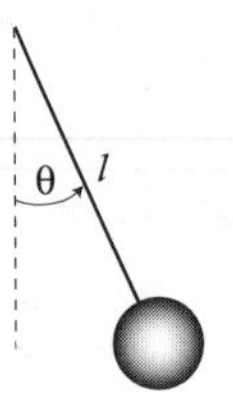

Fig. 6.3. The pendulum sways back and forth at a characteristic frequency f that depends on the length of the rod ℓ, the initial offset, and the strength of gravity g, but not on the mass or the size of the arc. With small displacements it moves slowly and traverses a small percentage of the arc. For larger displacements, it moves faster and swings further. The period remains the same.

There are two problems with the use of a pendulum as a timekeeper. First, friction will eventually cause the pendulum to stop swinging. This is not included in the model (6.2) and some method of regularly injecting energy into the pendulum is needed in order to counteract the inevitable decay. Second, oscillators like (6.1) and (6.2) are only "marginally" stable: there is no force that acts to restore them to their operating point once dislodged. Computer simulations of such oscillators eventually either die away to zero or explode in amplitude due to the accumulation of small errors. For example, using the ode23 command in MATLAB®, a linear oscillator at 1 Hz degenerates to an amplitude of 10^{-7} within about 5000 s of simulation time. This occurs because numerical roundoff errors aggregate over time. In order to create a pendulum clock that is unchanged by random perturbations, it is necessary to incorporate some kind of feedback mechanism.

In his 1658 book *Horologium*, Christian Huygens [B: 101] introduced an escapement mechanism, shown in Fig. 6.4, that made pendulum clocks practical. Such clocks are accurate to within a few minutes a day and were among the most precise ways to measure time until the invention of electric clocks in the early 20th century. Mathematically, the escapement can be modeled by

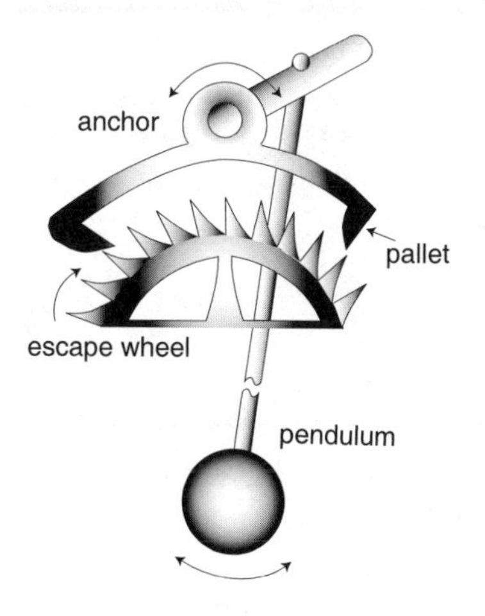

Fig. 6.4. Escapement mechanisms regulate the motion of the pendulum in a grandfather clock. As the pendulum rocks back and forth, the anchor seesaws, engaging and disengaging the pallets which constrain the gear. At each swing, the gear (called the escape wheel) ratchets one click. The sound of the escapement is the familiar tick and tock of a grandfather clock. At each swing, the mechanism transfers a small amount of energy ϵ from a descending mass that is attached to the escape wheel by a rope (not shown) to push the pendulum, counteracting the friction and stabilizing the oscillation.

adding two terms to (6.2). The first is a frictional term c proportional to the (negative) of the velocity of the angle $\dot{\theta}$. Friction opposes the motion of the pendulum in proportion to its speed. The second term represents the energy $g(t)$ added by the anchor-escapement mechanism. Thus (6.2) becomes

$$\ddot{\theta} - c\dot{\theta} + \omega^2 \sin(\theta) + g = 0. \tag{6.3}$$

For example, $g(t)$ might add a small pulse of energy ϵ each time the pendulum reaches the end of its arc, that is, whenever $\dot{\theta} = 0$. Unlike simulations of (6.1) and (6.2), numerical simulations of (6.3) do not decay (or explode) over time. A stable oscillator can be built by providing feedback around a linear oscillator.

Van der Pol's Oscillator: One of the most popular oscillators is named after Balthasar Van der Pol [B: 234] who modeled oscillating electronic circuits with the nonlinear differential equation

$$\ddot{x} - \epsilon(1 - x^2)\dot{x} + \omega^2 x = 0. \tag{6.4}$$

This can be interpreted as a linear oscillator ($\ddot{x} + \omega^2 x = 0$ when $\epsilon = 0$) surrounded by a small feedback proportional to ϵ that stabilizes the oscillation.

Accordingly, for small ϵ, the output $x(t)$ of the Van der Pol oscillator is nearly sinusoidal. Three typical trajectories are shown in Fig. 6.5.

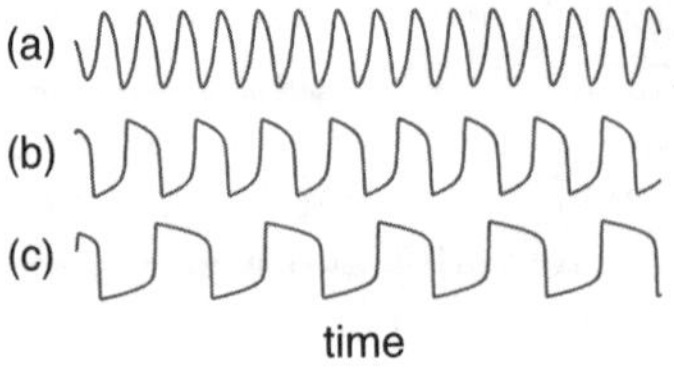

Fig. 6.5. Three typical trajectories of the Van der Pol oscillator with $\omega = 1$ and (a) $\epsilon = 1$, (b) $\epsilon = 5$, and (c) $\epsilon = 10$. Trajectories become more sinusoidal as $\epsilon \to 0$. The shape of the output waveform changes with the frequency.

The Fitzhugh–Nagumo Oscillator: This two-variable differential equation [B: 61] can be used to model the cyclical action potential of neurons. The model gradually accumulates energy until a threshold is reached; it then "fires" and releases the energy, beginning a new cycle. The model is given by the differential equation

$$\begin{aligned} \frac{dx}{dt} &= -x(x - \tau_1)(x - \tau_2) - y - \Omega \\ \frac{dy}{dt} &= \epsilon(x - \tau_3 y) \end{aligned} \tag{6.5}$$

where the τ_i are thresholds, ϵ defines the frequency of oscillation, and Ω is a driving term. A typical trajectory is shown in Fig. 6.6.

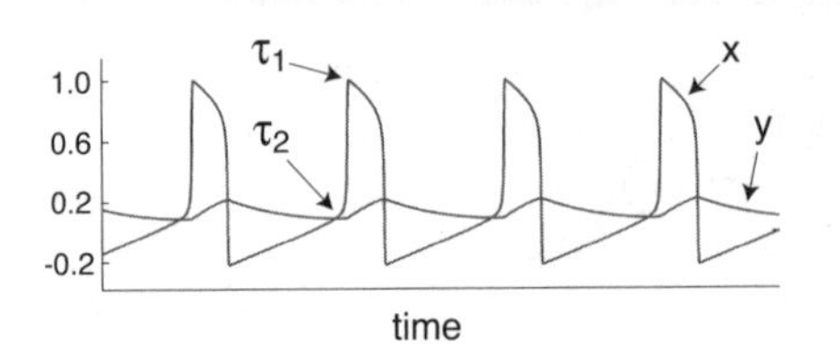

Fig. 6.6. A typical trajectory of the Fitzhugh–Nagumo oscillator. Parameters for this simulation are $\tau_1 = 1$, $\tau_2 = 0.2$, $\tau_3 = 1.2$, $\epsilon = 0.001$, and $\Omega = 0.1$.

The y variable rocks back and forth between modest growth and modest decline depending on whether $x > \tau_3 y$ or not. The interesting behavior is given by x, which is effectively a cubic. When the sign of the cubic term is positive, x experiences rapid growth; when the sign is negative, x decays rapidly. The period of growth occurs when $\tau_1 < x < \tau_2$, and the detailed shape of the waveform changes with the frequency. The Fitzhugh–Nagumo oscillator has been proposed for use in locating downbeats in rhythmic passages by Eck [B: 56].

Phase-reset Oscillators I: The stabilizing nonlinearities in the Van der Pol and Fitzhugh–Nagumo oscillators (6.4)–(6.5) are smooth and differentiable; other oscillators use more severe, discontinuous nonlinearities. For example the phase-reset methods, also called integrate-and-fire oscillators, contain an integrator (or a summer) that increases its value until it reaches a predefined positive threshold value τ. Whenever the output crosses the threshold, it is immediately reset to a fixed nonnegative value y_0. To write this mathematically, let $y[k]$ be the output of the oscillator at time k, let $b \geq 0$ be the amount of increase, and $a > 0$ the decay (or growth) factor. Then the oscillator is

$$y[k] = f(ay[k-1] + b) \tag{6.6}$$

where the nonlinearity $f(\cdot)$ is

$$f(s) = \begin{cases} s & 0 < s < \tau \\ y_0 & \text{otherwise} \end{cases} .$$

For example, if $a = 1$ and $y_0 = 0$, the output of the oscillator is a ramp or sawtooth wave with a period equal to $k = \frac{\tau}{b}$ as shown in Fig. 6.7(a). Typical trajectories for two $a \neq 1$ cases are shown in Fig. 6.7(b) and (c). The parameters have been chosen so that the period is $\frac{1}{2}$ s.

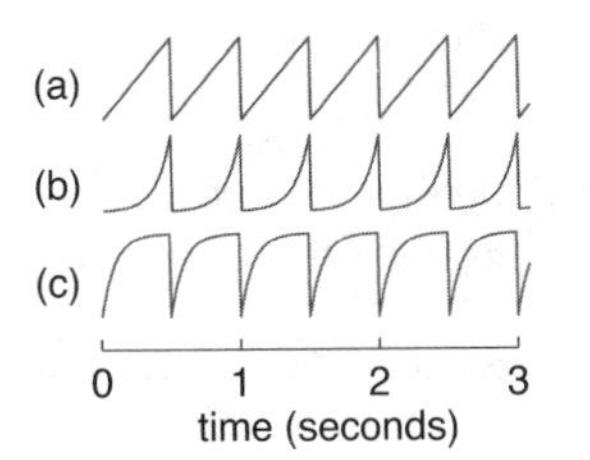

Fig. 6.7. The trajectories of three phase-reset oscillators: (a) a ramp oscillator with $a = 1$, (b) a growth oscillator with $a > 1$, and (c) a decay oscillator with $a < 1$. Specific parameter values used in these simulations are given in [B: 6].

Wavetable Oscillators I: A wavetable oscillator stores one or more periods of a signal in a table and reads values from the table to generate its output. In order to change frequency, the data is read at different speeds. This provides complete flexibility in the shape of the output of the oscillator because the data in the wavetable can be chosen arbitrarily. As a consequence, the shape of the output is independent of the frequency of the oscillator.

It is easiest to see how this works with continuous signals. Consider a periodic waveform $w(t)$ that is expanded or contracted by a factor of α; the result is $w(\alpha t)$. For example, $w(2t)$ has the same basic shape as $w(t)$, but all features are sped up by a factor of two. The result is a doubling of all frequencies. Similarly, $w(\frac{t}{2})$ stretches the waveform and moves all frequencies down an octave. This is demonstrated in Fig. 6.8.

Analogous transformations can be made directly on digital samples of a single period of $w(t)$. Suppose that $w(t)$ is sampled every T_s s and the samples are stored in a vector $w[k]$ for $k = 0, 1, 2, \ldots, N-1$. For example, Fig. 6.9(a) shows a single period of a waveform $w(t)$ and (b) shows a corresponding sampling that records 23 samples in the period. Using resampling techniques, these 23 samples can be transformed into (almost) any other number of samples; Fig. 6.9(c) shows a resampling into 28 samples. If played back at the same

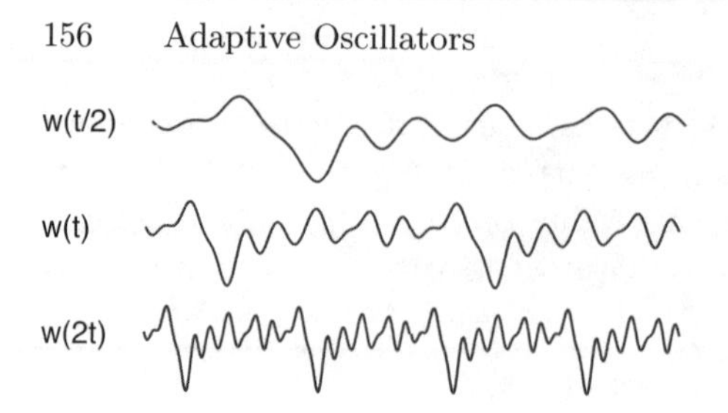

Fig. 6.8. Stretching and compressing a signal $w(t)$ by a factor of α changes all frequencies by a factor of α. The cases $\alpha = 2$ (up an octave) and $\alpha = \frac{1}{2}$ (down an octave) are shown.

rate, this would correspond to an α of $\frac{23}{28}$, a slowing down (lower pitched version) of the signal to about 82% of its original speed.

If the bandwidth of $w(t)$ is B and if the sampling interval is $T_s < \frac{1}{2B}$, it is possible to exactly reconstruct $w(s)$ at any point s using only the N samples. This *Shannon reconstruction*[2] is computationally intensive, but can be approximated using relatively simple interpolation methods. It is always possible to resample faster, to lower the pitch. But resampling slower (raising the pitch) can be done without distortion only as long as the resampled interval αT_s remains less than $\frac{1}{2B}$. For large α, this will eventually be violated. [B: 223] discusses this in more detail for a collection of standard waveshapes.

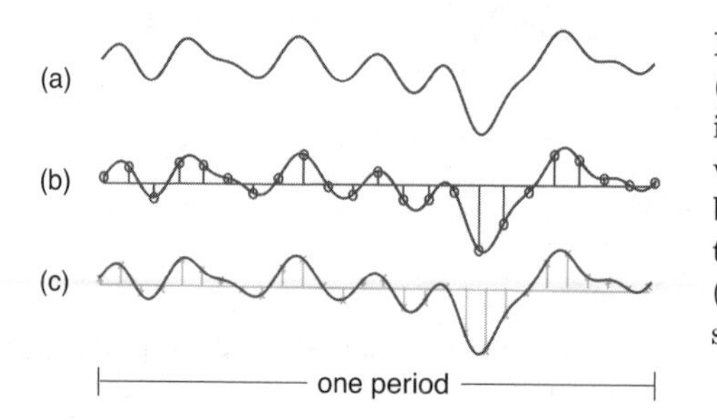

Fig. 6.9. A periodic waveform $w(t)$ in (a) is sampled at one rate in (b). Digital resampling of (b) results in (c), which represents the same waveform but at a different rate. According to the resampling theorem, the data in (c) is indistinguishable from a direct sampling of (a) at the new rate.

Indexing into a wavetable w of size N is accomplished using the recursion

$$s[k] = (s[k-1] + \alpha) \bmod N \tag{6.7}$$

where mod N is the remainder after division by N and where the starting value is $s[0] = 0$. The output of the oscillator at time k is

$$o[k] = w((s[k] + \beta) \bmod N). \tag{6.8}$$

Thus α specifies the frequency of the oscillation ($\frac{\alpha}{NT_s}$ Hz) and β specifies the phase.

If $s[k] + \beta$ happens to be an integer then the output of the oscillator is straightforwardly $w(s[k] + \beta)$. When $s[k] + \beta$ is not an integer, interpolation is needed to estimate an appropriate value. The simplest method is to use

[2] Details of the reconstruction of bandlimited signals along with MATLAB® code to carry out the required procedures can be found in [B: 102].

linear interpolation. For example, if $s[k]+\beta$ were 10.2 then the output would be $0.8w(10)+0.2w(11)$. More sophisticated interpolation methods would not be atypical.

Such wavetable synthesis is commonly used in sample playback electronic musical instruments [W: 42]. For example, the stored waveform might be a multi-period sampling of an acoustic instrument such as a flute playing C. Whenever a key is pressed on the keyboard, the recording of the flute note plays back but with a different α to transpose the note to the desired pitch. One advantage of the wavetable approach is that it allows straightforward control over the frequency (via α) and phase (via β), and cleanly separates these control parameters from the shape of the oscillatory waveform (the vector w).

Phase-reset Oscillators II: On the surface it may appear as if the wavetable equation (6.7) is the same as the phase-reset equation (6.6) with the mod function serving the same role as $f(\cdot)$. But it differs in two ways. First, there is no threshold in (6.7)–(6.8). Second, the output $y[k]$ of the phase-reset oscillator is the value of the oscillator at time k whereas $s[k]$ in the wavetable oscillator is the index at time k into the wavetable $w[k]$.

A threshold τ can be incorporated into the wavetable approach by indexing into w (an N-vector representing the waveshape) using the recursion

$$s[k]=\begin{cases}0 & o[k]\geq\tau\\ (s[k-1]+\alpha)\bmod N & \text{otherwise}\end{cases}\tag{6.9}$$

where $o[k]=w(s[k])$ is the output of the oscillator at time k. If w never exceeds the threshold, the period of the oscillator is $\frac{NT_s}{\alpha}$. If w does exceed the threshold, the period is $\frac{mT_s}{\alpha}$ where m is the (first) index at which $w(m)\geq\tau$. Phase-reset oscillators are important because they provide a different way to incorporate inputs.

6.4 Adaptive Oscillators

The oscillators of the previous section are unforced: they have no inputs other than the fixed parameters of the model. In order for an oscillator to respond to signals in the world, there must be a way for it to incorporate external signals into its operation: this is a *forced oscillator*. In order to be useful, there must be a way for the parameters of the oscillator to change in response to external events: this is an *adaptive oscillator*. There are four key issues in the design of an adaptive oscillator:

(i) the kind of oscillator (Van der Pol, phase-reset, wavetable, etc.)
(ii) the kind of inputs it is designed to accommodate (sinusoidal, impulsive, random, etc.)
(iii) which parameters will be adapted

(iv) the mechanism (or algorithm) implementing the adaptation.

The remainder of this section examines a number of different adaptive oscillators; subsequent sections examine their applicability to the rhythm-finding and beat tracking problems.

6.4.1 The Phase Locked Loop

Many modern telecommunication systems operate by modulating a sinusoidal "carrier" wave with a message sequence that is typically a long string of binary encoded data. In order to recover the data, the receiver must be able to figure out both the frequency and phase of this carrier wave. To accomplish this, a local oscillator at the receiver generates a sinusoid and adapts its frequency and phase to match the received (modulated) sine wave. The phase locked loop (PLL) is one way of accomplishing this goal.[3] The PLL addresses the four design issues of the adaptive oscillator in the following ways:

(i) the oscillator in a PLL is a wavetable with the shape of a sinusoid
(ii) the PLL is designed to operate with the modulated sinusoid (the received signal) as input
(iii) the phase of the local oscillator is adapted
(iv) the adaptation proceeds using a "hill climbing" strategy

The development of the PLL proceeds in two steps. First, the frequency is assumed known and an algorithm for finding the phase is derived. Next, both the phase and frequency are assumed unknown: a dual structure can estimate both.

Phase Estimation with the PLL: Suppose first that the input signal is a sinusoid $r(t) = \cos(2\pi ft + \phi)$ with known frequency f but unknown phase ϕ. One way to estimate the phase ϕ is shown in block diagram form in Fig. 6.10. The local oscillator (designated ∿) generates the sinusoidal signal $s(t) = \cos(2\pi ft + \theta)$. The multiplication of $r(t)$ and $s(t)$ shifts the energy to near zero where it is lowpass filtered (LPF) to remove high frequency components. The magnitude of the resulting low frequency term is adjusted by changing the phase θ. A bit of trigonometry (as shown below) demonstrates that the value of θ that maximizes the low frequency component is the same as the phase ϕ of $r(t)$.

To be specific, let

$$J(\theta) = \text{LPF}\{r(t)\cos(2\pi ft + \theta)\} \tag{6.10}$$

be a "cost" or "objective" function that represents the relationship between the signals in the PLL and the desired θ. Using the cosine product relationship,[4] this is

[3] This discussion draws heavily from [B: 102].
[4] i.e., the trigonometric identity $\cos(x)\cos(y) = \frac{1}{2}[\cos(x-y) + \cos(x+y)]$.

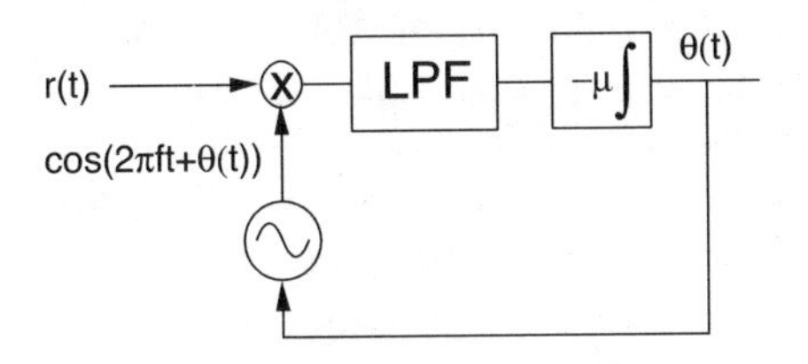

Fig. 6.10. The PLL can be viewed as a local oscillator that adapts its phase $\theta(t)$ over time so as to maximize the (averaged) zero frequency value of the demodulated signal. As time progresses, $\theta(t) \to \phi$, providing an estimate of the unknown phase of the input signal $r(t)$.

$$\begin{aligned} J(\theta) &= \text{LPF}\{\cos(2\pi ft + \phi)\cos(2\pi ft + \theta)\} \\ &= \frac{1}{2}\text{LPF}\{\cos(\phi - \theta)\} + \frac{1}{2}\text{LPF}\{\cos(4\pi ft + \theta + \phi)\} \\ &\approx \frac{1}{2}\cos(\phi - \theta) \end{aligned}$$

assuming that the cutoff frequency of the lowpass filter is well below $2f$. Thus, the values of θ that maximize $J(\theta)$ occur when $\theta = \phi + 2n\pi$ for any integer n.

One way to find the maximum of a function is to use a gradient strategy that iteratively moves the value of the argument in the direction of the derivative. Because the gradient always points "uphill," this is called a hill-climbing algorithm. The derivative of (6.10) with respect to θ can be approximated[5] as

$$\frac{dJ(\theta)}{d\theta} \approx -\text{LPF}\{r(t)\sin(2\pi ft + \theta)\}$$

and the corresponding hill-climbing algorithm is $\dot{\theta} = \mu\frac{dJ(\theta)}{d\theta}$ as depicted in block diagram form in Fig. 6.10. In discrete time this becomes

$$\theta[k+1] = \theta[k] + \mu\left.\frac{dJ(\theta)}{d\theta}\right|_{\theta=\theta[k]}.$$

Substituting for the derivative and evaluating at the appropriate θ yields

$$\theta[k+1] = \theta[k] - \mu\text{LPF}\{r(kT_s)\sin(2\pi fkT_s + \theta[k])\} \tag{6.11}$$

where T_s is the sampling interval and k is the iteration counter. The phase parameter θ in the PLL plays the same role that the phase parameter β plays in the wavetable oscillator (6.8). Using a wavetable w defined to be the N samples of a sinusoid, the evolution of θ acts to synchronize the phase of the local oscillator (that is, of w) to the phase of the input $r(t)$.

Frequency Estimation with the PLL: A standard way to adjust the frequency of a local oscillator to match the frequency of the input is based on the observation that frequency is equal to the integral of phase. For example,

[5] The LPF and the derivative commute because they are both linear operations. The approximation requires a small stepsize μ as described in Appendix G of [B: 102].

the sinusoid $\cos(2\pi f t + \theta(t))$ with time varying phase $\theta(t) = 2\pi ct + b$ is indistinguishable from $\cos(2\pi(f + c)t + b)$. Thus a "ramp" in phase is equivalent to a change in frequency that is proportional to the slope of θ.

One approach is to use a recursive lowpass filter that can track a ramp input. By linearizing, this can be shown to result in a $\theta(t)$ that converges to the ramp that properly accounts for the frequency offset. A conceptually simpler approach is based on the observation that the phase estimates of the PLL "converge" to a line. Since the slope of the line is proportional to the difference between the actual frequency of the input and the frequency of the oscillator, it can be used to make iterative corrections. This indirect method cascades two PLLs: the first finds the line and indirectly specifies the frequency. The second converges to a value appropriate for the phase offset.

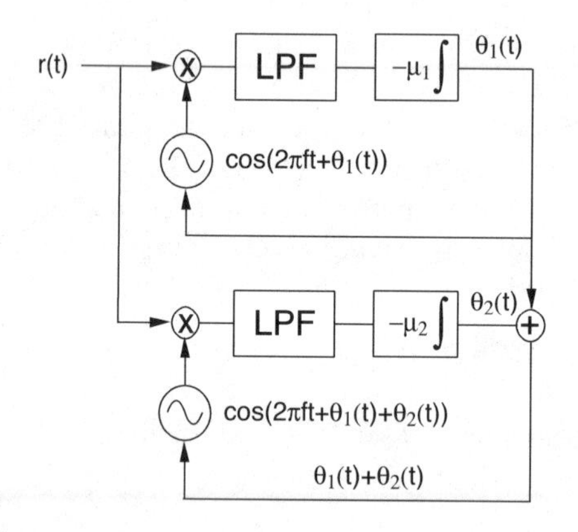

Fig. 6.11. A pair of PLLs can efficiently estimate the frequency offset at the receiver. The output θ_1 of the first PLL "converges" to a line with slope that estimates the true frequency of the input. Adding this line to the (phase input) of the second PLL effectively corrects its frequency and allows convergence of θ_2.

The scheme is pictured in Fig. 6.11. Suppose that the received signal has been preprocessed to form $r(t) = \cos(2\pi f_r t + \phi)$. This is applied to the inputs of two PLLs. The top PLL functions exactly as expected from previous sections: if the frequency of its oscillator is f_ℓ, then the phase estimates θ_1 converge to a ramp with slope $2\pi(f_r - f_\ell)$, that is,

$$\theta_1(t) \to 2\pi(f_r - f_\ell)t + b$$

where b is the y-intercept of the ramp. The θ_1 values are then added (as shown in Fig. 6.11) to θ_2, the phase estimate in the lower PLL. The output of the bottom oscillator is

$$\begin{aligned}\cos(2\pi f_\ell t + \theta_1(t) + \theta_2(t)) &= \cos(2\pi f_\ell t + 2\pi(f_r - f_\ell)t + b + \theta_2(t)) \\ &= \cos(2\pi f_r t + b + \theta_2(t)).\end{aligned}$$

Effectively, the top loop has synthesized a signal that has the "correct" frequency for the bottom loop since a sinusoid with frequency $2\pi f_r t$ and 'phase' $\theta_1(t)+\theta_2(t)$ is indistinguishable from a sinusoid with frequency $2\pi f t$ and phase

$\theta_2(t)$. Accordingly, the lower PLL acts just as if it had the correct frequency, and so $\theta_2(t) \to \phi - b$.

Variations on the PLL: The adaptation of the phase and frequency components of the PLL is based on the objective function (6.10) which essentially correlates the input with the waveshape of the oscillator. A variety of alternative objective functions can also be used. For example, the least squares objective

$$J_{LS}(\theta) = \text{LPF}\{(r(t) - \cos(2\pi f t + \theta))^2\} \tag{6.12}$$

leads to the algorithm

$$\theta[k+1] = \theta[k] - \mu\text{LPF}\{(r(kT_s) - \cos(2\pi f k T_s + \theta[k]))\sin(2\pi f k T_s + \theta[k])\}.$$

Similarly, the objective function

$$J_C(\theta) = \text{LPF}\{(r(t)\cos(2\pi f t + \theta))^2\}$$

leads to the "Costas loop"

$$\begin{aligned}\theta[k+1] = \theta[k] - \mu\ &\text{LPF}\ \{r(kT_s)\cos(2\pi f k T_s + \theta[k])\}\\ &\text{LPF}\ \{r(kT_s)\sin(2\pi f k T_s + \theta[k])\}.\end{aligned} \tag{6.13}$$

Analysis of these variations of the PLL (as in [B: 102]) show that they are more-or-less equivalent in the sense that they have the same answers in the ideal (no noise) case, though the algorithms may react differently to noises.

The algorithm (6.11) incorporates a lowpass filter. The requirements on this filter are mild, and it is common to simplify the iteration by removing the filter completely, since the integration with the small stepsize μ also has a lowpass character. Similarly, the lowpass filter may be absent from some implementations of (6.13).

6.4.2 Adaptive Wavetable Oscillators

A wavetable oscillator (as introduced in Sect. 6.3) consists of an array w containing N stored values of a waveform. The output of the oscillator at time k is

$$o_1[k] = w((s[k] + \beta_1) \bmod N)$$

where the indices into w are given by the recursion

$$s[k] = (s[k-1] + \alpha) \bmod N. \tag{6.14}$$

As before, α specifies the frequency of the oscillation while β_1 defines the phase. The oscillator can be made adaptive by adjusting the parameters to align the oscillator with an external input. This can be accomplished in several ways.

Suppose that the input to the oscillator is $i[k]$. An analogy with the PLL uses the correlation-style objective function

$$J(\beta_1) = \text{LPF}\{i[k]o_1[k]\}$$

that directly parallels (6.10). The β_1 that maximizes J provides the best fit between the input and the oscillator. It can be adapted using a hill-climbing strategy

$$\begin{aligned}\beta_1[k+1] &= \beta_1[k] + \mu\frac{dJ}{d\beta_1} \qquad (6.15)\\ &= \beta_1[k] + \mu\text{LPF}\{\ i[k]\ \left.\frac{dw}{d\beta_1}\right|_{\beta_1=\beta_1[k]}\}.\end{aligned}$$

Since w is defined by a table of values, $\frac{dw}{d\beta_1}$ is another table, the numerical derivative of w (the time derivative of w and the derivative with respect to β_1 are equal up to a constant factor). Several candidate wavetables and their derivatives are shown in Fig. 6.12.

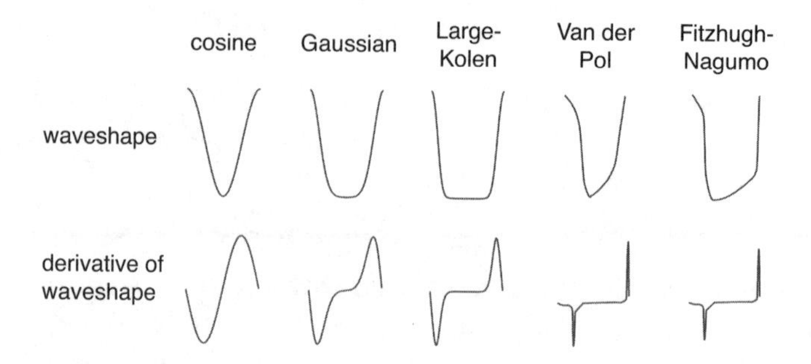

Fig. 6.12. Five common wavetables and their derivatives. The cosine wavetable is used in the PLL. The Gaussian shape is shifted so that the largest value occurs at the start of the table. The Large–Kolen oscillator wavetable is defined by $1 + \tanh(\gamma(\cos(2\pi f t) - 1))$. The Van der Pol and Fitzhugh–Nagumo waveshapes are defined using waveforms culled from the numerical simulations in Figs. 6.5(b) and 6.6.

As with the PLL, the update for the frequency can be done using the integral/derivative relationship between frequency and phase. To be specific, let

$$o_2[k] = w((s[k] + \beta_1[k] + \beta_2[k]) \bmod N)$$

be a second oscillator (corresponding to the bottom PLL in Fig. 6.11). A gradient update for β_2 proceeds as above by maximizing $J(\beta_2) = \text{LPF}\{i[k]o_2[k]\}$ with respect to β_2. The formula is the same as (6.15) with the appropriate substitutions.

While the two serial phase estimates of the dual PLL can track changes in frequency, they do not adapt the frequency parameter. Similarly, the adaptive oscillator using β_1 and β_2 can follow changes in the frequency of the input, but the parameter α in (6.14) is never changed. A consequence of this is that if the input ceases, the oscillator returns to its original frequency.

It is also possible to adapt the frequency parameter. This allows the oscillator to "learn" the frequency of the input and to continue at the new frequency even if the input stops. Perhaps the simplest technique for adapting the frequency is to use a gradient strategy that maximizes $J(\alpha) = \text{LPF}\{i[k]o_1[k]\}$. This is:

$$\begin{aligned}\alpha[k+1] &= \alpha[k] + \mu_\alpha \frac{dJ}{d\alpha} \qquad (6.16)\\ &= \alpha[k] + \mu_\alpha \text{LPF}\{\ i[k]\ \frac{dw}{ds}\frac{ds}{d\alpha}\bigg|_{\alpha=\alpha[k]}\ \}.\end{aligned}$$

Since $s[k]$ is defined by the recursion (6.14), the derivative with respect to α cannot be expressed exactly. Nonetheless, when the stepsizes are small, it can be approximated by unity,[6] and so the final form of the update is the same as the update for β. Because the frequency parameter is more sensitive, its stepsize μ_α is usually chosen to be considerable smaller than the stepsize used to adapt the phase.

In adapting the βs and αs of the adaptive wavetable oscillator, other objective functions may be used. For example, minimizing $\text{LPF}\{(i[k]-o[k])^2\}$ leads to an update that optimizes a least squares criterion while maximizing $\text{LPF}\{(i[k]o[k])^2\}$ leads to a "Costas loop" method of updating the oscillator parameters. These parallel directly the sinusoidal methods associated with the PLL in (6.13).

Besides the PLL, many common oscillators can be approximated using the adaptive wavetable oscillator structure. For example, for Large and Kolen's [B: 123] oscillator

(i) the wavetable is defined by $1 + \tanh\left(\gamma(\cos(2\pi f t) - 1)\right)$ as shown in Fig. 6.12
(ii) the input is optimized for pulses of the same shape as the input, but is also appropriate for a spike train
(iii) both the phase β and the frequency α of the local oscillator are adapted
(iv) the adaptation proceeds using a "hill-climbing" strategy.[7]

When attempting to locate the position of a train of spikes in time, oscillators that use pulses (such as Large and Kolen's or the Gaussian) are a good idea.

[6] In certain situations, this kind of approximation can lead to instabilities in the adaptation. See Sect. 10.6 of [B: 102] for an example using the PLL.

[7] The phase and period updates (equations (4) and (6) in [B: 123]) are the same as (6.15) and (6.16) but for the addition of some factors which scale the stepsizes.

The pulse can be thought of as providing a window of time over which the oscillator expects another spike to occur. If the spike occurs at exactly the right time, the derivative is zero and there is no change. If the spike occurs slightly early, then the derivative is positive and the phase increases. If the spike occurs late then the derivative is negative and the phase decreases. This process of adjustment actively aligns the oscillator with the spike train. Perhaps just as important, there is a zone between pulses where the waveshape and the derivative are small. This makes the oscillator insensitive to extraneous spikes or noisy data that occur far away from expected spike locations.

Figure 6.13 shows how the adaptive wavetable oscillator responds to an input

$$i(t) = \begin{cases} 1 & t = nT,\ n = 1, 2, \ldots, M \\ 0 & \text{otherwise} \end{cases} \tag{6.17}$$

that is a regular train of $M = 19$ spikes spaced $T = 500$ ms apart. This simulation used the Gaussian pulse oscillator with phase and frequency parameters adapted according to (6.15) and (6.16). The α parameter was initialized with period 550 ms, corresponding to a 10% error. The phase and frequency converge within a few seconds and the pulses align with the spikes. Observe that the oscillator continues at the adapted frequency even after the input ceases at around 10 s.

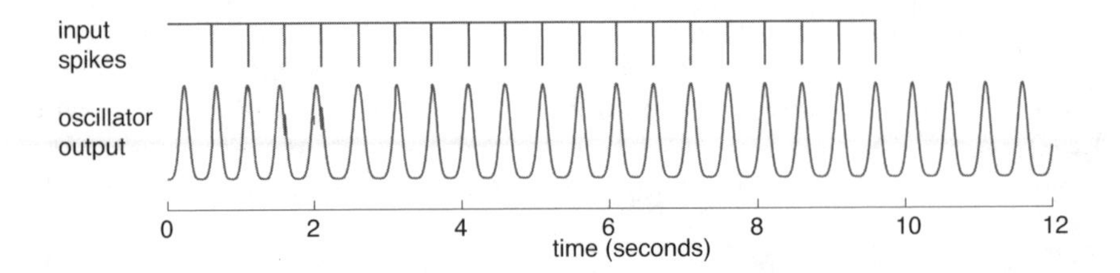

Fig. 6.13. A train of spikes are input into an adaptive oscillator using the Gaussian waveshape. Within a few periods, the oscillator synchronizes with the input. (The initial value was $\alpha = 550$ ms while the correct value was $\alpha = 500$ ms.) After synchronization, the input can cease (as occurred at 10 s) and the oscillator continues at the new rate.

Figure 6.14 shows that the same oscillator may synchronize in various ways to the same input depending on the initial values. The figure shows 1:2, 2:1, and 3:4 entrainments where $n{:}m$ means that n periods of the oscillator occur in the same time as m periods of the input. While such nonunity entrainments are common in the mode locking of oscillators, they are encouraged by the specifics of the waveshape. The dead (zero) region between pulses means that the adaptation is insensitive to spikes that occur far away from the expected location. For example, in the 1:2 entrainment, the "extra" spikes occur at precisely the point where they will have the least influence on the adaptation. On the other hand, in the 2:1 entrainment, the input is zero for the duration of the "off beat" pulse and hence does not affect adaptation. The final (bottom) simulation in Fig. 6.14 shows a stable 3:4 entrainment. These simulations (using the Gaussian pulse shape) are effectively the same as using the Large–Kolen oscillator. Using a cosine wave (as suggested in [B: 144], for instance) also converges similarly, suggesting that the details of the waveshape are not particularly crucial.

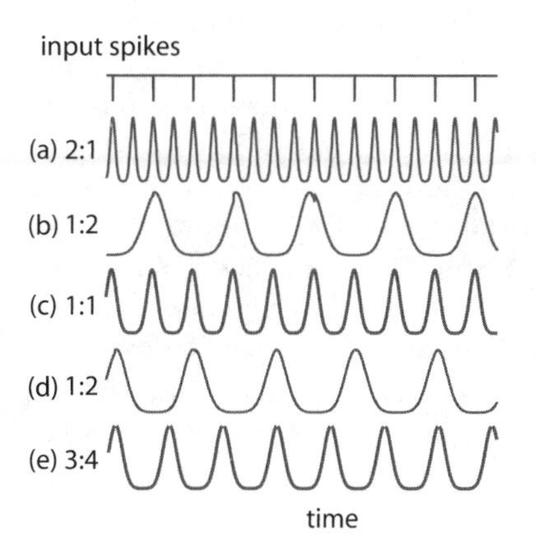

Fig. 6.14. The input spike train (6.17) excites the adaptive oscillator as in Fig. 6.13. The initial value in the top plot was $\alpha =$ 240 ms and the oscillator synchronizes to a 2:1 rate (two oscillator pulses occur for each input spike). The two middle plots were initialized at $\alpha = 1050$ ms. The oscillator synchronizes to a 1:2 rate (one oscillator output for every two input spikes). Depending on the initial value of β, the oscillator can lock onto either the odd or the even spikes. Other synchronizations such as 3:4 are also possible.

The stepsize parameters μ and μ_α affect the adaptation and synchronization in a direct way: larger stepsizes allow faster convergence but also may result in overshoot or unstable behaviors. Smaller stepsizes allow smoother adaptation at the cost of longer time to convergence. Finding useful values of the stepsizes is often a matter of trial and error. The behavior of adaptive wavetable oscillators will be discussed further in Sect. 6.5. The survey of kinds of adaptive oscillators turns next to consider ways that phase-reset oscillators may be made adaptive.

6.4.3 Adaptive Phase-reset Oscillators

The input to the adaptive wavetable oscillator is used only in the adaptation of the control parameters. The input to the adaptive phase-reset oscillator is incorporated directly into the thresholding process. In some cases, this allows the adaptive phase-reset oscillator to synchronize more rapidly to the input without suffering from the effects of a large stepsize.

As in (6.9), the threshold τ defines the indexing into a wavetable w using a recursion

$$s[k] = \begin{cases} 0 & o[k] + i[k] \geq \tau \\ (s[k-1] + \alpha) \bmod N & \text{otherwise} \end{cases} \tag{6.18}$$

where the input to the oscillator at time k is $i[k]$ and the output from the oscillator is $o[k] = w(s[k])$. The adaptation of the frequency parameter α can proceed in any of the usual ways. McAuley [B: 144] suggests[8] minimizing the least square error $J_{LS}(\alpha) = \text{LPF}\{i[k](\tau - o[k])^2\}$ using a gradient strategy. An alternative is to maximize the correlation objective $J_{cor}(\alpha) = \text{LPF}\{i[k]o[k]\}$. In either case, the update for α is

$$\alpha[k+1] = \alpha[k] \pm \mu \frac{dJ}{d\alpha}$$

where the $+$ is used to maximize and the $-$ is used to minimize. Both objective functions result in performance similar to that in Fig. 6.15. The top plot shows α converging to its final value of 1 within a few seconds. Observe that the convergence is not necessarily monotonic: the first step moves in the "wrong" direction because of an unfortunate initial relationship between the phase of the output of the oscillator and the onset of the input spikes. Despite the false start, convergence occurs within a few periods of the input. The bottom plot shows the output of the oscillator as it resets its phase with the onset of each input spike. Synchronization aligns the output with the input spikes. This simulation used the sinusoidal wavetable; others such as the Gaussian, the Large–Kolen, Van der Pol, and Fitzhugh–Nagumo shapes (recall Fig. 6.12) behave approximately the same.

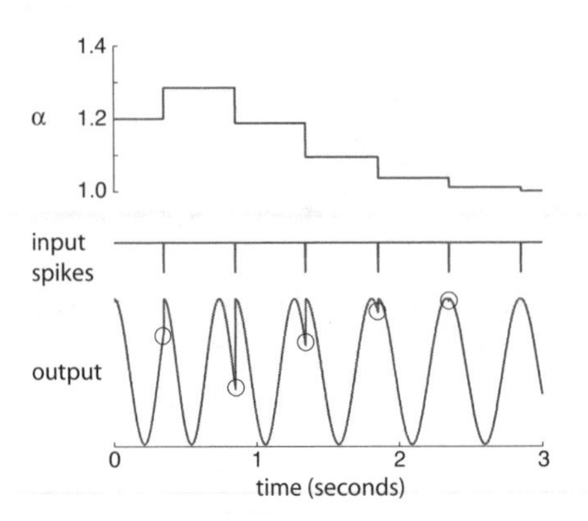

Fig. 6.15. The phase-reset oscillator (6.18) adapts to minimize the correlation between the input (6.17) and the output of the oscillator, shown in the bottom plot. The frequency parameter α converges rapidly to its desired value of 1, despite initially moving in the wrong direction. Each time an input spike arrives, the oscillator resets. The small circles annotate the output of the oscillator at the points of reset, which move to align the maximum points of the output with the input spikes.

6.4.4 Adaptive Clocking

The essence of a clocking device is that it must specify a sequence of equidistant time periods. This requires two parameters: one for the period T and one for the phase (or starting point) τ. The idea of adaptive clocking is to adjust these two parameters directly based on properties of the input. Thus the final kind of "adaptive oscillator" is not really an oscillator at all but a way of marking time so as to achieve synchronization with an input.

[8] McAuley's algorithm has other features (such as activation sharpening) that distinguish it from the present discussion.

The methods of the previous sections create an objective function by comparing the output of the oscillator to the input: the parameters are adjusted so as to make the output look more like the input. Since there is no wavetable and no function generating an output for the adaptive clocking, it must operate differently. One approach is to use the error in time between the predicted location of subsequent events and the actual location of those events. By adjusting the parameters, better predictions can be made. Thus the adaptive clocking uses an objective function that measures errors in time rather than errors in amplitude (like the least square and correlation objectives of the previous sections). This is analogous to one of the algorithms in [B: 48], though the inputs are quite different.

To be concrete, define the objective function

$$J(\mathcal{T},\tau) = \sum_{n=1}^{N} (I_n - (\tau + n\mathcal{T}))^2 \tag{6.19}$$

where the time index is suppressed and where I_n represents the (actual) location of the event n clock ticks into the future. This can be interpreted as minimizing the (squared) difference in time between N nearby events and the predictions $(\tau + n\mathcal{T})$ of when those events will occur. The parameters that minimize this J are (locally, at least) the best representation of the input. This is illustrated in Fig. 6.16 for $N = 2$.

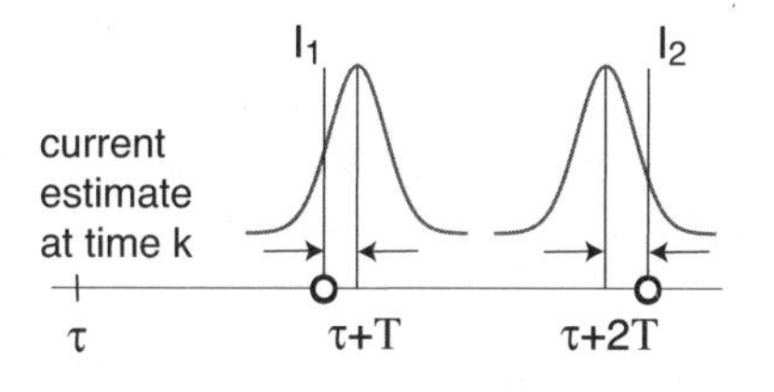

Fig. 6.16. In adaptive clocking, the actual locations of the input pulses (indicated by the I_n) are compared to the estimated locations $\tau + n\mathcal{T}$. The differences (indicated by the paired arrows) can be used to update the τ and $\mathcal{T}$ parameters to achieve synchronization.

As in the previous sections, optimal values can be found using a gradient strategy which leads to the iterations

$$\begin{aligned}\tau[k+1] &= \tau[k] + \mu_\tau\left((I_1[k] - \tau[k] - \mathcal{T}[k]) + (I_2[k] - \tau[k] - 2\mathcal{T}[k])\right) \quad (6.20)\\ \mathcal{T}[k+1] &= \mathcal{T}[k] + \mu_\mathcal{T}\left((I_1[k] - \tau[k] - \mathcal{T}[k]) + 2(I_2[k] - \tau[k] - 2\mathcal{T}[k])\right)\end{aligned}$$

for the $N = 2$ case. It is also possible to minimize J of (6.19) in one step using a matrix formulation since both parameters enter linearly. However, both of these approaches overlook a subtlety: the I values are also a function of the unknowns.

To see the issue, suppose that $i(t)$ is the input. Unless the locations of the events in the input are known beforehand (which is not a reasonable assumption since the goal of the adaptation is to locate the events that form the regular succession), the value of the I depends on the current estimates.

Let $g(t)$ be a function that is large near the origin and that grows small as t deviates from the origin. The Gaussian function $g(t) = e^{-t^2/\sigma^2}$ is one possibility, where σ is chosen so that the "width" of $g(t)$ is narrower than the time span expected to occur between successive events. The "actual" location of the nth event can then be defined as

$$I_n = \text{argmax}\{i[k]\ g(t - \tau[k] - n\mathcal{T}[k])\}. \tag{6.21}$$

To understand this, observe that $g(t-\tau[k]-n\mathcal{T}[k])$ is a (Gaussian) pulse with variance σ^2 and centered at $\tau[k] + n\mathcal{T}[k]$, the estimated location of the nth subsequent event. The product $i[k]g(t - \tau[k] - n\mathcal{T}[k])$ weights the input data so as to emphasize information near the expected event and to attenuate data far from the expected event. The argmax$\{\cdot\}$ function picks out the largest peak in the input lying near the expected location, and returns the location of this peak. This is (likely) the actual location of the event. The difference between the argmax (the likely location of the event as given in the data) and $\tau[k] + n\mathcal{T}[k]$ (the estimated location of the event) is thus the basis of the objective function. If the input were a regular succession of pulses and the estimates of τ and $\mathcal{T}$ were accurate, then the objective function would be zero. Appendix B of [B: 206] (which can be found on the CD) carries out an explicit calculation of the derivative of this argmax function and shows that it may be closely approximated by the simpler gradient method above.

One of the advantages of using a function like $g(t)$ to locate candidate event locations (instead of just taking the maximum in a window) is that it gives the algorithm a robustness to additive noise that some of the other methods may lack. Figure 6.17 shows the adaptive clock method locked onto an input that consists of a pulse train with a significant amount of noise. The vertical lines are the predictions of the adaptive clock after convergence.

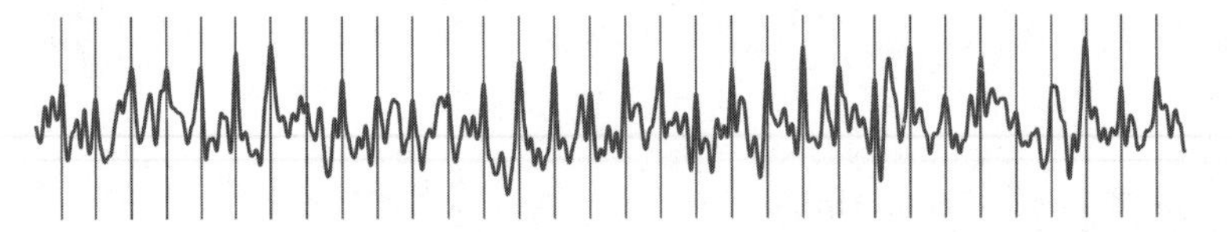

Fig. 6.17. The adaptive clocking algorithm can be applied in fairly noisy situations and still achieve synchronization

6.5 Behavior of Adaptive Oscillators

The behavior of an adaptive oscillator is dependent on several factors:

(i) the kind of oscillator
(ii) the waveshape
(iii) the fixed parameters (e.g., stepsizes, threshold values)
(iv) the method of adaptation

(v) the input

The discussion in the previous sections has focused on basic design choices within the adaptive oscillator. This section shows why the final factor (the input) is at the heart of the creation of useful oscillators for beat and rhythm tracking applications. When the character of the input is well modeled, oscillators can be specified to synchronize and track relevant features of the input. But when the input is too diverse or is poorly modeled, then the application of adaptive oscillator technology is likely to be problematic.

6.5.1 Regular Pulse Trains

Several adaptive oscillators are designed to synchronize to input sequences consisting of trains of isolated pulses such as (6.17). When the input is indeed of this form, it is reasonably straightforward to understand the convergence of the oscillator. A standard way to study the behavior of systems governed by an objective function is to plot the objective for all possible values of the parameters.[9] This "objective surface" shows how the algorithm behaves: when maximizing, the parameters climb the surface. When minimizing, the parameters descend the surface. The summits (or the valleys) are the values to which the algorithm converges.

For example, the objective function $J_{cor}(\alpha) = \text{LPF}\{i[k]o[k]\}$ for the phase-reset adaptive oscillator of Sect. 6.4.3 is shown in Fig. 6.18(a) and (b). The input is assumed to be a pulse train (6.17) with $T = 500$ ms, the threshold is $\tau = 1$, and the objective function $J(\alpha)$ is plotted for α between 0.1 and 5. Time is scaled so that a value of $\alpha = 1$ is the "correct" answer where one period of the oscillator occurs for each input spike. In Fig. 6.18(a), the waveshape is the cosine while (b) uses the Gaussian waveshape (recall Fig. 6.12). In both cases, if the α is initialized between 0.5 and 1.5 then it will climb the surface until it reaches the desired value of one. If it is initialized between 1.5 and 2.5, it converges to $\alpha = 2$, which is when two periods of the oscillator occur for each single input pulse. Similarly, it can also converge to any integer multiple of the input pulse train.

Figures 6.18(c) and (d) show the objective surfaces for the least squares $J_{LS}(\alpha) = \text{LPF}\{i[k](\tau - o[k])^2\}$. Part (c) uses the cosine waveshape and (d) uses the Gaussian waveshape; all other values are the same. In these cases, if the α is initialized between 0.5 and 1.5, it descends the surface (because it is minimizing rather than maximizing) until it reaches the bottom at the desired value of unity. If it is initialized between 1.5 and 2.5, it converges to $\alpha = 2$. In all cases, the adaptive parameter converges to an integer multiple of the input spike train.

Similar analysis can be carried out for other adaptive oscillators. For example, the adaptive wavetable oscillator of Sect. 6.4.2 adapts both the phase β

[9] For example, [B: 102] shows plots of the objective surfaces for the various forms of the PLL (6.11)–(6.13).

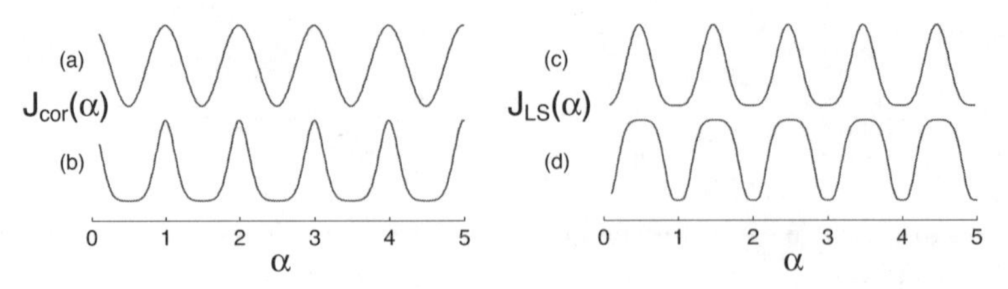

Fig. 6.18. Objective surfaces for the adaptive phase-reset oscillator. In all four cases (Gaussian and cosine waveshapes, least squares and correlation objectives) the α parameter is optimized at the "correct" value of 1, or at a simple integer multiple.

and the frequency α, and so the objective function $J(\alpha, \beta)$ is two dimensional. The surface for the correlation objective is shown in Fig. 6.19. The algorithm is initialized at some α, β pair and then evolves on this surface, climbing at each step. As before, α is normalized so that unity corresponds to one period of the oscillator for each input spike. As β ranges between zero and α, it covers all the possible phases.

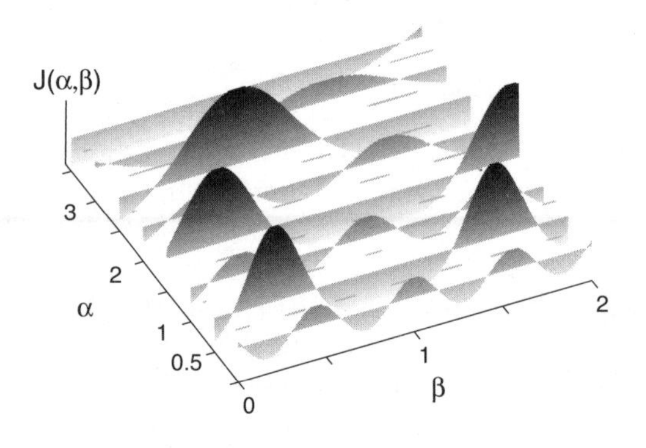

Fig. 6.19. The objective surface for the phase and frequency updates (6.15),(6.16) of the adaptive wavetable oscillator. Depending on the initial values, the period may converge to $\alpha = 1, 2, 3, \ldots$ or any of the other peaks at integer multiples. There are also stable regions of attraction surrounding $\alpha = 0.5, 1.5, 2.5$, and 3.5, which correspond to various $n{:}m$ synchronizations.

Observe that the oscillator may converge to different values depending on its initialization. If started near $\alpha = 1$, it inevitably synchronizes so that each period of the oscillator is aligned with an input spike. But other values are possible: α may converge to an integer multiple, or to a variety of $n{:}m$ synchronizations.

Another way to view the behavior of adaptive oscillators is to simulate a large number of oscillators, each with different initial conditions. Figure 6.20 shows the adaptive clocking algorithm as it adapts to a spike train with 500 ms between adjacent spikes. All possible initial periods between 200 and 2000 ms are used and each is initialized with a randomly chosen phase. The vast majority converge to one of the synchronization ratios shown, though several also converge to one of the intermediate limit cycles. Figure 6.20 also shows the approximate rate of convergence of the algorithm (though the time axis is not labeled, this typically occurs within 3 to 7 events (1.5 to 3.5 s).

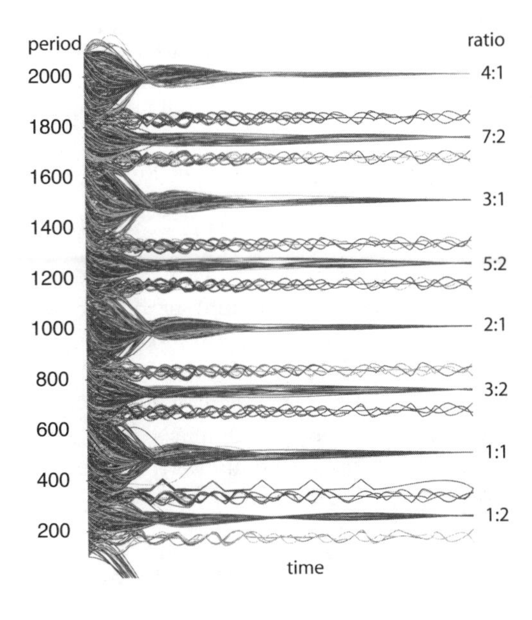

Fig. 6.20. The adaptive clocking algorithm is initialized with every possible period between 200 and 2000 ms in an attempt to lock onto an input spike train with period 500 ms. Those initialized between (about) 375 and 625 converge to the expected value of 500. Others converge to 2:1, 3:1, 1:2, etc. Initializations below (about) 175 diverge to zero. There are also several limit cycles observable at intermediate values which correspond to more complex synchronization ratios.

6.5.2 Irregular Pulse Trains

There are many ways that a regular pulse train may become irregular: spikes in the input may fail to occur, extra spikes may be present, or noise may be added to the input. This section compares the shape of the objective functions of the adaptive phase-reset oscillator and the adaptive wavetable oscillator under these kinds of irregularities. The discussion focuses on the Gaussian waveshape and the correlation objective function, but the general outlines of the results are the same for other wavetables and other objective functions.

First, adaptation is typically robust to the deletion of events. Converging takes longer and is more tenuous, but once in the neighborhood of the correct answer, the phase-reset algorithm can tolerate more than 90% deletions and still achieve synchronization. Figure 6.21 shows three cases where 50%, 80%, and 95% of the input spikes have been removed. In all three cases, there are still prominent peaks in the objective function at the expected integer values of α. The implication of the narrow peaks is that the region of convergence is reduced, that is, the range of initial α values that will converge properly is smaller than without the deletions. Initial values that are not near the peaks may converge to one of the local maxima, the smaller peaks that occur between the integer values. Similarly, the algorithms are fairly robust to the addition of randomly spaced spikes.

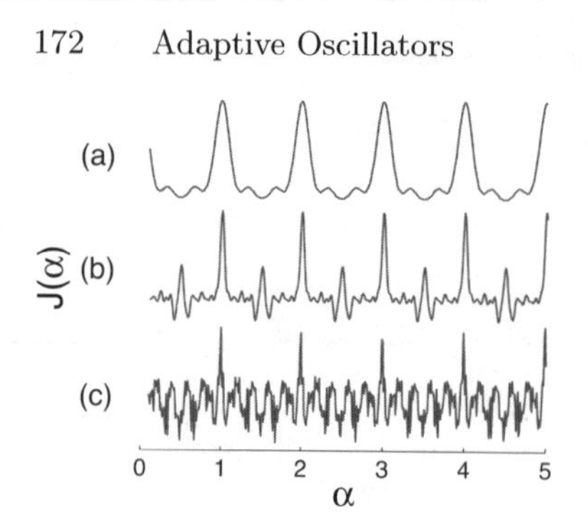

Fig. 6.21. The objective functions for the adaptive phase-reset oscillator when (a) 50% of the input spikes have been (randomly) deleted, (b) 80% have been deleted, and (c) 95% have been deleted. Despite the high percentages, all still retain their optima at $\alpha = n$ for integers n. As the number of deletions increases, the region of attraction of the optimal values shrinks and local maxima occur.

The Achilles heel of the adaptive phase-reset oscillator is its sensitivity to additive noise; even small amounts of noise can bias and/or destabilize the iteration. Figure 6.22 shows the objective function when Gaussian noise is added. In case (a), the standard deviation of the added noise is 0.004, in (b) it is 0.01 and in (c) it is 0.1. Even in (a), the convergent value for α is significantly biased away from one (the peak occurs at approximately $\alpha = 1.2$). In the modest noise case, the peak near one disappears completely, and in the high noise case, the algorithm is unstable for all initializations.

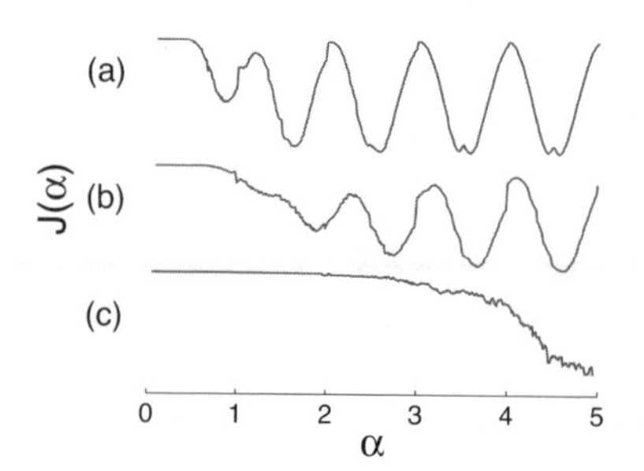

Fig. 6.22. The objective function for the adaptive phase-reset oscillator when (a) the input is subjected to additive Gaussian noise with standard deviation 0.004, (b) with standard deviation 0.01, and (c) with standard deviation 0.1. As the amount of noise increases, the optimal values cease to be maxima of the objective function and the oscillator cannot be expected to converge.

Why is the phase-reset adaptation so sensitive to such small additive perturbations? The heart of the problem lies in the thresholding. Any additive noise tends to cause the oscillator to fire early (because the threshold tends to be exceeded in each cycle when the noise adds to the oscillator output). Since the threshold is crossed early, the oscillator tries to compensate by increasing the period. This is why Fig. 6.22(a) and (b) shows such severe biasing of the α. Eventually, the noise gets large enough and the thresholding occurs irregularly enough that synchronization is lost completely.

The adaptive wavetable oscillator behaves similarly to the phase-reset oscillator when the input is a train of spikes, though its convergence tends to be slower. Both tend to be robust to random additions and deletions of spikes in the input. But the adaptive wavetable oscillator can withstand considerably larger additive noises without suffering biased estimates or losing synchronization. Figure 6.23, for example, shows the objective function when the input is corrupted by additive noise with standard deviation 0.04. The region of attraction of the humps is smaller than in the noise-free case and the noise floor of the figure is much higher. Nonetheless, the humps in the objective function occur at the desired integer locations and the oscillator can still achieve unbiased synchronization with the input. A more detailed comparison of the behavior of the various methods can be found in [B: 6].

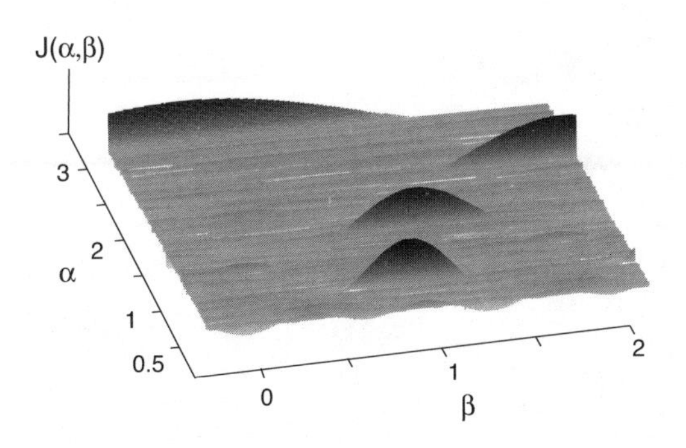

Fig. 6.23. The objective surface for the phase and frequency updates (6.15),(6.16) of the adaptive wavetable oscillator using the Gaussian waveshape. Additive noise with standard deviation of 0.04 is added to the input, which is a regular train of spikes. Depending on the initial values, the period may converge to $\alpha = 1, 2, 3, \ldots$

Different kinds of beat tracking problems encounter different kinds of noises. When tracking symbolic sequences such as MIDI, there may be extra pulses or missing pulses but there is no source of additive noise. In contrast, when tracking audio from a set of feature vectors, the noisy character of the signal assumes a major role. This dichotomy will restrict certain kinds of oscillators to certain problems. These issues are explored more fully in Chap. 8.

7

Statistical Models

The search for rhythmic patterns can take many forms. Models of statistical periodicity do not presume that the signal itself is periodic; rather, they assume that there is a periodicity in the underlying statistical distributions. In some cases, the randomness is locked to a known periodic grid on which the statistics are defined. In other cases, the random fluctuations may be synchronized to a grid with unknown period. In still other cases, the underlying rate or period of the repetition may itself change over time. The statistical methods relate the signal (for example, a musical performance) to the probability distribution of useful parameters such as the period and phase of a repetitive phenomenon. The models are built on joint work with R. Morris in [B: 151] and [B: 206].

There are two steps in the application of a statistical approach to the determination of periodic behavior. The most important is the creation of an appropriate model that specifies the unknown parameters in simple form. The sophisticated machinery of Bayesian analysis can then be applied to the model conditioned on the data. Typically, this involves writing a computer program to carry out the calculations required to find the probability distribution of the parameters representing the period and phase of the repetitive phenomenon.

The next section briefly reviews the basic probabilistic definitions that underlie the Bayesian approach and Sect. 7.2 discusses the modeling of various kinds of periodic phenomenon. The next several sections introduce a series of models each aimed at a particular task in the rhythm finding process. Section 7.3 considers the simplest setting in which the data consists of a sequence of binary events located on a fixed lattice of known duration. When the underlying pulse-rate is not known it must be inferred from the data, as suggested by the model of Sect. 7.4. Then Sect. 7.5 shows how the parameters of the model can be tracked through time recursively using a particle filter [B: 50, B: 75, B: 236]. This model can be applied to the beat tracking of MIDI performances. Section 7.6 then constructs a generative model of the probabilistic structure of feature vectors such as those of Sect. 4.4. The framework allows seamless and consistent integration of the information from multiple feature vectors into a single estimate of the beat timing. This final model will be used in Chap. 8 for the beat tracking of audio performances.

7.1 Probability and Inference

Let $P(A)$ be the probability that event A is true and $P(B,A)$ be the probability that both A and B are true. The conditional probability that B is true given that A is true is designated $P(B|A)$ and is defined implicitly by

$$P(B,A) = P(B|A)P(A)$$

when $P(A) \neq 0$. Similarly, the conditional probability $P(A|B)$ is defined by

$$P(A,B) = P(A|B)P(B).$$

Since $P(A,B)$ and $P(B,A)$ are the same,

$$P(B|A)P(A) = P(A|B)P(B)$$

which is usually written

$$P(B|A) = \frac{P(A|B)P(B)}{P(A)}. \tag{7.1}$$

This is known as Bayes' rule, and it relates the conditional probability of A given B to the conditional probability of B given A.

Books on probability are filled with problems where conditioning plays an important role. A typical example involves a hapless student who dutifully draws balls from an urn containing black and white balls. Suppose that black balls are drawn with some probability q. If the value of q happened to be known, it would be possible to calculate many interesting facts such as the probability distribution of the number of black balls drawn, the mean and variance of this distribution, and the chance that five black balls are chosen sequentially. The existence of the urn is the model on which the problems are based; it is the context or the background information. The particular value (in this case the percentage q of balls that are black) can be thought of as a known parameter of the model. The problems require calculation of various functions of the distribution given the value of the parameter. That is, they require calculation of the probability of some set of data $\mathcal{D}$ given the value of q. In symbols, this is $P(\mathcal{D}|q)$. The calculations may be easy or difficult, but they are straightforward because the problem statement contains enough information so that a unique solution is possible.

Sometimes the model may be given but some of the parameters within the model may not be known. Perhaps there is an urn containing an unknown number of balls of different colors. Certainly it must be true that it is possible to infer information about the contents of the urn by experimenting. Balls that are drawn from the urn provide data $\mathcal{D}$ that helps to pin down reasonable values for the percentages of the various colors. To be explicit, let q be the (unknown) percentage of black balls. The goal is to learn as much as possible about the distribution of q given the data and the prior information. That is, the goal is to find the distribution of $P(q|\mathcal{D})$.

Bayes' rule is perfect for this task because it relates the probability of the data given the parameter (one of the straightforward problems above) to the probability of the parameter given the data. That is, it relates $P(\mathcal{D}|q)$ to $P(q|\mathcal{D})$. To be explicit, (7.1) can be rewritten

$$P(q|\mathcal{D}) = \frac{P(\mathcal{D}|q)P(q)}{P(\mathcal{D})}, \tag{7.2}$$

though many books on Bayesian statistics (such as [B: 212, B: 137]) explicitly distinguish the background information from the parameters q of the model by writing all quantities as conditioned on the background information. For ease of reference, each term in (7.2) is given a name:

$$\text{posterior probability} = \frac{\text{likelihood} \times \text{prior information}}{\text{evidence}}.$$

Bayes' rule shows how the conditional probability of the data given the parameters (and the background information) is related to the conditional probability of the parameters given the data (and the background information). That is, it relates the likelihood (which is often easy to calculate) to the posterior. In many situations, this can be rewritten as the proportionality

$$P(q|\mathcal{D}) \propto P(\mathcal{D}|q)P(q)$$

which emphasizes that the denominator in (7.2) is a normalization term that is the same for all possible values of q. In practical terms, this means that it is possible to compare the posteriors for different candidate values of q (say q_1 and q_2) by directly comparing $P(\mathcal{D}|q_1)P(q_1)$ and $P(\mathcal{D}|q_2)P(q_2)$.

7.2 Statistical Models of Periodic Phenomenon

The kinds of models used in this chapter to represent periodic or repetitious behavior are only slightly more complicated than the ball and urn model of the previous section. Imagine that each urn in the carousel shown in Fig. 7.1 contains some percentage of black and white balls. Each time a ball is drawn, a new urn rotates into place. The chances of drawing black balls changes periodically with the number of urns and the problem of inferring the number of urns is related to the problem of locating regularities in the draws of the balls.

This illustrates that models of statistical periodicity do not presume that the data itself is periodic; rather, they assume that there is a periodicity in the underlying statistical distributions. Thus the draws from the urns of Fig. 7.1 do not lead to periodic patterns of black and white balls; rather, there is

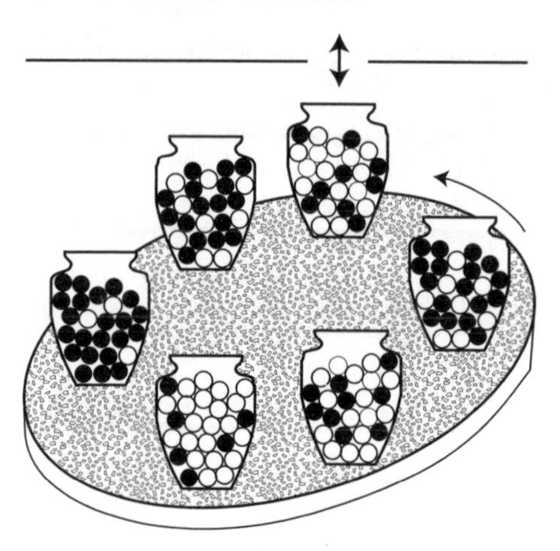

Fig. 7.1. Each time a ball is removed from one of the N urns (indicated by the arrow), the platform rotates, bringing a new urn into position. If the number of urns were known, this would be the same as N copies of the original ball and urn problem. But when N is unknown, it becomes another parameter (the periodicity) that must be inferred from the experiments. In terms of the periodicity-finding goals of *Rhythm and Transforms*, inferring N is often more important than inferring the individual percentages of black or white balls.

a tendency for the statistics of the balls to fluctuate in synchrony with the rotation of the urns.

Such combinations of repetition and randomness are important in many fields. For example, the message in telecommunications is effectively random while the modulation, synchronization, and frame structure impose periodic fluctuations. In mechanics, rotating elements provide periodicity while cavitation, turbulence, and varying loads impose randomness. Rhythmic physiological processes such as the heartbeat and brainwaves are clearly repetitive but are neither completely periodic nor fully predictable.

In some cases, the randomness is locked to a known grid on which the statistics are defined by an underlying periodicity. For example, a record of temperature versus time might experience cycles that are locked to the daily and yearly cycles of the Earth and Sun. In other cases, the random fluctuations might be synchronized with a grid, but the period of the grid might be unknown. This is common in astrophysics where the rotation of a celestial object can be assumed, but the period is unknown. In still other cases, the underlying rate or period of the repetition may itself change over time. Heartbeats speed up or slow down depending on the activity level of the animal.

Such processes have been studied extensively in the mathematical literature. A discrete-time stochastic process[1] x_k is called *wide sense stationary* if both the expectation $E\{x_k\}$ and the autocorrelation $R_x(k+\tau, k) = E\{x_{k+\tau}x_k\}$ are independent of k. Stationarity captures the idea that while a process may be random from moment to moment, it has an underlying unity in that the distribution of the process remains fixed through time. Only slightly more complex is the idea that the mean and autocorrelation may be periodic in time. A process x_k is called *wide sense cyclostationary* if both the expectation $E\{x_k\}$ and the autocorrelation $R_x(k+\tau, k)$ are periodic functions of k [B: 58]. If a_k is a periodic sequence and x_k is a stationary process, the

[1] Analogous definitions apply to continuous-time stochastic processes.

sum $a_k + x_k$ and the product $a_k x_k$ are cyclostationary. Many of the models considered in this chapter are (wide sense) cyclostationary.

Statistical models of periodicity need not be much more complex than the ball and urn problem illustrated in Fig. 7.1 and this chapter presents several models that generate repetitive behavior.[2] The simplest model, in Sect. 7.3, considers sequences of binary random variables lying on a grid that is fixed and known. When the grid is not known, the model of Sect. 7.4 can be used for the detection of pulse trains in symbolic sequences. This is generalized in Sect. 7.5 to a method that can track changes in the underlying periodicities of symbolic sequences. These models are not chosen arbitrarily; they mimic three levels of rhythmic processing in musical sequences. The simplest situation parallels the search for regularities in a musical score. The second parallels the finding of the pulse in a musical performance. The third tracks changes in the pulse and is applied to MIDI performances in Chap. 8. The final model, in Sect. 7.6, can be applied to search for statistical periodicities in feature vectors such as those of Sect. 4.4. This is applied to the beat tracking of audio, also in Chap. 8.

7.3 Regularities in Binary Sequences

The numerical notations for musical sequences from Sect. 2.1.4 are built on a fixed time grid in which each location represents a possible event. A "1" in the ith position indicates that an event occurs at grid point i while a "0" indicates that no event occurs. Thus the data $\mathcal{D}$ is a binary sequence with elements d_i that indicate when events occur. The time base of the sequence is completely regular: the interval between adjacent grid points is fixed and known. This section shows a way to apply the statistical approach to locate repetitive behavior in such binary sequences.

One simple model presumes that there is a basic underlying periodicity in the statistics of the data. Every $\mathcal{T}$ timesteps, events occur with some large probability q_L while at all other times, events occur with some small probability q_S. This is the carousel model of Fig. 7.1 with $\mathcal{T}$ urns: one urn has a high percentage of black balls (ones) while the rest have mostly white balls (zeroes). This model generates binary sequences that are repetitive but not periodic.

Let I_L be the set of indices $\{\tau, \tau + \mathcal{T}, \tau + 2\mathcal{T}, \tau + 3\mathcal{T}, \ldots\}$ where the q_L random variables occur and let I_S be the complement, where the q_S random variables occur. The model contains two kinds of Bernoulli random variables:

$$x_i^L = \begin{cases} 1 & \text{with probability } q_L \\ 0 & \text{with probability } 1 - q_L \end{cases}$$

which occurs whenever $i \in I_L$ and

[2] See [B: 203] for a variety of other models aimed at similar goals.

$$x_i^S = \begin{cases} 1 & \text{with probability } q_S \\ 0 & \text{with probability } 1 - q_S \end{cases}$$

for $i \in I_S$. The parameters of the model are the period $\mathcal{T}$, the phase τ, and the probabilities q_L and q_S, which are gathered into the vector

$$\mathbf{t} = [\tau, \mathcal{T}, q_L, q_S].$$

There are two kinds of parameters: the *structural* parameters define the probabilities of events and the *timing* parameters specify the periodicities. As will be shown, it is often possible to assume that the structural parameters q_L and q_S remain the same across many pieces and so these may be estimated off-line from training data. For example, when considering binary representations of musical scores, the average number of notes per second translates (roughly) into reasonable values for the probabilities. The timing parameters τ and $\mathcal{T}$ lie at the heart of the problem and each piece of music will have its own periodicities.

Since the model assumes that the data is constructed from 0–1 Bernoulli random variables, the likelihood $p(\mathcal{D}|\mathbf{t})$ is straightforward. In a data sequence of x_i^L variables, the probability that a specific sequence of n_L ones and m_L zeroes occurs is $q_L^{n_L}(1 - q_L)^{m_L}$. Similarly, in a data sequence of x_i^S variables, the probability that there are n_S ones and m_S zeroes is $q_S^{n_S}(1 - q_S)^{m_S}$. Since the x_i^L and x_i^S are assumed independent, the likelihood $p(\mathcal{D}|\mathbf{t})$ is proportional to[3]

$$q_L^{n_L}(1 - q_L)^{m_L} q_S^{n_S}(1 - q_S)^{m_S} \tag{7.3}$$

where the parameters can be written directly in terms of the data $\mathcal{D}$ as $n_L = \sum_{i \in I_L} d_i, m_L = \sum_{i \in I_L}(1 - d_i), n_S = \sum_{i \in I_S} d_i$, and $m_S = \sum_{i \in I_S}(1 - d_i)$. For numerical reasons, it is often advantageous to use the log of the likelihood, which has the same set of maxima and/or minima. This is

$$n_L \log(q_L) + m_L \log(1 - q_L) + n_S \log(q_S) + m_S \log(1 - q_S). \tag{7.4}$$

If the values of the parameters are unknown, Bayes' theorem can be used to relate the probability of the parameters given the data to the probability of the data given the parameters

$$p(\mathbf{t}|\mathcal{D}) \propto p(\mathcal{D}|\mathbf{t})\; p(\tau)\; p(\mathcal{T})\; p(q_L)\; p(q_S).$$

The prior probabilities $p(\tau)$, $p(\mathcal{T})$, $p(q_L)$, and $p(q_S)$ are fixed with respect to the length of the data record and are assumed independent of each other.[4] The

[3] Strictly speaking, $p(\mathcal{D}|\mathbf{t})$ contains terms that count up the number of ways that the n ones and m zeroes can occur. Since these are independent of the parameters of interest (the q_i), they do not change the maxima or minima.

[4] In reality, the priors cannot be truly independent. For example, the structure of the model dictates that $q_L > q_S$ and that $\tau \leq \mathcal{T}$.

likelihood $p(\mathcal{D}|\mathbf{t})$ is strongly dependent on the data; it is a distribution that concentrates mass at the most probable values of $\mathbf{t}$ and vanishes at improbable values. Thus the likelihood, which is easy to calculate from (7.3) and (7.4), dominates the right hand side and can be used to estimate the unknown parameters.

To see how this works, suppose that the data has a periodic component with period $\mathcal{T} = 8$ and phase offset $\tau = 3$. The parameters $q_L = 0.5$ and $q_S = 0.2$ generate data sequences where it is hard to see the underlying periodicity by eye. Assuming that the structural parameters are known but the timing parameters are unknown, $p(\mathcal{D}|\mathbf{t})$ can be calculated (up to a normalization constant) for all possible periods and all possible phases directly from (7.4). The result is plotted in Fig. 7.2. The maximum (the most probable value of

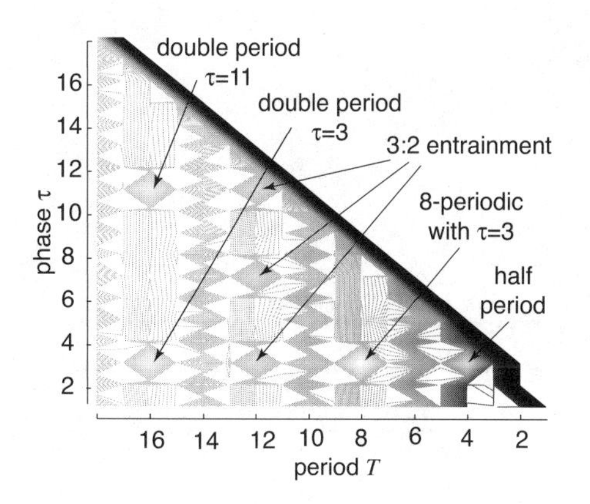

Fig. 7.2. The distribution of the unknown period $\mathcal{T}$ and phase τ can be calculated using Bayes' theorem. The largest values on this contour plot indicate the most probable values of the parameters. The actual period and phase coincide with the maximum, and some simple integer multiples feature prominently.

the parameters given the data) occurs at the correct period and phase values $\mathcal{T} = 8$ and $\tau = 3$. There are also peaks at the double period, which reflects the fact that if a sequence is periodic with period 8 then it is also periodic with period 16. There is a small peak at the half period $\mathcal{T} = 4$ and another at $\mathcal{T} = 12$, which represents a 3:2 synchronization.

A more complete approach would also assume that the structural parameters are unknown, but this may not be necessary because the maxima are fairly insensitive to the particular numbers chosen, providing q_L remains sufficiently larger than q_S. This reinforces the observation that the structural parameters can be safely estimated off-line. More care would also need to be taken with the prior probabilities if the data record were short. For large data records, the priors may be safely ignored.

This approach is not limited to finding a single periodicity. For example, data was generated with two periods: $\mathcal{T} = 7$ with $\tau = 4$ and $\mathcal{T} = 12$ with $\tau = 5$. The same procedure was followed (i.e., estimating $p(\mathbf{t}|\mathcal{D})$ via $p(\mathcal{D}|\mathbf{t})$ of (7.4)) and the distribution of the unknown period(s) $\mathcal{T}$ and phase(s) τ are shown in Fig. 7.3. Both periods are located with their correct phases. There are also a number of double and triple-period maxima and a 3:2 synchronization.

The unlabeled maxima at $\mathcal{T} = 28$ are four times the $\mathcal{T} = 7$ fundamental and the small maxima at $\mathcal{T} = 30$ represent a 5:2 synchronization with the $\mathcal{T} = 12$ period.

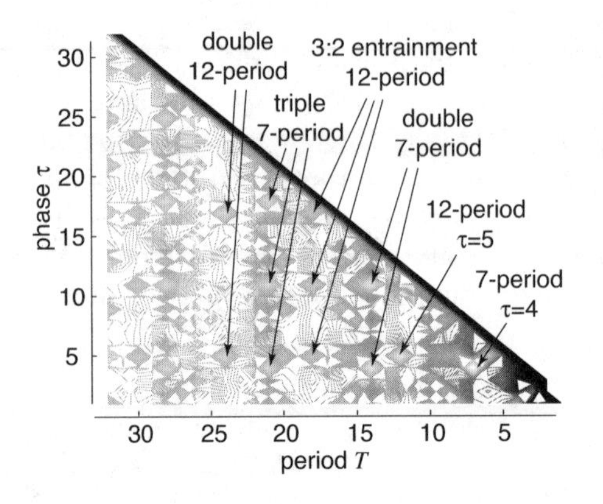

Fig. 7.3. The largest values on this contour plot indicate the most probable values of the parameters. The actual periods and phases coincide with the two largest maxima, and some simple integer multiples feature prominently.

7.4 A Model for Symbolic Pulse Detection

The model of the previous section presumes that the data sequence lies on a fixed grid defining the interval between adjacent elements. In automated musical processing, this grid typically represents the tatum, the underlying regular succession, and it is likely to be unknown. The tatum must therefore be inferred from the data, and the model must be able to automatically detect the grid timepoints.

One approach is to represent time more finely (at a faster rate) than the tatum by using a pulse train instead of a list of event times. Let t_i be a sequence of numbers specifying when events occur. For instance, the sequence might be derived from a MIDI performance and the t_i may be a list of times when MIDI notes occur. The function

$$d(t) = \begin{cases} 1 & \text{if } t = t_i \\ 0 & \text{otherwise} \end{cases} \tag{7.5}$$

is a pulse train with "1" at each event and "0" between. A perfectly regular grid would be a spike train with ones at times $\tau, \tau + \mathcal{T}, \tau + 2\mathcal{T}, \ldots$ and zeroes elsewhere.

In real data, the times are not exact: events may occur slightly before or slightly after the specified grid, occasional spurious events might occur between the grid points, and some events might fail to occur when expected. This uncertainty can be modeled statistically by presuming that events are highly probable at (or near) grid points and unlikely (but not impossible) at points between. Thus there is a probability q_L that events will occur at the grid times (as expected) and a smaller probability q_S that events may occur in between grid points. In order to handle the uncertainty in time, the function $q(t)$ is defined to move smoothly between q_L to q_S (and back), as shown in Fig. 7.4. Thus there is a reasonable chance that events cluster near lattice points even if they do not occur precisely on time. The transitions between q_L and q_S are governed by a width parameter ω, which defines the standard deviation of a Gaussian pulse centered at each grid point.

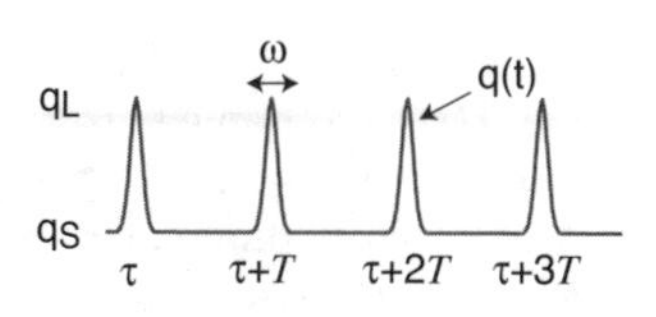

Fig. 7.4. The function $q(t)$, bounded above by q_L and below by q_S, defines the probability that an event will occur at time t. $q(t)$ is large near the lattice with period $\mathcal{T}$ and phase τ, and small otherwise, implying that events are likely to occur near the lattice and less probable elsewhere.

The model is based on the Bernoulli random variables

$$x(t_i) = \begin{cases} 1 & \text{with probability } q(t_i) \\ 0 & \text{with probability } 1 - q(t_i) \end{cases} \tag{7.6}$$

where $q(t)$ is a function of the timing parameters (the period $\mathcal{T}$ and the phase τ) and of the structural parameters (the width ω and the two extreme probabilities q_L and q_S), which are gathered into the vector

$$\mathbf{t} = [\tau, \mathcal{T}, \omega, q_L, q_S].$$

Since the data is constructed from independent Bernoulli random variables, the likelihood $p(\mathcal{D}|\mathbf{t})$, analogous to (7.3), is proportional to

$$\prod_i q(t_i)^{d(t_i)} (1 - q(t_i))^{(1-d(t_i))} \tag{7.7}$$

where $d(t_i)$ is the data at time t_i arising from $x(t_i)$. The log of the likelihood

$$\sum_i d(t_i) \log(q(t_i)) + (1 - d(t_i)) \log(1 - q(t_i)) \tag{7.8}$$

has the same set of maxima and/or minima and may be preferred for numerical reasons.

As before, when the values of the parameters are unknown, Bayes' theorem can be used to relate the probability of the parameters given the data to the probability of the data given the parameters

$$p(\mathbf{t}|\mathcal{D}) \propto p(\mathcal{D}|\mathbf{t})\ p(\tau)\ p(\mathcal{T})\ p(\omega)\ p(q_L)\ p(q_S). \tag{7.9}$$

As before, when there is a large data set, the likelihood dominates the calculation because it concentrates mass at the most probable values of $\mathbf{t}$. Since the priors are fixed with respect to the data, the maxima of the posterior $p(\mathbf{t}|\mathcal{D})$ occur at the same values as the maxima of the likelihood $p(\mathcal{D}|\mathbf{t})$ and (7.8) can be used to estimate the unknown parameters.

Of course, other statistical models are possible. One alternative, suggested by Cemgil in [B: 27], preprocesses the data by convolving with a Gaussian window to create a pulse train. A collection of varying width spike trains then attempts to locate the grid using the "tempogram" and a Kalman filter as in Fig. 7.5.

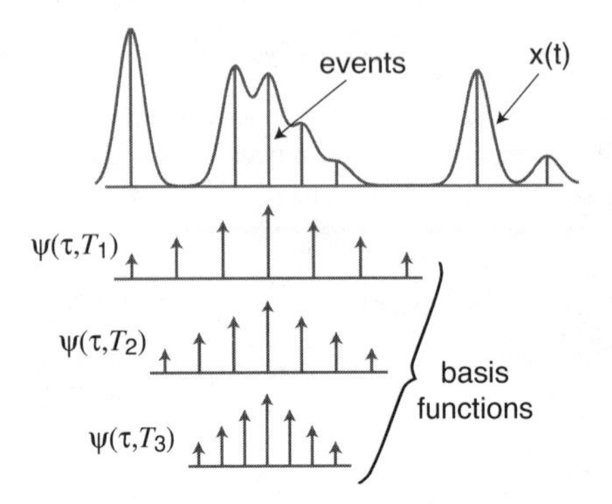

Fig. 7.5. In Cemgil's model [B: 27], the continuous signal $x(t)$ is obtained from the event list by convolution with a Gaussian function. Below, three different basis functions $\psi(\tau, \mathcal{T}_i)$ are shown. All are localized at the same τ but with different $\mathcal{T}_i$. The "tempogram" is calculated by taking the inner product of $x(t)$ and $\psi(\tau, \mathcal{T}_i)$ and the parameters can be tracked using a Kalman filter.

7.5 A Model for Symbolic Pulse Tracking

The data in the sequences of Sect. 7.3 was assumed to lie on a grid of time-points that is both fixed and known. The model was generalized in Sect. 7.4 to detect the time interval between successive elements of the grid, that is, the grid could be unknown but it was still required to be fixed. In musical applications, the tatum grid may change as the music progresses and the model must track these changes.

The calculation of $p(\mathbf{t}|\mathcal{D})$ in (7.9) provides a way of finding a fixed tatum-rate of a performance encoded into symbolic events. One way that the pulse can vary is that it might speed up or slow down. Fortunately, it is easy to modify the probabilities $q(t)$ by adding a derivative term $\delta\mathcal{T}$ to the timing parameters that allows the grid to smoothly contract or expand, as shown in Fig. 7.6. But music (typically) does not change tempo in such a simple fashion;

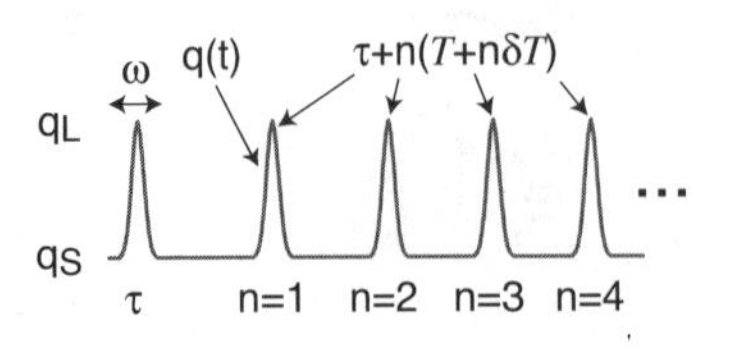

Fig. 7.6. The function $q(t)$ defines the probability that an event will occur at time t. The derivative term $\delta\mathcal{T}$ compresses (or expands) the rungs of the lattice to account for increasing (or decreasing) tempos.

rather, it may speed up or slow down many times over the course of a performance. Perhaps the simplest approach to tracking arbitrary time variations is to apply the model sequentially over short segments of the performance.

Partition the spike train into blocks of N samples.[5] Collect the timing parameters, τ, $\mathcal{T}$ and $\delta\mathcal{T}$ into a state vector $\mathbf{t}$, and let $p(\mathbf{t}_{k-1}|\mathcal{D}_{k-1})$ be the distribution of the parameters given the data in block $k-1$. The goal is to recursively update this to estimate the distribution over the parameters at block k given the new data in the kth partition, that is, to estimate $p(\mathbf{t}_k|\mathcal{D}_k)$. For example, Fig. 7.7 shows a spike train divided into partitions. If the param-

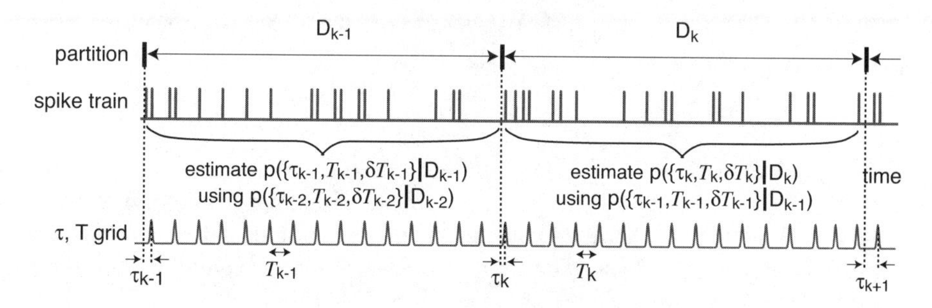

Fig. 7.7. The spike train is segmented into equal-time partitions D_k and the distribution of the parameters $p(\mathbf{t}_{k-1}|\mathcal{D}_{k-1})$ is calculated for the $k-1$st partition. This is used to estimate the distribution $p(\mathbf{t}_k|\mathcal{D}_k)$ in partition D_k.

eters entered the system linearly and if the noises were Gaussian, this could be optimally solved using the Kalman Filter [B: 92]. However, the timing parameters $\mathbf{t}$ enter into (7.9) in a nonlinear manner, and so the Kalman Filter is not directly applicable.

A naive approach would estimate the parameters anew at each partition after directly moving forward in time. For example, the timing parameter $\mathcal{T}_k$ would be assumed to be equal to $\mathcal{T}_{k-1}$ and the phase parameter would be updated[6] as

$$\tau_k = \tau_{k-1} + \mathcal{T}_{k-1}\lceil \frac{N}{\mathcal{T}_{k-1}} \rceil - N$$

where $\lceil z \rceil$ is the first integer larger than z. Such an approach effectively ignores the distributional information about the parameters at time k that is implied by previous calculations at time $k-1$.

A way to make this information explicit is to use a technique called *particle filtering* which divides the tracking into two stages, *prediction* and *update.*

[5] Somewhere between 4 and 16 seconds appears to be a reasonable time span. Shorter partitions allow the parameters to change faster, longer partitions are more resistant to portions of the performance where the spike train fails to reflect the underlying pulse.

[6] This formula presumes $\delta\mathcal{T} = 0$.

Because the pulse period does not remain precisely fixed, knowledge of the timing parameters becomes less certain over the partition and the distribution of $\mathbf{t}$ becomes more diffuse. The update step incorporates the new block of data from the next partition and provides new information to lower the uncertainty and narrow the distribution.

The predictive phase details how $\mathbf{t}_k$ is related to $\mathbf{t}_{k-1}$ in the absence of new information. In general this is a diffusion model

$$\mathbf{t}_k = f_{k-1}(\mathbf{t}_{k-1}, w_{k-1})$$

where f_{k-1} is some known function and w_{k-1} is a vector of random variables. The simplest form is to suppose that as time passes, the uncertainty in the parameters grows as in a random walk

$$\mathbf{t}_k = \mathbf{t}_{k-1} + w_{k-1}$$

where the elements of w_{k-1} have different variances that reflect prior information about how fast the particular parameter is likely to change. For pulse tracking, these variances are dependent on the style of music; for instance, the expected change in $\mathcal{T}$ for a dance style would generally be smaller than for a style with more rubato.

At block k a noisy observation is made, giving the signal

$$d_k = h_k(\mathbf{t}_k, v_k)$$

where h_k is a measurement function and v_k is the measurement noise. To implement the updates recursively requires expressing $p(\mathbf{t}_k|\mathbf{R}_k)$ in terms of $p(\mathbf{t}_{k-1}|\mathbf{R}_{k-1})$, where $\mathbf{R}_{k-1}$ represents all the data in the partitions up to block $k-1$.

This can be rewritten

$$p(\mathbf{t}_k|\mathbf{R}_{k-1}) = \int p(\mathbf{t}_k|\mathbf{t}_{k-1}) p(\mathbf{t}_{k-1}|\mathbf{R}_{k-1}) d\mathbf{t}_{k-1} \tag{7.10}$$

as the product of the predictive distribution $p(\mathbf{t}_k|\mathbf{t}_{k-1})$ (which can be calculated from the diffusion model) and the posterior distribution at time $k-1$ (which can be initialized using the prior), then integrated over all possible values $\mathbf{t}_{k-1}$. Bayes' theorem asserts that

$$p(\mathbf{t}_k|\mathbf{R}_k) = p(\mathbf{t}_k|d_k, \mathbf{R}_{k-1}) = \frac{p(d_k|\mathbf{t}_k) p(\mathbf{t}_k|\mathbf{R}_{k-1})}{p(d_k|\mathbf{R}_{k-1})} \tag{7.11}$$

where the term $p(d_k|\mathbf{t}_k)$ is a simplification of $p(d_k|\mathbf{t}_k, \mathbf{R}_{k-1})$ as the current observations are conditionally independent of past observations given the current parameters. The numerator is the product of the likelihood at block k and the predictive prior, and the denominator can be expanded

$$p(d_k|\mathbf{R}_{k-1}) = \int p(d_k|\mathbf{t}_k) p(\mathbf{t}_k|\mathbf{R}_{k-1}) d\mathbf{t}_k.$$

In a one shot estimation (as in Sects. 7.3 and 7.4), this normalization can be ignored because it is constant. In the recursive form, however, it changes in each partition.

This method can be applied to the model of (7.6) by writing the predictive distribution

$$p(\mathbf{t}_k|\mathbf{t}_{k-1}) \sim e^{-\frac{(\mathcal{T}_k-\mathcal{T}_{k-1})^2}{2\sigma_{\mathcal{T}}^2}} e^{-\frac{(\delta\mathcal{T}_k-\delta\mathcal{T}_{k-1})^2}{2\sigma_{\delta\mathcal{T}}^2}} e^{-\frac{(\tau_k-\tau_{k-1})^2}{2\sigma_{\tau}^2}}$$

where $\sigma_{\mathcal{T}}^2$, σ_{τ}^2, and $\sigma_{\delta\mathcal{T}}^2$ are the variances of the diffusions of the period, phase, and derivative, which are assumed independent. The likelihood is given by (7.7) and (7.8). Assuming an initial distribution $p(\mathbf{t}_1|\mathbf{R}_0)$ is available, these equations provide a formal solution to the estimate through time of the distribution of the timing parameters [B: 13].

In practice, however, for even moderately complex distributions, the integrals in the above recursion are analytically intractable. Particle filters [B: 50, B: 236] overcome this problem by approximating the (intractable) distributions with a set of values (the "particles") that have the same distribution, and then updating the particles over time. More detailed presentations of the particle filter method can be found in [B: 51, B: 75].

Applied to the pulse tracking problem, the particle filter algorithm can be written succinctly in three steps. The particles are a set of M random samples, $\mathbf{t}_k(i), i = 1, 2, \ldots, M$ distributed as $p(\mathbf{t}_{k-1}|\mathbf{R}_{k-1})$.

(i) **Prediction:** Each sample is passed through the system model to obtain samples of

$$\mathbf{t}_k^{\dagger}(i) = \mathbf{t}_{k-1}(i) + w_{k-1}(i) \quad \text{for} \quad i = 1, 2, \ldots, M,$$

which adds noise to each sample and simulates the diffusion portion of the procedure, where $w_{k-1}(i)$ is assumed to be a three-dimensional Gaussian random variable with independent components. The variances of the three components depend on how much less certain the distribution becomes over the block.

(ii) **Update:** With the new block of data values d_k, evaluate the likelihood for each particle using Equation (7.7). Compute the normalized weights for the samples

$$g_i = \frac{p(d_k|\mathbf{t}_k^{\dagger}(i))}{\sum_j p(d_k|\mathbf{t}_k^{\dagger}(j))}.$$

(iii) **Resample:** Resample M times from the discrete distribution over the $\mathbf{t}_k^{\dagger}(i)$s defined by the g_is to give samples distributed as $p(\mathbf{t}_k|\mathbf{R}_k)$.

To initialize the algorithm, draw M samples from the prior distribution $p(\tau, \mathcal{T})$, which is taken as uniform over some reasonable range. If more information is available (as studies such as Noorden and Moelants on preferred rates of tapping [B: 156] suggest), then better initializations may be possible. A number of alternative resampling schemes [B: 51, B: 52] with different numerical properties could be used in the final stage of the algorithm.

7.6 A Model for Audio Feature Vectors

The models of the previous sections operate with inputs composed of event lists and binary spike trains. To deal with audio, the model must operate with real-valued feature vectors meant to represent the rhythmic structure of the piece. Each feature vector is a method of data reduction that uses a different method of (pre)processing to extract different low level audio features, and so provides a (somewhat) independent representation of the beat. Feature vectors based on auditory models and on the detection of various kinds of auditory boundaries are described in detail in Sect. 4.4.

Inspection of the feature vectors such as those in Figs. 4.17 and 4.20 (on pp. 104 and 107) reveals that, to a first approximation, they are composed primarily of large values at (or near) the beat timepoints and smaller values away from the beat points. This suggests that they may be modeled as a collection of random variables with changing variances: small variance when "between" the beats and large variance when "on" the beat. A simple model that captures this structure supposes that the feature vectors are formed from realizations of an independent Gaussian noise process, where the variance of the noise "on" the beat is larger than the variance "off" the beat.

The simplest variance model assumes an underlying T-periodic sequence $\sigma_0^2, \sigma_1^2, \sigma_2^2, \ldots, \sigma_{T-1}^2$ that defines the variance of a zero mean cyclostationary process s_i with T-periodic distribution

$$s_i \sim \mathcal{N}(0, \sigma^2_{i \bmod T})$$

where $\mathcal{N}(\cdot, \cdot)$ denotes the Gaussian distribution. The $T = 3$ case is illustrated in Fig. 7.8, and the model is fully specified by the distribution of the s_i, the variances σ_i^2, and the starting time τ.

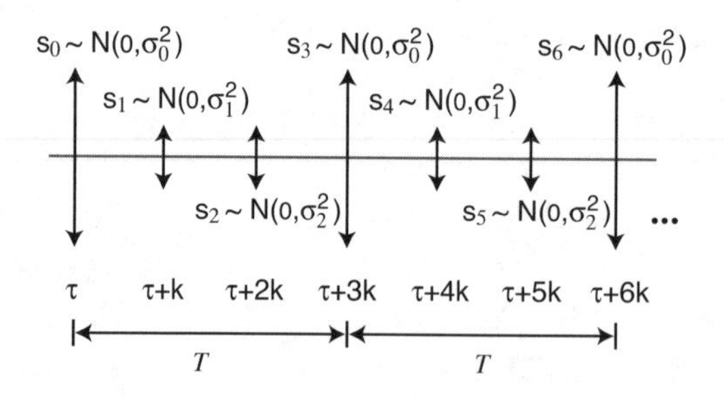

Fig. 7.8. Outputs of the variance model are generated from a stochastic process s_i defined by a periodic pattern of variances σ_i^2. A three-periodic example is shown.

Much as the model of regularities in binary sequences (of Sect. 7.3) requires that the data lie on a fixed and known grid, the variance model requires a rigid T-periodicity of the statistics. This can be relaxed by creating a smooth function that varies between a large variance σ_L^2 near the beat and a small variance σ_S^2 far from the beat. Figure 7.9 shows the parameters and the structure of the model, along with a typical realization. It is a zero mean cyclostationary process with parameters defined by a function σ_t that looks much like the $q(t)$ of Fig. 7.4 but where the values represent the instantaneous variance of a Gaussian process rather than the Bernoulli probability of events.

Such a model ignores much of the structure that is present. For example, the group delay feature vector (in Fig. 4.20(c)) often shows larger positive peaks than negative peaks. The tracks also regularly display oscillatory behavior, mirroring the intuitively obvious idea that the samples cannot be truly independent (this is most obvious in the feature vector of Fig. 4.14). However, this model is shown in the experiments to capture enough of the structure to allow reliable beat extraction in a variety of musical situations. While it is in principle possible to derive the distribution of the samples in the feature vectors from a probabilistic model of the original audio, this is too complex to result in a feasible algorithm. Also, it is no more obvious how to construct a model for the audio than for the feature vectors directly.

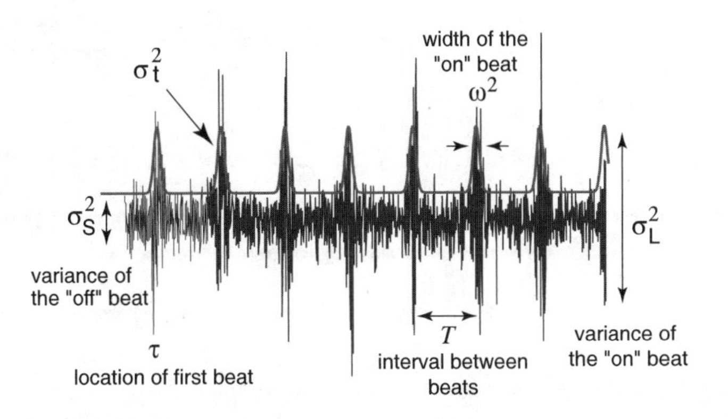

Fig. 7.9. Parameters of the feature vector model are $\mathcal{T}$, τ, ω, σ_S, σ_L and $\delta\mathcal{T}$ (not shown). The dark curve σ_t shows the periodic pattern of variances that defines the cyclostationary process. The jagged curve shows a typical realization.

The parameters can be divided into two sets. The structural parameters remain essentially constant and are estimated off-line from training data:

- σ_S^2 is the "off the beat" variance,
- σ_L^2 is the "on the beat" variance, and
- ω^2 is the beatwidth, the variance of the width of each set of "on the beat" events. For simplicity, this is assumed to have Gaussian shape.

The timing parameters lie at the heart of the beat extraction:

- τ is the time of the first beat,
- $\mathcal{T}$ is the period of the beat, and
- $\delta\mathcal{T}$ is the rate of change of the beat period.

Gather these parameters into a single vector

$$\mathbf{t} = [\tau, \mathcal{T}, \delta\mathcal{T}, \sigma_S, \sigma_L, \omega].$$

Given the signal $\mathbf{r}$ (the feature vector), Bayes' theorem asserts that the probability of the parameters given the signal is proportional to the probability of the signal given the parameters multiplied by the prior distribution over the parameters. Thus

$$p(\mathbf{t}|\mathbf{r}) \propto p(\mathbf{r}|\mathbf{t})\ p(\tau)\ p(\mathcal{T})\ p(\delta\mathcal{T})\ p(\sigma_S^2)\ p(\sigma_L^2)\ \ p(\omega^2)$$

where the priors are assumed independent.[7] Each of the prior probabilities on the right hand side is fixed with respect to the length of the data record, while the first term becomes more concentrated as a function of the length of the data. Accordingly, the first term dominates. Let $t_i = i\mathcal{T} + \sum_{j=0}^{i} j\delta\mathcal{T} + \tau$ be the time of the ith beat and let $\lambda_t = \sum_{i=-\infty}^{\infty} e^{-\frac{(t-t_i)^2}{2\omega^2}}$ be a sum of shifted Gaussian functions. The variance

$$\sigma_t^2 = \lambda_t \sigma_L^2 + (1 - \lambda_t)\sigma_S^2 \tag{7.12}$$

specifies the likelihood of the feature vector model as

$$p(r_t|\tau, \mathcal{T}, \delta\mathcal{T}, \sigma_S, \sigma_L, \omega) \sim \mathcal{N}(0, \sigma_t^2) \tag{7.13}$$

where t is the (positive) time at which the sample r_t is observed. This periodic variance function σ_t^2 defines the cyclostationary statistics of the process and is illustrated by the dark curve in Fig. 7.9.

Because the feature vector values r_t are assumed independent, the probability of a block of values $\{r_t, t_1 < t < t_2\}$ is simply the product of the probability of each value. Thus σ_t^2 is a combination of the variances on and off the beat, weighted by how far t is from the nearest (estimated) beat location. While this may appear noncausal, it is not because it only requires observations up to the current time. Also note that the summation in the definition of λ_t can in practice be limited to nearby values of j.

To see the relationship between experimentally derived feature vectors and the model, the data in Fig. 7.9 is an "artificial" feature vector constructed from alternating small and large variance normally distributed random variables. Qualitatively, this provides a reasonable model of the various feature vectors in Fig. 4.20.

Much as the model of pulse detection (of Sect. 7.4) requires that the data lie on a fixed grid, the feature vector model of Fig. 7.9 does not directly track changing parameters. The symbolic pulse tracker of Sect. 7.5 used recursive particle filtering [B: 75] to follow changes in the pulse and the same idea can be applied to the tracking of audio feature vectors by operating over successive blocks using the output distribution at one block as the prior distribution (initialization) for the next.

---

[7] In reality, the priors cannot be truly independent, for example, the structure of the model dictates that $\sigma_L > \sigma_S$.

Divide each feature vector into blocks, typically about 400–600 samples long.[8] Collect the timing parameters, $\tau, \mathcal{T}$ and $\delta\mathcal{T}$ into a state vector $\mathbf{t}$, and let $p(\mathbf{t}_{k-1}|\cdot)$ be the distribution over the parameters at block $k-1$. The goal of the (recursive) particle filter is to update this to estimate the distribution over the parameters at block k, that is, to estimate $p(\mathbf{t}_k|\cdot)$.

Again, the tracking can be divided into two stages, *prediction* and *update.* The predictive phase details how $\mathbf{t}_k$ is related to $\mathbf{t}_{k-1}$ in the absence of new information. The simplest form is to suppose that as time passes, the uncertainty in the parameters grows as in a random walk

$$\mathbf{t}_k = \mathbf{t}_{k-1} + w_{k-1} \tag{7.14}$$

where the elements of w_{k-1} have different variances that reflect prior information about how fast the particular parameter is likely to change. At block k a noisy observation is made, giving the signal

$$r_k = h_k(\mathbf{t}_k, v_k)$$

where h_k is the measurement function and v_k is the noise. To implement the updates recursively requires expressing $p(\mathbf{t}_k|\mathbf{R}_k)$ in terms of $p(\mathbf{t}_{k-1}|\mathbf{R}_{k-1})$, where $\mathbf{R}_{k-1}$ represents all the feature vector samples up to block $k-1$.

The same logic as in Sect. 7.5, equations (7.10) and (7.11), shows that the predictive distribution is

$$p(\mathbf{t}_k|\mathbf{t}_{k-1}) \sim e^{-\frac{(\mathcal{T}_k - \mathcal{T}_{k-1})^2}{2\sigma_{\mathcal{T}}^2}} e^{-\frac{(\delta\mathcal{T}_k - \delta\mathcal{T}_{k-1})^2}{2\sigma_{\delta\mathcal{T}}^2}} e^{-\frac{(\tau_k - \tau_{k-1})^2}{2\sigma_\tau^2}}$$

where $\sigma_{\mathcal{T}}^2$, $\sigma_{\delta\mathcal{T}}^2$ and σ_τ^2 are the variances of the diffusions on each parameter, which are assumed independent. In this case, the likelihood is

$$p(r_k|\mathbf{t}_k) \sim \mathcal{N}(0, \sigma_k^2) = e^{-\frac{r_k^2}{\sigma_k^2}} \tag{7.15}$$

where σ_k^2 is the value of the function σ_t^2 in (7.12) evaluated at the time of the kth sample. Assuming an initial distribution $p(\mathbf{t}_1|\mathbf{R}_0)$ is available, these equations provide a formal solution to the estimate through time of the distribution of the timing parameters. As before, the integrals in the recursion are analytically intractable, and so the particle filtering algorithm can be used.

A major advantage of the Bayesian approach is its ability to incorporate information from multiple feature vectors in a straightforward manner. Assuming that the various feature vectors provide independent measurements of the underlying phenomenon (a not unreasonable assumption given that the tracks measure different aspects of the input signal), the likelihood for a set of feature vectors is simply the product of the likelihood for each track. Thus the numerical complexity of estimating the optimal beat times from a collection

[8] At an effective sampling rate of 60 Hz, this is in the middle of the same 4–16 second range suggested in Sect. 7.5. The same tradeoffs apply.

of four feature vectors is only four times the complexity of estimating from a single feature vector. Examples of the application of the models of this chapter to the beat tracking problem appear in Chap. 8.

8

Automated Rhythm Analysis

Just as there are two kinds of notations for rhythmic phenomenon (the symbolic and the literal), there are two ways to approach the detection of rhythms; from a high level symbolic representation (such as an event list, musical score, or standard MIDI file) or from a literal representation such as a direct encoding in a `.wav` *file. Both aspire to understand and decompose rhythmic phenomena, and both exploit a variety of technologies such as the transforms, adaptive oscillators, and statistical techniques of Chaps. 5–7. This chapter begins with a discussion of the rhythmic parsing of symbolic sequences and then incorporates the perceptually motivated feature vectors of Chap. 4 to create viable beat detection algorithms for audio. The performance of the various methods is compared in a variety of musical passages.*

Listeners can easily identify complex periodicities such as the rhythms that normally occur in musical performances, even though these periodicities may be distributed over several interleaved time scales. At the simplest level, the pulse is the basic unit of temporal structure, the foot-tapping beat. Such pulses are typically gathered together in performance into groupings that correspond to metered measures, and these groupings often cluster to form larger structures corresponding to musical phrases. Such patterns of grouping and clustering can continue through many hierarchical levels, and many of these may be readily perceptible to an attentive listener.[1] An overview of the problem and a taxonomy of beat tracking methods can be found in [B: 194], and a review of computational approaches for the modeling of rhythm is given in [B: 81].

Attempts to automatically identify the metric structure of musical pieces often begin with a symbolic representation such as a musical score. This simplifies the rhythmic analysis in several ways: the pulse is inherent in the score, note onsets are clearly delineated, multiple voices cannot interact in unexpected ways, and the total amount of data to be analyzed is small. Three levels of difficulty are contrasted in Table 8.1. At the simplest level, Sect. 8.1 applies the various methods (the transforms of Chap. 5, the adaptive oscillators of Chap. 6, and the statistical methods of Chap. 7) of identifying

[1] Recall the discussion in Sect. 3.2, especially Figs. 3.1 and 3.2 on pp. 55 and 56.

Table 8.1. The three levels of rhythmic analysis in order of increasing difficulty

	Score	MIDI	Audio
Tatum	given	inferred	inferred
Note Events	given	given	inferred
Section/Page	8.1/194	8.2/201	8.3/209

repetitions to a simple symbolic representation of *La Marseillaise.* All of the methods are capable of detecting the kinds of regularly recurring rhythmic structures associated with the beat and/or higher levels of metrical structure.

Finding repetitive phenomena in a MIDI sequence is, in general, a more complex task because the underlying pulse must be inferred from the signal. Section 8.2 discusses how the various methods can be applied and an essential dichotomy emerges. Techniques (such as transforms and autocorrelations) that require a steady underlying pulse to locate rhythmic behavior fail when the tatum of the MIDI signal is variable. Techniques (such as the adaptive oscillators and statistical methods) that are able to track changes in the underlying pulse continue to perform well.

But listeners are not presented with patterns of symbols when attempting to understand a musical passage. Discovering repetitive behavior in an audio signal is tricky, and there are two basic approaches. In the first, the audio is parsed to identify note boundaries and interonset intervals. Since MIDI representations are defined directly in terms of note-onset times, this effectively reduces the audio problem to an event list, which can then be solved by any of the techniques that work with MIDI sequences. The second approach reduces the audio to a set of feature vectors and then parses the feature vectors for the existence of a regular succession. This bypasses the difficulties of extracting individual notes from the audio signal. Section 8.3 uses the feature vectors of Chap. 4 to preprocess the audio, and demonstrates the strengths and weaknesses of the various technologies for detecting rhythm. The rhythm finding approaches are then examined in a number of musical examples in Sect. 8.3.1, beginning with a simple polyrhythm and proceeding through a variety of musical styles. Sound examples demonstrate the proper (and improper) functioning of the methods.

8.1 Analysis From a Musical Score: *La Marseillaise*

Many of the earliest attempts at automated rhythm and meter detection begin with a musical score. This simplifies the problem in several ways: the data is already segmented into notes with known pitches and the basic beat of the music is known, since all notes of equal duration are represented identically. Perhaps the simplest case is to use a single (usually melodic) line from a musical score.

8.1.1 Rule-based Approaches

Several researchers have created programs to automatically parse a musical score in order to generate a "high-level" explanation of the grouping of musical rhythms. Steedman [B: 222] proposes a theory of how listeners might infer the meter of a passage by comparing note lengths. Using only the interonset intervals within a single melodic line, the goal is to try to explain why a given rhythm is heard to be in a particular meter. This involves building a metrical hierarchy from the given set of interonset durations. The method is applied to the pieces of Bach's *Well Tempered Clavier* and the meter is often correctly found by following the suggested set of rules. Similarly, Longuet-Higgins [B: 132] creates a computer program to parse monophonic musical scores. The meter is regarded as a tree-like structure that can be derived from a kind of generative grammar, though the grammar must be augmented by the incorporation of various aspects of perception. A second study [B: 133] augments this with higher level phrasing rules which attempt to parse the rhythm so as to minimize the number of syncopations. One conclusion is that "any given sequence of note values is in principle infinitely ambiguous, but this ambiguity is seldom apparent to the listener." This underscores the difficulty of capturing musical meter and phrasing in a simple set of grammatical rules.

Given that the inputs are monophonic, pitchless, and uninflected, even the limited success of such rule-based programs is encouraging. The researchers report their results in glowing terms, saying [B: 222], "this program is intended to constitute a psychological theory of our perception of meter in melody." Yet there is no "perception" involved in this work. The program operates on high-level symbolic data (the interonset intervals taken from a musical score) that is far removed from the act of perception. To explain the cases where the program fails to discover the correct metrical structure, Steedman writes: "... where the program does not infer the meter that the score indicates, then either the melody is ambiguous, or the composer has exercised the artist's privilege to break the rules." There is also a third possibility: that the ideas used in the program are not sufficiently developed to explain the metrical structure of the pieces.

Rosenthal [B: 184] created a rhythm parsing program called `Fa` which searches for regularly spaced onset times in a MIDI data stream. The program forms and ranks several hypotheses "according to criteria that correspond to ways in which human listeners choose rhythmic interpretations." These criteria include quite sophisticated ideas such as having accented notes fall on "strong" beats, noticing motivic repetitions, and measuring salience. An example is given of the best rhythmic parsing found by `Fa` for the song *La Marseillaise*, which is reproduced here as Fig. 8.1. The four measure phrase can be heard in [S: 61].

Such rule-based approaches view the list of note events as a string of symbols to be parsed and makes decisions based on properties of the sequence.

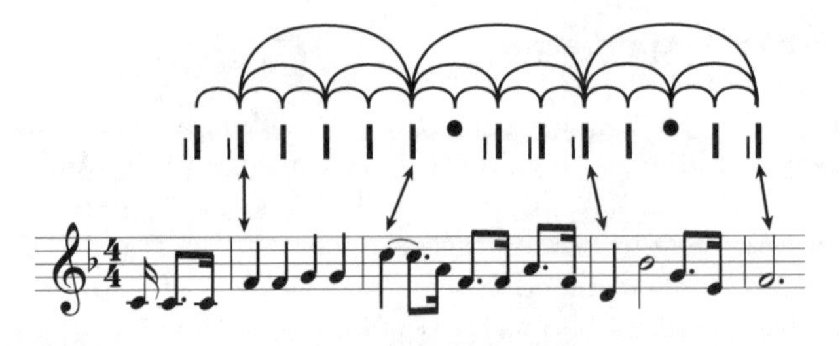

Fig. 8.1. The first four measures of *La Marseillaise* are encoded into the regular dark lines (representing notes that occur on the beat) and the short light lines that encode off-beat sixteenth notes. The three sets of arches show the rhythmic groupings calculated by Rosenthal's program `Fa`, which discerns three levels of metrical structure corresponding to the beat, the half measure and the full measure. The two dark circles • are "ghost tones" that occupy a place in the metric hierarchy but are unsounded. The oblique arrows indicate how the metrical analysis lines up with the musical score.

This problem of finding rhythmic parsings can also be viewed as a search for periodicity or frequency data from a time domain signal.

8.1.2 Transform Techniques

Autocorrelation is one way to look for periodicities and to detect non-randomness in a data sequence. Brown [B: 21] extracts a single voice from a musical score, codes it into a modified form of the binary notation, and then applies autocorrelation. In many cases, metrical features of the piece (such as the beat and the length of the measure) can be correlated with peaks of the autocorrelation function. In some cases, the half-measure was located at the peak of the autocorrelation function. This was also considered successful since listeners may also choose this level of the metric hierarchy as among the most significant.

For example, the four measures of *La Marseillaise* shown in Fig. 8.1 can be coded into the binary string $\mathcal{A}$:

11001100010001000100010000001100110011000100000000100110000000000

Each digit represents a time equal to that of one sixteenth note. The symbol 1 indicates that a note event occurred at that time, while a 0 means that no new note event occurred. Using the modified form of binary notation (as described in Sect. 2.1.4 on p. 29, this replaces each "1" with a number that specifies the length of the note), this becomes the string $\mathcal{B}$:

1 3 0 0 1 4 0 0 0 4 0 0 0 4 0 0 0 4 0 0 0 7 0 0 0 0 0 0 1 3 0 0
1 3 0 0 1 4 0 0 0 8 0 0 0 0 0 0 0 3 0 0 1 11 0 0 0 0 0 0 0 0 0 0

The autocorrelation of $\mathcal{B}$ is shown in Fig. 8.2, which plots the magnitude of the correlation coefficient at every possible delay. Large values indicate that the sequence is similar to itself at that delay. The curve shows several temporal groupings that correspond to musically significant features. The peak at delay 4 represents the quarter note pulse, the peak at delay 8 is the half note, and the peak at delay 16 corresponds to the measure. Similarly, the peak at delay 32 shows a strong correlation at two measures. Other peaks do not have clear musical interpretations.

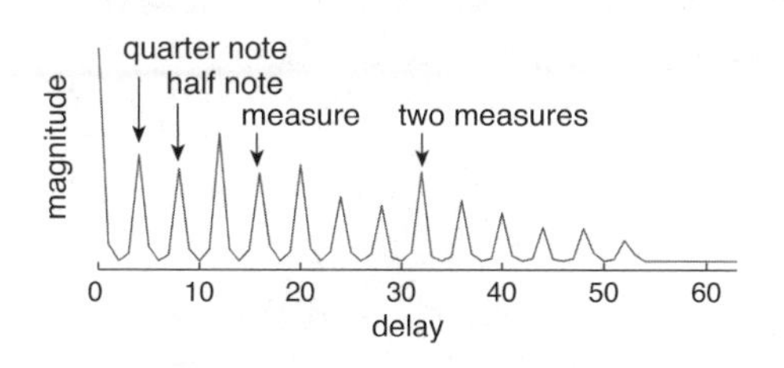

Fig. 8.2. The autocorrelation of the string $\mathcal{B}$ derived from the score to *La Marseillaise* shows several musically significant features (the quarter and half notes and the measure). Other peaks in the autocorrelation do not have an obvious interpretation.

For comparison, the binary sequence $\mathcal{A}$ (concatenated with itself four times) was transformed using the DFT. Assuming a sampling rate of 16 Hz, the duration of each quarter note is 1/4 second, and the resulting magnitude spectrum is shown in Fig. 8.3. In this figure, the largest peak at 4 Hz represents the quarter note pulse, while the peak at 1 Hz corresponds to the measure. It is unclear how to interpret the remainder of the information in this spectrum. In both this and the autocorrelation, the use of the binary notation $\mathcal{A}$ or the modified version $\mathcal{B}$ makes little difference.

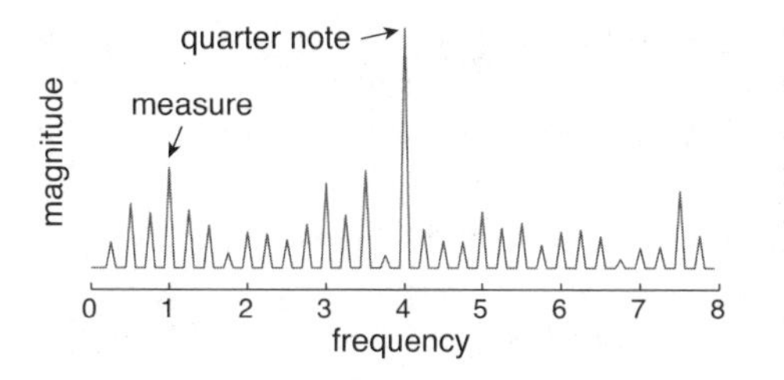

Fig. 8.3. The DFT of the binary sequence $\mathcal{A}$ derived from the score to *La Marseillaise*, assuming a sampling rate of 16 Hz (one measure per second). The peak at 4 Hz represents the quarter note pulse, and the peak at 1 Hz corresponds to the measure.

Both the autocorrelation and the DFT show significant musical features: the quarter note beat is prominent and other musical durations are apparent after inspection. Unfortunately, each is also riddled with extra peaks that have no obvious interpretation. In the autocorrelation of Fig. 8.2, for instance, the largest peak occurs at a delay of 12 which corresponds to a dotted half note (or a "measure" of three quarter notes). There are also large peaks at delays of 20, 24, and 28, etc. that do not directly correspond to sensible durations. Similarly, the DFT of Fig. 8.3 consists mostly of peaks that are integer harmonics of a very low frequency that corresponds to the length of the sequence. Most of these are unrelated to useful features of the musical score.

The periodicity transform of Sect. 5.5 fares somewhat better since it locates many of the most significant musical durations without clutter. Figure 8.4 shows the output of the M-Best periodicity transform with $M = 5$; the transform searches for the five best periodicities that occur within the sequence $\mathcal{B}$. The largest peak at period 4 represents the quarter note pulse. Also appearing are the eighth note (period 2) and the half note (period 8), as well as longer structures at the measure with period 16 and the two measure phrase with period 32. The appearance of the eighth note periodicity is interesting because there are no eighth notes in the sequence; rather, the algorithm has inferred this level of the metrical hierarchy from the sixteenth and dotted eighth notes.

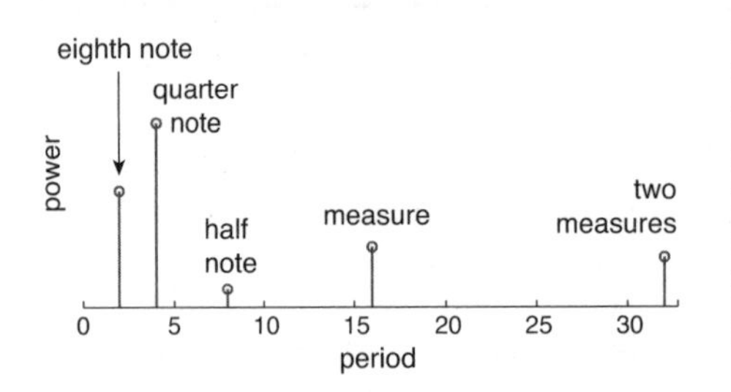

Fig. 8.4. The Periodicity Transform of the sequence $\mathcal{B}$ derived from the score to *La Marseillaise* shows five major periodicities: the quarter note beat (with period 4), the eighth note, the half note, and larger structures at the measure and at the two measure phrase

8.1.3 Statistical Methods

Another way to search for regularities in binary data is to use the Bayesian model of Sect. 7.3. Given the encoding of *La Marseillaise* into binary notation $\mathcal{A}$, the probability of the sequence $\mathcal{A}$ given the values of the parameters is the likelihood $P(\mathcal{A}|\{\tau, \mathcal{T}, q_L, q_s\})$. This can be calculated using (7.4), where q_L and q_S are the probabilities of events at the grid points and away from the grid points respectively. Using the default values of $q_L = 0.5$ and $q_S = 0.2$ for the structural parameters, the likelihood is plotted in Fig. 8.5 over a range of period $\mathcal{T}$ and phase τ. The contour plot displays the most prominent features such as the quarter note pulse, the half note, and the measure, though there are also smaller peaks (such as the one at three times the quarter note) that do not correspond to musically sensible intervals. In principle it would be proper to estimate the probabilities q_L and q_s rather than simply assuming them. But running simulations with a variety of values provides essentially the same output, suggesting that the peaks of the likelihood are fairly immune to the particular choice of values for the structural parameters. This is helpful because it reduces the problem from four dimensions to two.

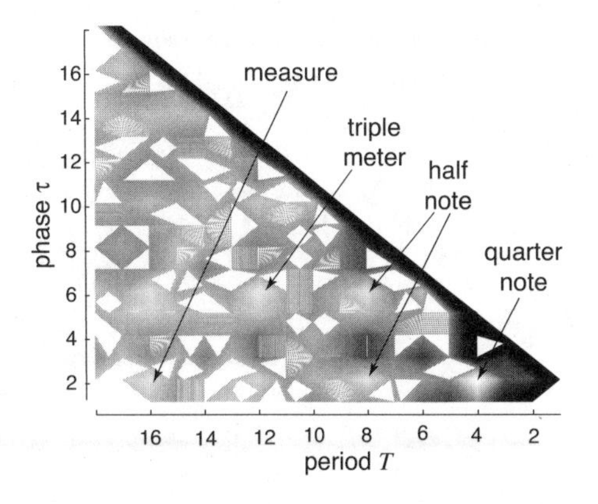

Fig. 8.5. The likelihood function for the binary AM model of Sect. 7.3 applied to *La Marseillaise* (string $\mathcal{A}$) has its largest peak at the quarter note pulse. The second largest peak occurs at the half note, and the third largest at the measure. Smaller peaks also appear at three times the quarter note.

8.1.4 Adaptive Oscillators

All of the methods of the previous sections (whether rule-based, transform, or statistical) assume that the tatum (the basic underlying pulse) of the piece is known beforehand. For example, in *La Marseillaise*, all durations are defined as simple multiples of the sixteenth note. This assumption underlies the musical score and numerical representations such as the strings $\mathcal{A}$ and $\mathcal{B}$. Because all intervals between adjacent note events are specified in terms of this idealized tatum, the interonset intervals represent idealized values, not actual values as might occur in performance.

When the tatum is not known beforehand, or where it might reasonably be expected to change throughout the development of a piece, time must be represented with a finer granularity than the tatum-rate. The simplest method is to create a function that represents the event list $\mathcal{A}$. Suppose that the ith element of the list occurs at time t_i. The function

$$x(t) = \begin{cases} 1 & \text{if } t = t_i \\ 0 & \text{otherwise} \end{cases} \tag{8.1}$$

is a pulse train that can be input to an adaptive oscillator. Figure 8.6 shows the spike train representation for *La Marseillaise*, which has the same generic appearance as the spike trains in Figs. 6.13–6.15 on pp. 164–166. Applying these inputs to adaptive oscillators provides a way to locate and track changes in the tempo of a sound pattern (as long as the change is not too rapid). By using several adaptive oscillators with different starting values, it is possible to locate more than one level of the metrical hierarchy simultaneously since the oscillators operate independently.[2]

Figure 8.6 shows the output of three adaptive phase-reset oscillators (defined in Sect. 6.4.3) when initialized near three different values corresponding to the tatum, the quarter note and the half note. All three synchronize within

[2] While some authors have suggested that better performance might result by coupling the oscillators in some way, it is not obvious how to implement the coupling.

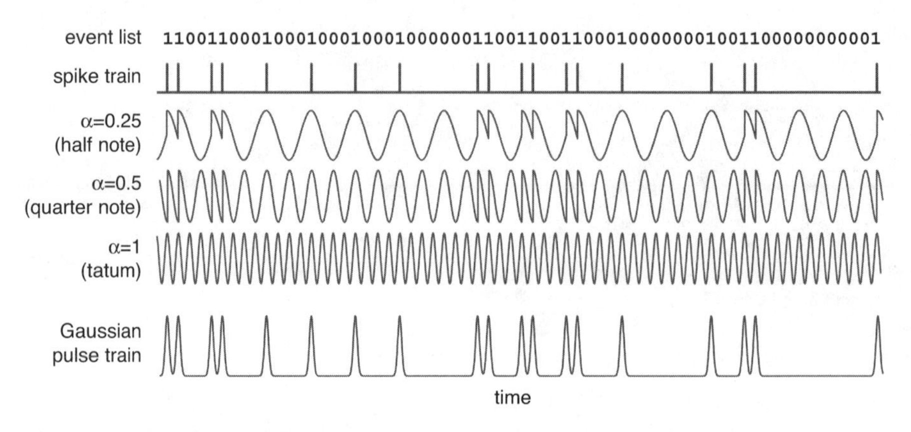

Fig. 8.6. The binary event list $\mathcal{A}$ for *La Marseillaise* is represented as a time-based pulse train. It is input to an adaptive phase-reset oscillator initialized near three different rates ($\alpha = 1, 0.5$, and 0.25) corresponding to the tatum, the quarter note and the half note. The output of the oscillators synchronize to the three different levels of the metric hierarchy. The bottom signal shows the spike train after convolution with a bell-shaped pulse, which produces an input useful for the adaptive clocking method.

a few spikes. The simulation shows the output signal during the second pass through the spike train; subsequent passes are identical. When the underlying rate of the input spikes is changed (up to about 10% per repetition) the output flawlessly tracks the input.

The bottom signal in Fig. 8.6 shows the spike train $x(t)$ convolved with a Gaussian pulse. This reshapes each spike in the binary representation into a smooth pulse that can be input into an adaptive clocking oscillator (given by (6.20) on p. 167). This input was applied to a bank of 500 independent oscillators that were identical except for their initial values. Most of the oscillators converged within a few repetitions, and the final (converged) values are displayed in the histogram Fig. 8.7, which shows the number of oscillators that converged to each period $\mathcal{T}$. About half of the oscillators converged to one of the musically sensible periods, with the simpler periods attracting the most oscillators. About one-fifth of the oscillators failed to converge (for example, the clusters of values near $\mathcal{T} = 450$ and $\mathcal{T} = 720$). The remaining oscillators appeared to be trapped in complex limit cycles with no obvious (musical) significance.

At first glance it may seem that a situation where one-half of the oscillators fail to converge to useful values is excessive. But it is easy to understand. The data contains several different near-periodic features. When initialized near a period of the data, an oscillator will most likely converge to that period. The objective surface for the wavetable oscillator (recall Fig. 6.19) shows the various stable points and the size of the regions of attraction. Thus an oscillator initialized close to the $\mathcal{T} = 100$ period will, with high probability,

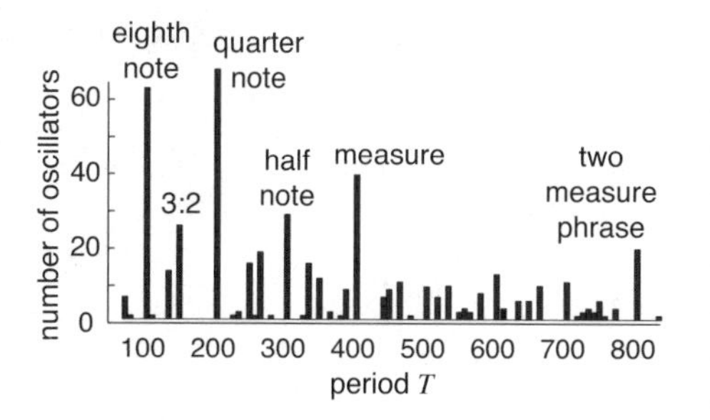

Fig. 8.7. The Gaussian pulse train representation of *La Marseillaise* drives a collection of 500 adaptive clock oscillators initialized with random starting values between 40 and 900 ($\mathcal{T} = 50$ was the tatum rate). The oscillators are allowed to converge through repeated presentations, and the resulting converged values are displayed in this histogram.

converge to $\mathcal{T} = 100$ while an oscillator initialized at $\mathcal{T} = 700$ may be too far from any actual period of the data to be attracted. Thus adaptive oscillators may be most useful when considered in aggregate. This lends plausibility to analogies with biological systems in which neuronal elements are typically considered in aggregate rather than in isolation.

For example, in a pair of studies Eck [B: 55, B: 56] used a collection of Fitzhugh–Nagumo oscillators (recall Fig. 6.12 on p. 162) to locate downbeats within the Povel and Essens patterns [B: 174, B: 202]. The frequency of the oscillators was fixed at the correct value and the phases were initialized randomly. The oscillators were run for several cycles until the outputs settled into a steady state. The behavior of the oscillators was then examined in aggregate. Though the adaptive oscillators of Chap. 6 are different in detail from Eck's oscillators, they have similar dynamics and behaviors.

8.2 MIDI Beat Tracking

All of the technologies for finding repetitive behavior can be successfully applied to symbolic sequences. The various transforms, the statistical methods, the adaptive oscillators, and the rule-based approaches can all locate the underlying periodicities when the pulse is known. But MIDI performance data is more complex in two ways. First, the time between successive events is uniform and idealized in symbolic sequences and musical scores. When data is derived from a musical performance, the interonset intervals (IOIs) need not be exactly equal. One pair of notes may be 150 ms apart and another 160 ms, and these are typically perceived as "the same duration." One task required for rhythmic understanding is to classify such durations together. Second, the tempo of a musical piece may change over time. Thus the pair of notes that are 150 ms apart at the start of a piece may be 140 ms apart when the "same" section is repeated at the end.

The classification problem can be seen in Fig. 8.8, which shows the IOIs in a MIDI performance of the *Maple Leaf Rag*. The data is displayed event-by-event in the left hand plot and as a histogram on the right. The values cluster around the tatum, around the eighth note, and around zero (representing notes that are played simultaneously).

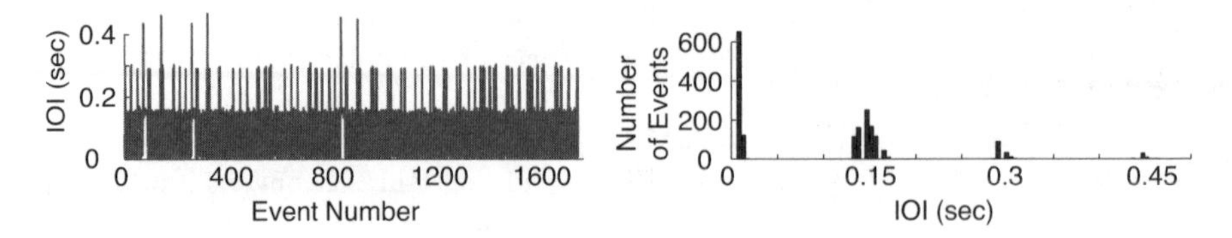

Fig. 8.8. The interonset intervals of Trachtman's [W: 51] rendition of the *Maple Leaf Rag* cluster in a few groups including 0.15 s (the tatum) and 0.3 s (the eighth note) that correspond to the surface structure

Desain [B: 43] attacks the classification problem directly using an "expectancy" function. Suppose that an interval A is perceived. What intervals are likely to follow? The most probable is another A, though simple integer multiples and divisors are also quite likely. Unrelated intervals such as πA and $\frac{6A}{7}$ are unlikely, but not impossible. The expectancy curve is shown in Fig. 8.9. When a sequence of IOIs occurs, the expectancies can be added, perhaps with a decay for those furthest back in time. Desain creates a network of nodes that represent the IOIs. These interact and alter their values until they approach simple rational multiples. Thus the expectancy model can be used to adjust the underlying tatum to the current IOIs and to classify all the intervals into a small number of uniform times.

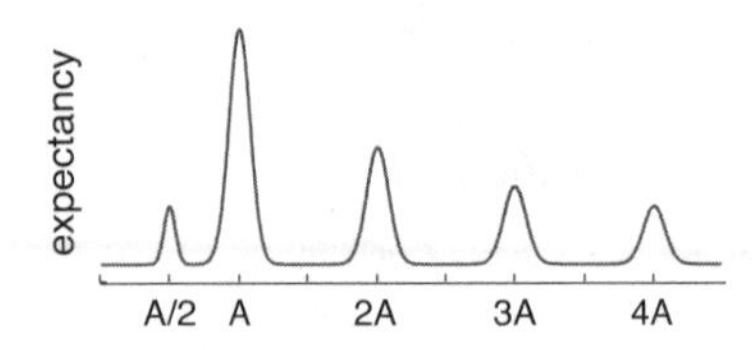

Fig. 8.9. The basic expectancy curve for a single time interval A. The most likely interval is another A, followed by simple integer multiples $2A$, $3A$, and $4A$ and simple divisors such as $\frac{A}{2}$.

Such a direct ad hoc approach may not be necessary. The transform methods, for instance, implicitly accomplish such a clustering. MIDI data can be converted into a time function using (8.1) which translates an event list into a spike train. (This was illustrated in Fig. 8.6.) The transforms can then be applied to the spike train. Figure 8.10 shows three analyses of the same MIDI version of the *Maple Leaf Rag*. Despite the fluctuations in the IOIs, the DFT and PT identify the required tatum, and the PT locates two higher levels of metric structure.

Unfortunately, none of the transform methods are able to handle tempo variation gracefully. Typical failures appear in Fig. 8.11, which shows the "same" MIDI performance of the *Maple Leaf Rag* played with a modest time variation (the piece slows down throughout the first half and speeds

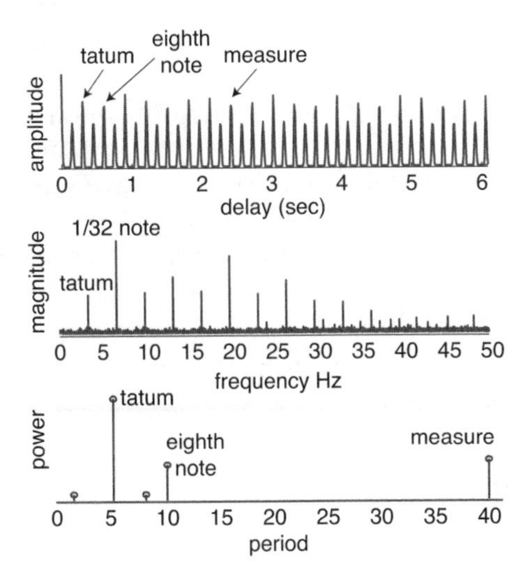

Fig. 8.10. A MIDI rendition of the *Maple Leaf Rag* is analyzed by autocorrelation, the DFT, and the Periodicity Transform. While peaks that correspond to the tatum and certain other features can be identified in the autocorrelation and DFT, they are overwhelmed by a maze of irrelevant peaks. The Periodicity Transform does better, but requires care in choosing the effective sampling rate.

up throughout the second half). None of the rhythmic features of the piece are visible and there is no obvious indication of the tempo variation.

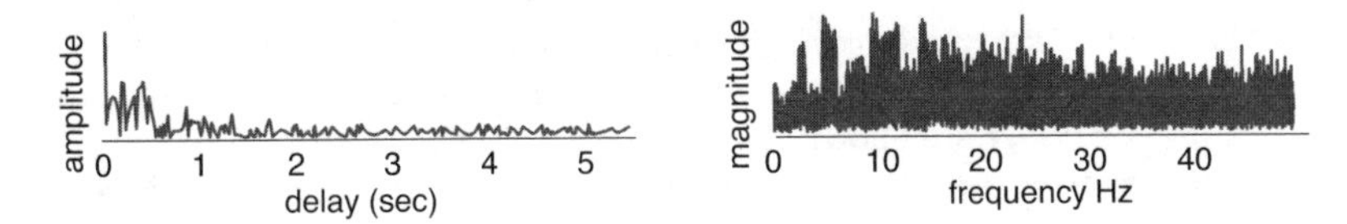

Fig. 8.11. When the *Maple Leaf Rag* changes tempo, none of the transform methods give reliable indications of the tatum or other rhythmic features. Shown are the autocorrelation and the DFT, which may be contrasted to Fig. 8.10. The PT fails similarly.

Thus neither the transform methods nor the rule-based approaches can be expected to be generally successful when parsing data from musical performances. Fortunately, this pessimistic assessment does not apply to all of the technologies, and the next two sections show how both adaptive oscillators and statistical techniques can be applied to track time variations in MIDI performances. Looking ahead, Sect. 8.3 considers the (yet more complex) problem of tracking the beat directly from an audio source.

8.2.1 Adaptive Oscillators

A MIDI sequence defines a set of musical events and specifies the times when they occur. An adaptive oscillator defines a kind of clock with an adjustable period, and the goal of the adaptation is to synchronize the oscillator to the events in the list. As suggested in Sect. 8.1.4, the time base for the oscillator is unknown and varying, and this makes oscillators ideal candidates for tracking MIDI performance data.

For example, Fig. 8.12 shows the trajectories of a collection of adaptive phase-reset oscillators applied to a MIDI rendition of the *Maple Leaf Rag*. The MIDI event list is first translated into a spike train[3] and the oscillators adjust their period throughout the performance. In the left hand plot, the tempo is held constant (as it was performed). Most of the oscillators converge to rates that correspond to the tatum (the sixteenth note) or to twice the tatum. In the right hand plot, the piece was artificially slowed down and then sped up (using sequencing software to manipulate the MIDI file). The phase-reset oscillators track the changes without problem.

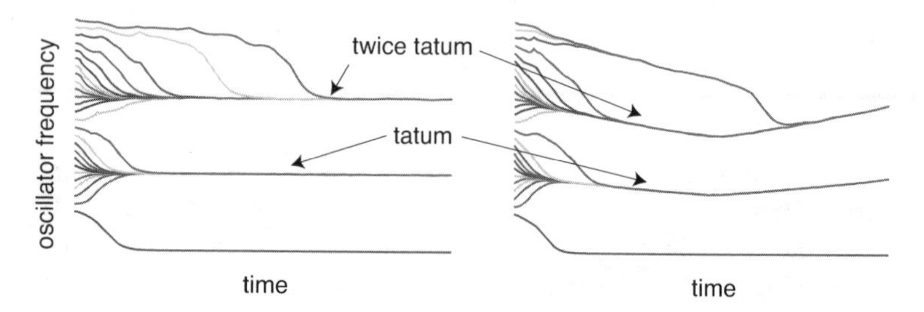

Fig. 8.12. A collection of adaptive phase-reset oscillators are used to find rhythmic features in a rendition of the *Maple Leaf Rag*. Trajectories of 26 oscillators (with different initializations) are shown. All but one converge to either the eighth note or the tatum (sixteenth note). The tempo is fixed in the left plot, but varies (first slowing and then speeding up) throughout the right plot. The oscillators track the changes. In both cases, the bottom oscillator converges to zero, a degenerate stable point. Other initializations (larger than those shown) do not converge.

To demonstrate beat tracking in a more realistic scenario, a bank of adaptive oscillators is applied to a MIDI rendition of the Beatles' song *Michelle*. This MIDI file is from a data set containing expressive polyphonic piano performances by twelve pianists (four "classical," four "jazz," and four "amateurs") with considerable fluctuation in the tempo. Cemgil et al. [B: 27] write that each arrangement of the Beatles' song was played at "normal, slow and fast tempo (all in a musically realistic range and all according to the judgment of the performer)" and the files are available at the *Music, Mind, Machine* website [W: 35].

Each oscillator contains an adjustable parameter α that specifies the frequency.[4] One trajectory is shown in the left hand plot of Fig. 8.13, which

[3] in the same way that the event list for *La Marseillaise* was translated into a spike train in Fig. 8.6.

[4] The period of the oscillator is $\frac{N}{\alpha T_s}$ where N is the number of samples in the wavetable and T_s is the time base on which the spike train is sampled.

shows the tempo as it increases throughout the first 30 s and then slows towards the end. The output of the oscillator is shown in the right hand plot, which also superimposes the MIDI spike train. Observe that the output of the oscillator resets to one each time a new spike occurs. Thus Fig. 8.13 gives a visual indication of the functioning of the beat tracking.

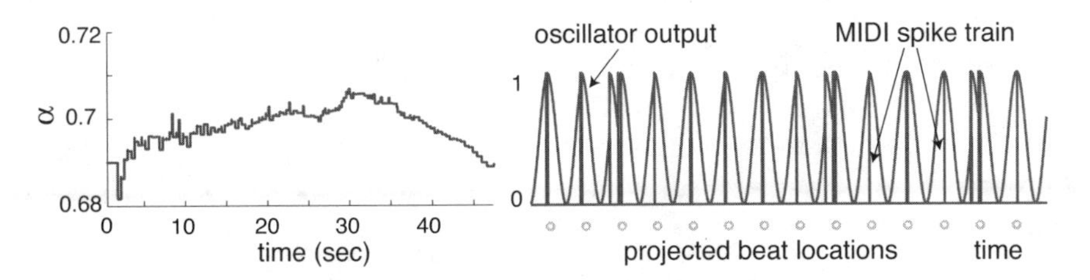

Fig. 8.13. A MIDI rendition of the Beatles' *Michelle* is beat tracked using an adaptive phase-reset oscillator. The α parameter is proportional to the frequency of the adaptive oscillator, which tracks the tempo of the piece. The right hand plot shows how the MIDI spike train resets the output of the oscillator at each event. A noise burst imposed at each projected beat location allows the ear to judge the accuracy of the beat tracking in examples [S: 62]– [S: 64].

More insightful, however, is to listen. One approach is to superimpose a noise burst at each beat location. These are calculated by projecting one period into the future: at the kth beat, the time of the $k+1$st noise burst is estimated based on the current period. The output of this procedure can be heard in [S: 62]. The noise bursts lock onto the beat rapidly and follow changes. Careful listening reveals a glitch at around 29 s, which is caused by a rapid succession of note events that momentarily increase α. By about 37 s, the pulse is regained. Because of the projection, the process is causal and can be implemented in real time (though the simulations reported here are implemented offline in MATLAB®). The second example in [S: 62] replaces the phase-reset oscillator with a wavetable oscillator using a Gaussian wavetable: this avoids the glitch and gives smoother results.

To investigate further, the adaptive wavetable oscillator is applied to three different MIDI renditions of the *Maple Leaf Rag* in sound example [S: 63]. In each case the oscillator is initialized far enough from the actual pulse rate that it is possible to hear the oscillator converging. Synchronization occurs within three or four beats, and the oscillators subsequently track the tatum of each of the performances. Two seconds of a typical run in Fig. 8.14 show the three generic situations that the beat tracker encounters. Often, the MIDI events occur at the times of the beats and the oscillator output achieves its peak value at these times. Many MIDI events occur between beat locations. The beat tracker must not be unduly influenced by such misaligned events since they are common even in metrically simple music. Finally, the beat tracker must locate some beats where there are no events. All three situations are illustrated in Fig. 8.14.

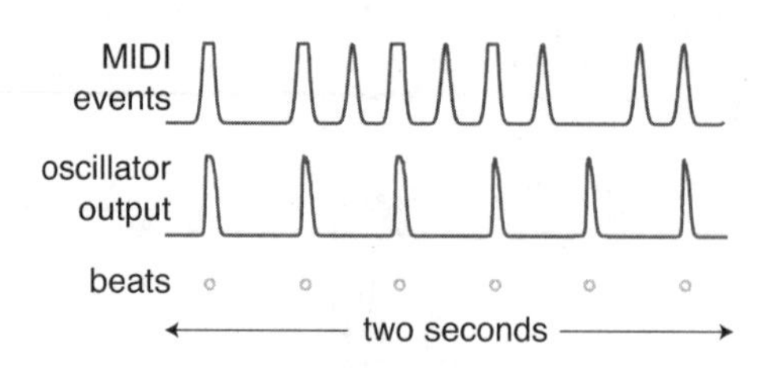

Fig. 8.14. The output of the adaptive wavetable oscillator tracks MIDI events. In many places, events coincide with the output of the oscillator. In others, events occur between or off the detected beat locations. In yet others, the beat continues even when there are no events.

By initializing the oscillator at a slower rate, it is often possible to locate other levels of the metric hierarchy. For example, when the same wavetable oscillator is initialized at half speed, it converges to the quarter-note rate of the *Maple Leaf Rag* rather than to the eighth-note pulse. This can be heard in [S: 64].

These examples give something of the flavor of what is possible when using adaptive oscillators to track MIDI performances. A number of researchers have developed this idea in various directions. One of the earliest uses of adaptive oscillators as a model of metric perception is the work of Large and Kolen [B: 123] which uses an "integrate-and-fire" oscillator with phase and frequencies that adapt via a gradient-like strategy. Toiviainen [B: 229] uses a similar bank of adaptive oscillators for the recognition of meter and applies this to the automatic accompaniment of piano playing in [B: 230]. Toiviainen expertly demonstrates his system by playing in real time to an accompaniment that follows his rhythmically sophisticated performance.

Each of the three kinds of adaptive oscillators has its own idiosyncrasies. The phase-reset oscillator has the fastest convergence. But it is not robust to additive noises and hence cannot be sensibly applied to audio feature vectors. The phase-reset oscillator also tends to be limited to only the fastest levels of the metrical hierarchy since it does not have a good mechanism for "ignoring" large numbers of events that occur between successive beats. For example, all initializations at frequencies below those shown in Fig. 8.12 converge to the degenerate oscillator with frequency zero. The adaptive clock method has the slowest convergence, though it is also perhaps the most resistant to noises. The adaptive wavetable converges quickly, and can be adjusted for different rates of synchronization by choosing the shape (and particularly the width) of the wavetable. Sharply peaked wavetables help the algorithm ignore unexpected intermediate events while wider wavetables increase the robustness to noises.

8.2.2 Statistical Methods

The model for symbolic pulse tracking of Sect. 7.5 is ideal for beat tracking MIDI performances. The timing parameters τ (phase), $\mathcal{T}$ (period), and $\delta\mathcal{T}$

(change of period) are gathered into a state vector $\mathbf{t}$ and the input is a set of time stamped MIDI events. The procedure is:

(i) Change the MIDI events into a spike train using the technique of (8.1) (as shown in the first part of Fig. 8.6).
(ii) Partition the data into frames $\mathcal{D}_k$, each representing 2 to 6 seconds (as in Fig. 7.7).
(iii) Apply the particle filter method of Sect. 7.5 to estimate the distribution of the parameters $p(\mathbf{t}_k|\mathcal{D}_k)$ in frame $\mathcal{D}_k$ from the previous distribution $p(\mathbf{t}_{k-1}|\mathcal{D}_{k-1})$.

An audible output can be created by superimposing a percussive sound at the detected beat locations. This requires choosing a single "best" value for the parameters in each frame. This can be the mean, the median, the mode, or any other single "most likely" point value.

(iv) Choose the most likely values of τ, $\mathcal{T}$, and $\delta\mathcal{T}$ from $p(\mathbf{t}_k|\mathcal{D}_k)$ and use these as the "best estimate" of the beat within frame k.

This procedure is applied to two MIDI renditions of the *Maple Leaf Rag* in [S: 65] and to the Beatles' *Michelle* in [S: 66]. The detected beat follows the pulse of the music flawlessly even as the tempo changes. The performance by Roache contains numerous ornaments and flourishes, and the algorithm is able to "ignore" these (metrically extraneous) events.

One way to control the algorithm is via the initialization, the prior distribution $p(\mathbf{t}_0|\mathcal{D}_0)$ of the first frame. In the simulations [S: 65] and [S: 66], the initial distribution of the period $\mathcal{T}$ was chosen to be uniform over some reasonable range (between $\underline{t} = 0.2$ and $\overline{t} = 0.4$ s per beat for the *Maple Leaf Rag* and between $\underline{t} = 1.0$ and $\overline{t} = 2.0$ s for the slower *Michelle*). The initial distribution of the phase τ was chosen to be uniform over one period $(0, \overline{t} - \underline{t})$ and the derivative $\delta\mathcal{T}$ was initialized to zero.

Other initial distributions also lead to useful behaviors. For example, if the initial range of the period for the *Maple Leaf Rag* is doubled to $\underline{t} = 0.4$ and $\overline{t} = 0.8$, the output settles to the next slowest (quarter note) level of the metric hierarchy as in [S: 67]. If nothing is known about the true tempo, a good strategy is to run the algorithm several times (or to run several versions in parallel[5]) with different initial distributions. If the initializations differ by factors of two or three, they will often converge to different levels of the metric hierarchy. For example, for the *Maple Leaf Rag*, each of the $\mathcal{T}$ initializations $(\underline{t}, \overline{t}) = (0.1, 0.2)$, $(0.2, 0.4)$, $(0.4, 0.8)$, and $(0.8, 1.6)$ leads to a description of the piece at a different level. The fastest of these (see [S: 68]) locates the tatum at every sixteenth note. Even faster synchronizations are possible, though they become increasingly difficult to listen to.

The slowest example of the *Maple Leaf Rag* in [S: 68] beats steadily once per $\frac{2}{4}$ measure. In this example, the beat tracker does not typically lock onto

[5] The method operates considerably faster than real time even in MATLAB®.

the first beat of each measure. Rather, it locks onto the first "and" where the meter is counted "one-and-two-and." While it is possible to force the algorithm to synchronize to the "one" by suitably restricting the initial values of the phase τ, the syncopation is strong enough that the algorithm consistently prefers the "and" to the "one." This is diagrammed in Fig. 8.15 where the underlying beats shown as "1+2+" correspond to the metric notation in the musical score. Four different synchronizations are shown, with periods equal to the quarter note and to the measure. Which synchronization will occur depends primarily on the initialization.

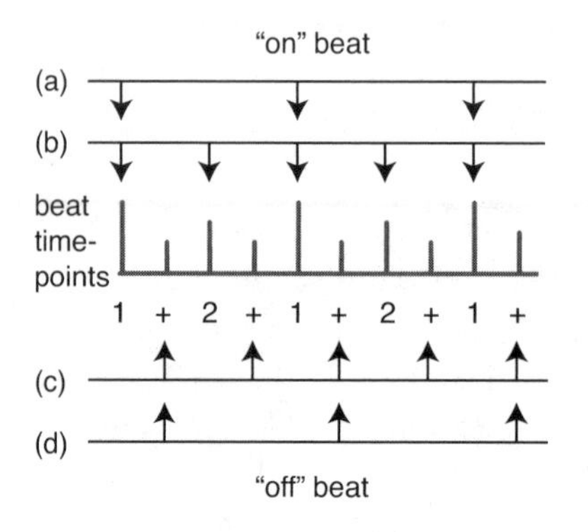

Fig. 8.15. Four possible synchronizations to the beats of the *Maple Leaf Rag*: (a) and (b) show the beat tracker synchronized to the "on" beat at rates of the measure and the quarter note. (c) and (d) show the same two rates synchronized to the "off" beat. When initialized near the measure, the statistical approach prefers the off-beat synchronization (d) to (a), in accordance with the idea that the syncopation in the *Maple Leaf Rag* is significant.

The diffusion parameters give another dimension of control over the algorithm by specifying the anticipated change in the parameter values between successive frames. Where modest changes are likely (such as a rendition of the *Maple Leaf Rag*) these variances may be chosen small. Where larger changes can be expected (as in *Michelle*) the variances must be chosen larger.

In the course of a musical performance, parameters may change at any time. By processing the data in discrete frames, the algorithm can only report changes at the frame boundaries. This is shown in Fig. 8.16 for a phase change τ occurring in the midst of frame B. The statistical method can retain multiple estimates of the parameters since the distribution $p(\mathbf{t}_k|\mathcal{D}_k)$ in step (iii) may contain many possible values, each with their own probability. However, in step (iv), when it is necessary to choose a particular estimate for output (for instance, a set of times at which to strike the percussion within frame B) only a single set of parameters is possible within a given frame. This problem is

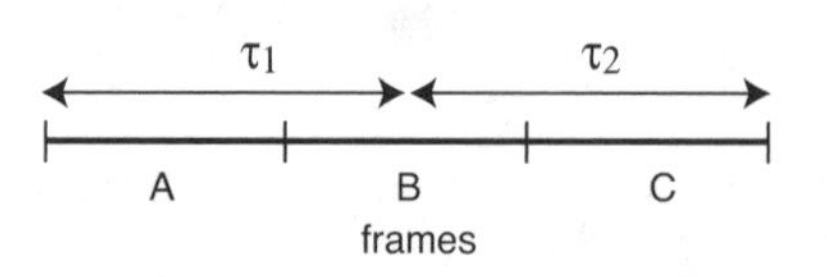

Fig. 8.16. Parameter values may change at any time; the phase τ is shown changing in the middle of frame B. A frame-based analysis may locate the actual values in frames A and C, but it must choose a compromise value within frame B.

inherent in any method that uses frame-based processing, and is one of the reasons that the frame size should be kept as small as possible.

The method (i)–(iv) is one way to exploit probabilistic reasoning in the context of MIDI beat tracking. Other researchers have proposed a variety of methods including the Kalman filter [B: 27], MCMC methods and particle filters [B: 28] (based on Cemgil's model of Fig. 7.5 on p. 184), and a Bayesian belief network [B: 179]. Cemgil and Kappen [B: 28] compare two algorithms for the beat tracking of MIDI data and conclude that the particle filter methods outperform iterative methods.

8.3 Audio Beat Tracking

There are two approaches to the beat tracking of audio signals. The first locates interonset intervals and note events, gathers them into a list, and applies one of the MIDI beat tracking techniques. Since MIDI beat trackers can operate quite effectively (as shown in the previous sections), this method is limited primarily by the accuracy of the detection of the note events. The second approach uses feature vectors to reduce the audio data to a manageable size and the beat tracker searches the feature vectors for evidence of a regular succession. This bypasses the need to accurately detect individual note events, but is limited by the fidelity of the feature vectors to the underlying pulse of the music.

Because "notes" feature so prominently in the human conception of music, a commonsense perspective suggests that note detection is a prerequisite to rhythm identification. Indeed, the majority of beat tracking algorithms (see for example the overview of automated rhythm description systems by Gouyon and Dixon [B: 78]) begin with the detection of note events. But the difficulties in transcribing complex polyphonic music are well established, and things may not be as straightforward as they appear. Section 4.3.8 observed that any factors that can create auditory boundaries can be used to create rhythmic patterns (recall that sound examples [S: 47]–[S: 50] demonstrate rhythms without loudness contours or individually identifiable notes). This suggests that it may sometimes be easier to identify a regular succession of auditory boundaries than to reliably identify the individual note events that make up the succession. This is a kind of chicken-and-egg problem discussed further in Sect. 12.1.

The bulk of this section focuses on the parsing of feature vectors and examines the ability of the pattern-finding techniques of the previous chapters (transforms, oscillators, and statistical) to solve the beat tracking and rhythm detection problems.

8.3.1 Transform Techniques

Many psychoacoustically based models of the auditory system begin with a set of feature vectors created from a bank of filters that divide the sound into a number of frequency regions. For example, the first two minutes of the *Maple Leaf Rag* are shown in a spectrogram-like display in Fig. 4.5 on p. 82. Such feature vectors have been used as a first step in beat tracking in [B: 208], in the wavelet approach of [B: 232] and in Gouyon's subband decomposition [B: 79].

This section presents several examples where the transforms of the feature vectors correlate closely with various levels of the rhythmic structure. These begin with a simple three-against-two polyrhythm, which is explored in depth to show how various parameters of the model effect the results. A series of musical examples are taken from a variety of sources with a steady tempo: dance music, jazz, and an excerpt from a Balinese gamelan piece. The transforms are able to discern the pulse and several "deeper" layers of rhythmic structure with periodicities corresponding to measures and musical phrases. When the beat is steady and unchanging, transform techniques can be successful at locating rhythmic phenomenon even on multiple metrical levels. As might be expected from the MIDI results, however, when the underlying tempo of the audio changes, the transform methods fail.

Three Against Two

The *Three Against Two* polyrhythm familiar from Sect. 3.9 is notated in Fig. 8.17. It was played at a steady tempo of 120 beats per minute using two different timbres, "wood block" and "stick," for 15 seconds (see [S: 69]). This was recorded at 44.1 kHz and then downsampled using the data reduction technique of Sect. 4.4.1 to an effective sampling rate of 140 Hz. The downsampled data was divided into 23 frequency bands (resulting in 23 feature vectors) which can be pictured as a spectrogram-style plot (such as Fig. 4.5) in which each row represents one of the frequency bands and each column represents a time instant of width $\frac{1}{140}$ s. The rows can then be searched for patterns and periodicities.[6]

A standard signal processing approach to the search for periodicities is to use the DFT. Figure 8.18 superimposes the magnitudes of the DFT of all 23 rows. While not completely transparent, this plot can be meaningfully interpreted by observing that it contains two harmonic series, one based at 2 Hz (which represents the steady quarter note pulse) and the other at 3 Hz (which represents the steady triplet rhythm). These two "fundamentals" and their "harmonics" are emphasized by the upper and lower lattices which are superimposed on the plot. However, without prior knowledge that the spectrum of Fig. 8.18 represents a three against two polyrhythm, this structure would not

[6] This effective sampling rate e_f was derived using the technique of Sect. 5.5.5. An initial periodicity analysis used an effective sampling rate of 132.43 Hz (corresponding to a FFT overlap of 333 samples). The strongest periodicities detected occurred at 227 samples, corresponding to a periodicity at 1.714 seconds. A new sampling interval was chosen so that 240 samples occurred within the same time span; this corresponds to the effective sampling rate of 140 Hz.

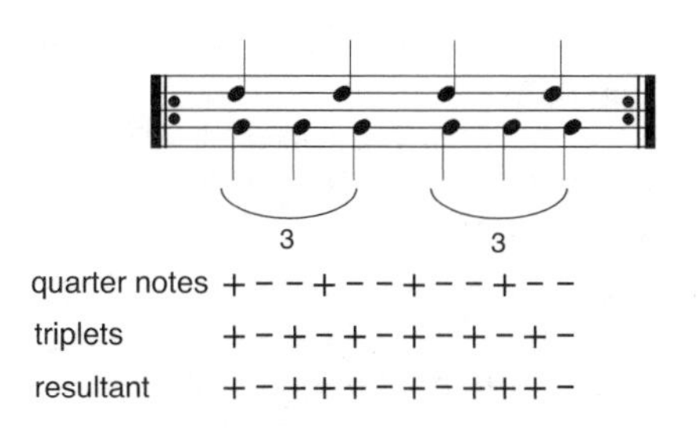

Fig. 8.17. The *Three Against Two* polyrhythm is written in standard musical notation. At a tempo of 120 beats per minute, two quarter notes occur in each second; three triplets occupy the same time span. This can be heard in [S: 69].

be obvious. This result directly parallels the earlier results in Figs. 8.3 and 8.10 where the DFT was applied to symbolic sequences and to MIDI files.

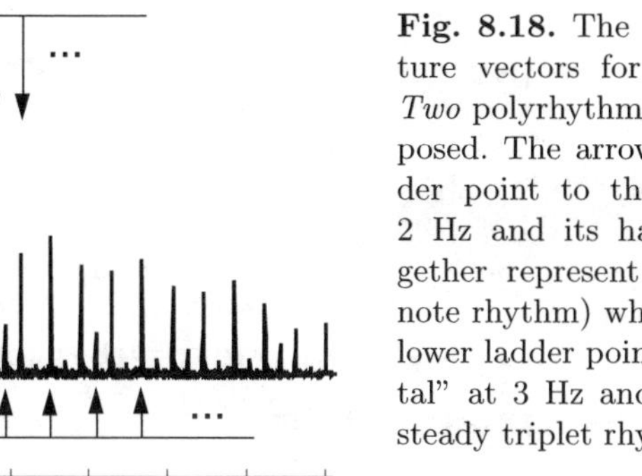

Fig. 8.18. The DFTs of the 23 feature vectors for the *Three Against Two* polyrhythm [S: 69] are superimposed. The arrows of the upper ladder point to the "fundamental" at 2 Hz and its harmonics (which together represent the steady quarter note rhythm) while the arrows of the lower ladder point to the "fundamental" at 3 Hz and its harmonics (the steady triplet rhythm).

Applying the periodicity transforms to the feature vectors (in place of the DFT) leads to plots such as Fig. 8.19. Here the Best-Correlation method detects three periodicities, at $p = 72$ (which corresponds to the quarter note pulse), at $p = 48$ (which corresponds to the triplets), and at $p = 24$ (which corresponds to the speed of the resultant in Fig. 8.17). Clearly, it is much easier to interpret the periodicities in Fig. 8.19 than to locate the harmonic templates in the spectrum of Fig. 8.18.

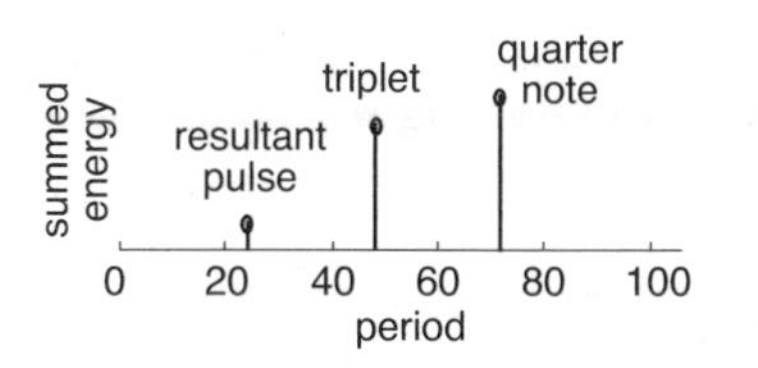

Fig. 8.19. The Best-Correlation algorithm is used to detect periodicities in the *Three Against Two* polyrhythm [S: 69]. The horizontal axis is the period (each sample represents $\frac{1}{140}$ second). The vertical axis shows the amount of energy detected at that period, summed over all the frequency bands.

Accelerating Three Against Two

A significant limitation of the transform methods is that they require a steady tempo; pieces which change speed lack the kinds of periodicities that are easily detected. To investigate the effect of unsteady pulses, the "same" *Three Against Two* polyrhythm was re-recorded in [S: 70], but the tempo was increased by 5% every eight beats. This resulted in an increase in tempo of more than 25% over the course of the 15 seconds. As expected, the transforms are unable to cope gracefully with the time variation. Figure 8.20 shows the DFT spectrum and the detected periodicities for the Small-to-Large, Best-Correlation, and M-Best algorithms. Each of the algorithms has its own peculiarities, but none of the periodicities accounts for a significant percentage of the energy from the signal. The Small-to-Large algorithm detects hundreds of different periods, most far longer than the "actual" signals of interest. Both the Best Correlation and M-Best algorithms detect clumps of different periodicities between $33 < p < 40$ and $53 < p < 60$ which are in the midrange of the changing speed. One can view these clumps as decreasing in period, as should be expected from an increase in tempo, but it would be hard to look at these figures and to determine that the piece had sped up throughout. Similarly, the harmonic templates of the DFT (the top plot in Fig. 8.20) are smeared beyond recognition and give little useful information.

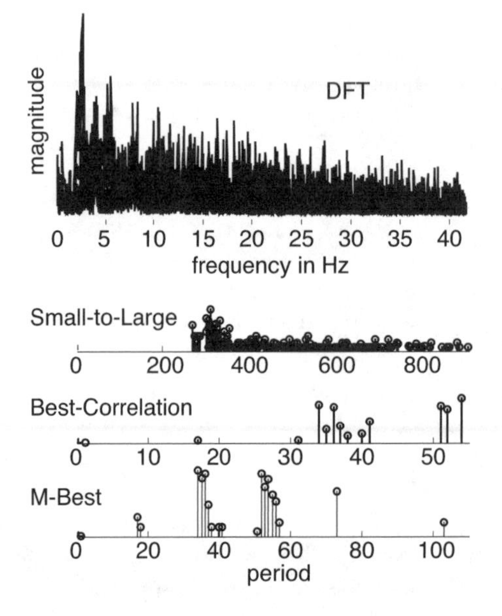

Fig. 8.20. The *Three Against Two* example is performed with the tempo increasing 5% after every eight beats in [S: 70], for a total increase of about 25% over the 15 second duration. None of the periodicity algorithms show significant rhythmic features of the signal and the DFT also fails to reveal any significant structure. The autocorrelation (not shown) is similarly uninformative.

This highlights the greatest limitation to the use of transforms in the detection of rhythm; when the underlying pulse rate changes, the transforms are unable to follow. Nonetheless, when the beat is steady, the transform techniques can be used quite effectively, as the following examples show.

True Jit

The first analysis of a complete performance is of the dance tune *Jit Jive* performed by the Bhundu Boys [D: 4]. Though the artists are from Zimbabwe, the recording contains all the elements of a dance tune in the Western "pop" tradition, featuring singers, a horn section, electric guitar, bass and drums, all playing in a rock steady 4/4 beat. A preliminary analysis was performed at a convenient effective sampling rate (in this case 100.23 Hz, an overlap factor of 440), and there was a major periodicity at $p = 46$ samples. The analysis was then redone[7] at a sampling rate so that this same time interval was divided into 60 samples giving an effective sampling rate of 130.73 Hz. This was used to generate Fig. 8.21, which compares the outputs of the Best-Correlation algorithm, the M-Best algorithm, and the DFT. In all cases, the transforms

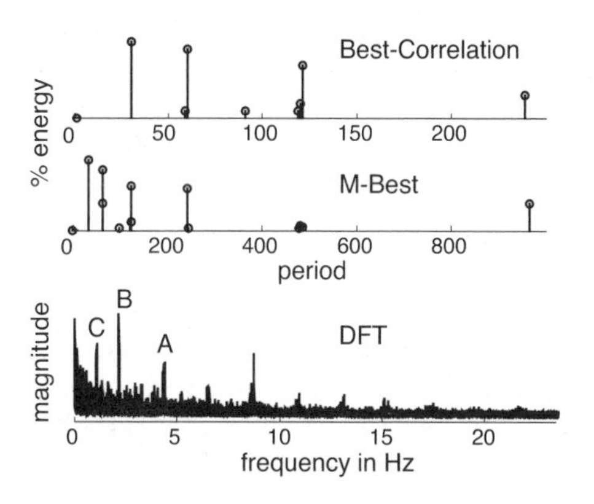

Fig. 8.21. Periodicity analysis of *Jit Jive* by the Bhundu Boys [D: 4] clearly reveals structural aspects of the performance. The major periodicities occur at 30 (which represents the pulse at 230 ms), 60 (two beats), 120 (the four beat measure), 240 (the two measure phrase) and 960 (an eight bar phrase). The DFT shows three meaningful peaks: peak A represents the 230 ms pulse, and peak C represents the four beat measure. The intermediate peak B corresponds to the half measure.

are conducted on each of the 23 feature vectors independently, and then the results are added together so as to summarize the analyses in a single graph. Thus the figure labeled DFT is actually 23 DFTs superimposed, and the figure labeled M-Best represents 23 independent periodicity analyses. The vertical axis for the DFT is the (summed) magnitude, while the vertical axes on all the periodicity transform figures is the amount of energy contained in the basis functions, normalized by the total energy in the signal. Hence it depicts the percentage of energy removed from the signal by the detected periodicities.

There are several major peaks in the DFT analysis, of which three are readily interpretable in terms of the song. The peak marked A represents the basic beat or pulse of the song which occurs at 230 ms (most audible in

[7] as suggested in Sect. 5.5.5 for the determination of a good effective sampling rate.

the incessant bass drum) while peak C describes the four beat measure. The intermediate peak B occurs at a rate that corresponds to two beats, or one half measure.

The periodicity analysis reveals even more of the structure of the performance. The major periodicity at 30 samples corresponds to the 230 ms beat. The periodicities at 60 and 120 (present in both periodicity analyses) represent the two beat half note and the four beat measure. In addition, there is a significant periodicity detected at 240, which is the two bar phrase, and (by the M-Best algorithm) at the eight bar phrase, which is strongly maintained throughout the song. Thus the transforms, in conjunction with the feature vectors defined by the critical band filters, can be easily interpreted in terms of a hierarchical rhythmic structure for the performance. Plots for a variety of musical pieces that are qualitatively similar to the DFT and PT in Fig. 8.21 can also be found in [B: 63] and [B: 109].

Take Five

Both of the previous pieces were rhythmically straightforward. Brubeck's *Take Five* [D: 9] is not only in the uncommon time signature $\frac{5}{4}$, but it also contains a drum solo and complex improvisations on both saxophone and piano. Figure 8.22 shows the periodicity analysis by the Best-Correlation, Best-Frequency, and M-Best algorithms. All show the basic five to one structure (the period

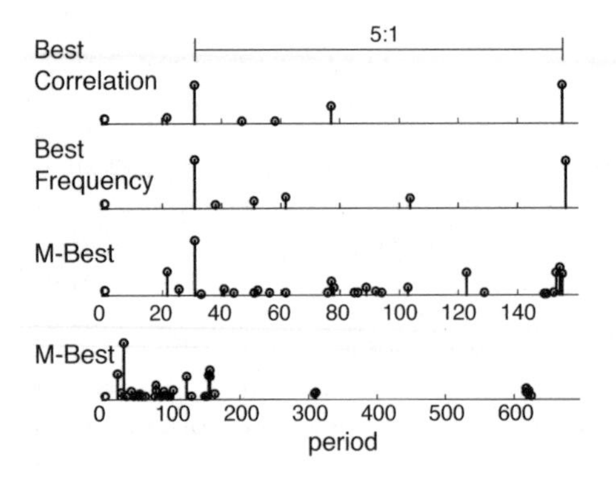

Fig. 8.22. Periodicity analysis of Brubeck's *Take Five* [D: 9] clearly reveals the "five" structure. The periodicity at 31 represents the beat, while the periodicity at (and near) 155 represents the five beat measure. The lower two plots show the periodicities detected by the M-Best algorithm: the top is an expanded view of the bottom, which shows the larger periodicities near 310 (two measures) and near 620 (four measures). The piece is often notated in eight bar phrases.

at 31 represents the beat, while the period at 155 corresponds to the five beat measure). In addition, the M-Best algorithm finds periodicities at 310 (two measures) and at 620 (the four bar phrase). The piece would normally be notated in eight bar phrases, which is the length of both of the melodies. As is clear, there are many more spurious peaks detected in this analysis than in the previous two, and likely this is due to the added complexity of the performance. Nonetheless, Fig. 8.22 displays several of the major structural levels.

Baris War Dance

The *Baris War Dance* is a standard piece in the Balinese Gong Kebyar (gamelan) repertoire [D: 15]. The performance begins softly, and experiences several changes from soft to loud, from mellow to energetic, but it maintains a steady underlying rhythmic pulse throughout. This beat is alternately carried by the drum and the various kettle instruments (the bonangs and kenongs), while the bass gong is struck steadily every eight beats.

These rhythmic elements are clearly reflected in the periodicity analysis. Figure 8.23 shows the pulse at period 45, and other periodicities at two, four, and eight times the pulse. The striking of the large gong every eight beats is shown clearly by the M-Best analysis. Such regular punctuation appears to be a fairly generic character of much of the Gong Kebyar style [B: 221]. Thus the periodicity analysis is applicable cross culturally to musics with a steady pulse.

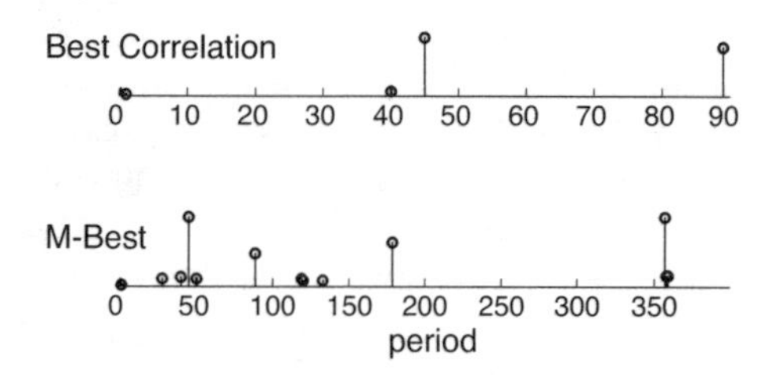

Fig. 8.23. Periodicity analysis of the *Baris War Dance* [D: 15] clearly reflects structural aspects of the performance. The major periodicities occur at the beat $p = 45$ and at two, four, and eight times the beat. The largest of these is most prominent in the bass, and the periodicity at 360 is the rate at which the largest gong is struck.

Periodicity Transforms are designed to locate periodicities within a data set by projecting onto a set of (nonorthogonal) periodic subspaces. This can be applied to musical performances through the intermediary of feature vectors, and examples show that the method can detect periodicities that correspond to the tatum, the beat, the measure, and even larger units such as phrases, as long as the basic pulse is steady. The major weakness of the transform approaches is the inability to gracefully account for time variations in the musical pulse. Fortunately, other techniques do not have the same limitation.

8.3.2 Statistical Beat Tracking

The statistical approach to audio beat tracking operates on feature vectors designed to capture and condense relevant characteristics of the music. This section presents a number of examples that demonstrate the functioning of the algorithm. In all cases, the output is a sequence of times designed to represent when beat timepoints occur, when listeners "tap their feet." To make this accessible, an audible burst of noise is superimposed over the music at the predicted time of each beat. By listening, it is clear when the algorithm has "caught the beat" and when it has failed. Please listen to the `.mp3` examples from the CD to hear the algorithms in operation; graphs such as Fig. 8.24 are a meager substitute.

The statistical approach of Sect. 7.6 partitions the feature vectors into frames and sequentially estimates the distribution of the model's parameters. As indicated in Fig. 7.9 (on p. 189) there are six parameters in the model. The three structural parameters are fixed throughout all the examples while the three timing parameters estimate the temporal motion of the feature vectors. To be specific, the particle filter method outlined in Sect. 7.5 is used to implement the prediction and update phases using (7.14) and (7.15), that is, the method estimates the distribution of the timing parameters in the k+1st frame of the feature vector based on the distribution in the kth frame.

Initial values for the structural parameters σ_S^2, σ_L^2, and ω (the variance of the off-beat, the variance of the on-beat, and the width of the Gaussian pulse) were chosen by hand from an inspection of the feature vectors. Using these values, the algorithm was run to extract the beats from several pieces. These results were then used to re-estimate the parameters using the entire feature vectors, and the values of the parameters from the different training tracks were then averaged.[8] These averaged values were then fixed when estimating the beats in subsequent music (i.e., those not part of the training set). The nominal values were $\omega = 0.02$, $\sigma_S^2 = 0.14$, and $\sigma_L^2 = 0.2$, while the initialization of $\mathcal{T}$ was uniform in the range $[\underline{t}, \overline{t}] = [0.2, 0.4]$ s, $\delta\mathcal{T}$ was uniform $[-0.0001, 0.0001]$ and τ was uniform $[0, \overline{t} - \underline{t}] = [0, 0.2]$.

The statistical method of beat tracking has been applied to about 300 different pieces in a variety of styles and a representative sample appear in Table A.1 on p. 289. The first (approximately) thirty seconds of each are excerpted in the corresponding sound examples [S: 71] which demonstrate the beat tracking by superimposing a burst of white noise at each detected beat. In each case, the algorithm locates a (slowly changing) regular succession that corresponds to times when a listener might plausibly tap the foot.

In some cases, the detected beat rate feels uncomfortably fast. For example, using the default values, the algorithm locks onto a pulse near $\mathcal{T} = 0.24$ s when beat tracking *Howell's Delight* [S: 71](12). Because the piece is rhythmically fluid and moves slowly in a stateful $\frac{6}{8}$, this feels frenetic. Doubling the initial range of the period $\mathcal{T}$ to $[\underline{t}, \overline{t}] = [0.4, 0.8]$ allows convergence to a more reasonable $\mathcal{T} \approx 0.72$ s, which taps twice per $\frac{6}{8}$ measure. This can be heard in [S: 72]. Taken together, these have located the fastest two levels of the metrical hierarchy. Observe how much musical activity occurs between each detected timepoint. Similarly, the default values applied to *Julie's Waltz* [S: 71](15) lock onto the eighth-note tatum at $\mathcal{T} \approx 0.30$. Doubling the initial range of the period $\mathcal{T}$ allows it to lock onto the quarter note beat at $\mathcal{T} \approx 0.61$, and this can be heard in [S: 73].

[8] It is important to scale the feature vectors so that they have approximately equal power. This allows use of one set of parameters for all feature vectors despite different physical units.

Similarly, the detected period in *Lion Says* [S: 71](14) is the tatum at $\mathcal{T} \approx 0.21$ when using the nominal parameters. By doubling the initial period to $[\underline{t}, \overline{t}] = [0.4, 0.8]$, the algorithm locates the quarter note beat at $\mathcal{T} \approx 0.42$ s. Interestingly, the phase can lock onto either the on-beat or the off-beat (depending on the exact initialization). These can be heard in [S: 74]. This is the same kind of synchronization issue as in Fig. 8.15.

Thus, depending on the range of the initial timing parameter $\mathcal{T}$, the algorithm may lock onto a pulse rate that corresponds to the tatum, the beat, or to some higher level of rhythmic structure. Given that reasonable people can disagree by factors of two[9] on the appropriate beat (one clapping hands or tapping feet at twice the rate of the other), and that some people tend to clap hands on the on-beat, while others clap on the off-beat, such effects should be expected in a beat tracking algorithm.

While searching for regularities in symbolic sequences using the statistical method, the likelihood function often has several peaks. For example, the maxima in Figs. 7.2 and 7.3 (on pp. 181 and 182) occur at the basic pulse and at various simple integer ratios. Similarly, in audio beat tracking, there are rare cases that may lock onto two equally spaced taps for every three beats or to three equally spaced taps for every two beats. These can typically be "fixed" by running the algorithm again and/or by increasing the number of particles. Since the initial particles of the algorithm (the initial guesses) are chosen randomly, there is no guarantee of finding the best possibilities and unlikely answers (such as the 3:2 or 2:3 synchronizations) may occasionally occur. Moreover, once the algorithm becomes entrained, it can be stable and persist throughout the piece. While such cases are rare (occurring in perhaps two percent of the cases), it is easy to force such synchronizations by suitably restricting the initial values of the period. For example, if the period estimates of the *Maple Leaf Rag* are constrained to $[\underline{t}, \overline{t}] = [0.5, 0.53]$, the algorithm identifies a 3:2 synchronization at $\mathcal{T} \approx 0.52$ which is $\frac{3}{2}$ the actual beat rate of $\mathcal{T} = 0.34$ (as in [S: 71](5)). This can be heard in [S: 75].

In typical operation, the output of the algorithm can be pictured as in Fig. 8.24, which shows the four feature vectors of Sect. 4.4.3 at the start (between 2 and 6 seconds) of *Pieces of Africa* by the Kronos Quartet [S: 71](8). The smooth curves σ_t define the variance at each time t which is approximately constant ($= \sigma_S^2$) between beats and larger ($\approx \sigma_L^2$) near the bumps. Spaced $\mathcal{T}$ seconds apart, the bumps represent the estimated beat times. Some of the feature vectors show the pulse nicely, and the algorithm aligns itself with this pulse. Feature vector (c) provides the cleanest picture with large spikes at the beat locations and small deviations between. Similarly, feature vector (a) shows the beat locations but is quite noisy (and temporally correlated) between spikes. Feature vector (d) shows spikes at most of the beat locations, but also has many spikes in other locations, many at twice the tap rate. Feature vector (b) is unclear, and the lattice of times found by the algorithm

[9] And also factors of three for music in a triple meter.

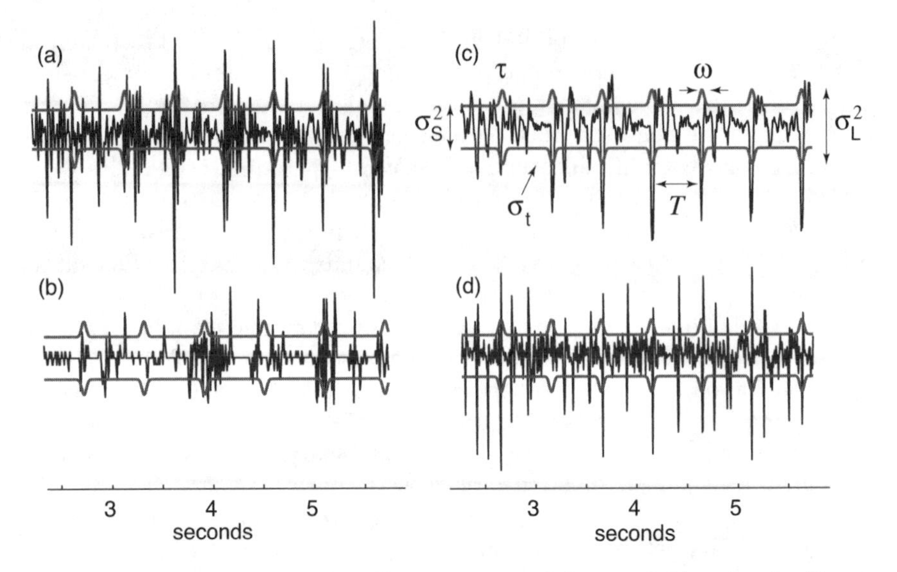

Fig. 8.24. A few seconds of four feature vectors of *Pieces of Africa* by the Kronos Quartet [S: 71](8) are shown. The estimated beat times (which correctly locate the pulse in cases (a), (c), and (d)) are indicated by the bumps in the curve σ_t that are superimposed over each vector. The three structural parameters (ω, σ_S^2, and σ_L^2) are fixed while the three timing parameters $\mathcal{T}$, τ, and $\delta\mathcal{T}$ (not shown) are estimated from the feature vectors. See also Fig. 7.9.

when operating only on this track is unrelated to the real pulse of the piece. In operation, the algorithm derives a distribution of samples from all four feature vectors that is used to initialize the next block. The algorithm proceeds through the complete piece block by block.

An audio file is a record of a performance of a piece. Therefore, a rhythmic analysis of an audio file directly reveals information about the performance and only indirectly about the underlying music. The enduring popularity of the *Maple Leaf Rag* makes it an ideal candidate for a multi-performance rhythmic analysis because of the many artists, working in a variety of musical styles, who have interpreted it. Chapter 11 conducts a detailed comparison of the beat structure of the different performances. To lay the groundwork for this investigation, Table A.2 (on p. 290) lists 27 renditions of the rag that have been beat tracked using the technique of this section. About half are piano renditions, the instrument the rag was originally composed for. Other versions are performed on solo guitar, banjo, and marimba. Renditions are performed in diverse styles: a klezmer version, a bluegrass version, three different big band versions (by Sidney Bechet, Tommy Dorsey, and Butch Thompson), one by the Canadian brass ensemble, two symphonic versions and an *a capella* version.

The beat was correctly located in all 27 versions shown in Table A.2 using the default values in the algorithm, as were all but 8 of 67 versions I have located. Six of these eight were correctly identified by increasing or decreasing the time window over which the algorithm operates (the length of a frame) and by increasing the number of particles. The remaining two are apparently beyond the present capabilities of the algorithm. One is bathed in reverberation and the other (by Sindel) is a solo electric guitar with a substantial delay-feedback effect. In both of these, the feature vectors appear to be smeared by the effect and fail to respect the underlying pulse. It is worth noting (parenthetically) that versions by Hyman [S: 76](7) and Dorsey [S: 76](14) which had been problematic in our earlier report [B: 206], can now be tracked successfully using nominal parameters. The improvement is due to small changes in the algorithm, better choice of parameters, and the use of more particles (typically 1500 rather than 500).

The statistical beat tracking method is generally successful at identifying the initial tempo parameters and at following tempo changes. One mode of failure is when the tempo changes too rapidly for the algorithm to track, as might occur in a piece with extreme rubato. It should be possible to handle abrupt changes by including an additional parameter that represents the (small) probability of a radical change. The price of this would be that more particles would be required. Perhaps the most common mode of failure is when the feature vectors fail to have the hypothesized structure (rather than a failure of the algorithm in identifying the structure when it exists). Thus a promising area for research is the search for better feature vectors. There are many possibilities: feature vectors could be created from a subband decomposition, from other distance measures in either frequency or time, or using probabilistic methods. What is needed is a way of evaluating the efficacy of a candidate feature vector. Also at issue is the question of how many feature vectors can be used simultaneously. In principle there is no limit as long as they remain "independent." Given a way of evaluating the usefulness of the feature vectors and a precise meaning of independence, it may be possible to approach the question of how many degrees of freedom exist in the underlying rhythmic process. Some progress on these issues can be found in [B: 204, B: 205].

Finally, it should be noted that the algorithms are capable of real time operation because they process a single frame at a time, though the simulations reported here are not real time because they are implemented in MATLAB®.

8.3.3 Beat Tracking Using Adaptive Oscillators

The promise of the adaptive oscillator approach is its low numerical complexity. That the oscillators are in principle capable of solving the beat tracking problem is indicated in Fig. 8.25 which shows 70 different runs of the adaptive clocking algorithm of Sect. 6.4.4 applied to the *Theme from James Bond*

[S: 71](9). Each run initializes the algorithm at a different starting value between 20 and 90 samples (between 0.11 and 0.52 seconds). In many cases, the algorithm converges nicely. Observe that initializations between 35 and 45 converge to the eighth-note beat at 0.23 seconds per beat, while initializations between 75 and 85 converge to the quarter-note beat at about 0.46 seconds. Other initializations do not converge, at least over the minute analyzed. This generic behavior should be expected from the analogous results for symbolic sequences as in depicted in Fig. 8.7.

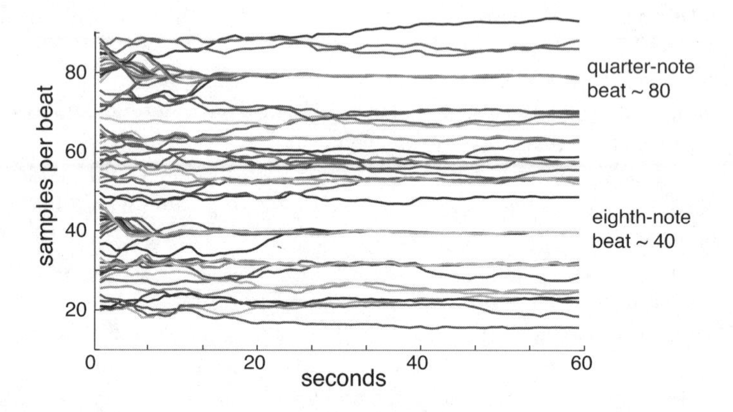

Fig. 8.25. Estimates of the beat period for the *Theme from James Bond* using the adaptive clocking algorithm. Depending on the initial value it may converge to the eighth-note beat at about 40 samples per period (about 0.23 seconds) or to the quarter-note beat near 80 samples (about 0.46 seconds).

When first applying the algorithm, it was necessary to run through the feature vectors many times to achieve convergence. By optimizing the parameters of the algorithm (stepsize, number of beats examined in each iteration, etc) it was possible to speed converge to within 30 or 40 beats (in this case, 15 to 20 seconds). What is hard to see because of the scale of the vertical axis in Fig. 8.25 is that even after the convergence, the estimates of the beat times continue to oscillate above and below the correct value. This can be easily heard as alternately rushing the beat and dragging behind. The problem is that increasing the speed of convergence also increases the sensitivity.

In order to make the oscillator method comparable to the Bayesian approach, the basic iteration (6.20) needed to be expanded to use information from multiple $\mathcal{T}$ intervals simultaneously and to use information from multiple feature vectors simultaneously. Both of these generalizations are straightforward in the sense that predictions of the beat locations and deviations can follow the same method as in (6.20) (from p. 167) whether predicting one beat or n beats into the future and whether predicting from a single feature vector or many. What is new is that there must be a way of combining the multiple estimates. There are several possibilities including averaging the updates from all n beats and all feature vectors, using the median of this value, or weighting the estimates. The most successful of the schemes (used to generate the examples such as Fig. 8.25) weight each estimate in proportion to $r(t_k^*)\, g(t_k^* - \tau_k - T_k)$ (using the notation from (6.21)) since this places more emphasis on estimates which are "almost" right.

Overall, the results of the adaptive oscillators were disappointing. By hand tuning the windows and stepsizes, and using proper initialization, it can often find and track the beat. But these likely represent an unacceptable level of user interaction. Since the Bayesian algorithm converges rapidly within a few beats it can be used to initialize the oscillators, effectively removing the initial undulations in the timing estimates. Of the 27 versions of the *Maple Leaf Rag*, this combined algorithm was able to successfully complete only twelve: significantly fewer than the particle filter alone.

8.4 Summary

This chapter has applied each of the three technologies for locating patterns (transforms, adaptive oscillators, and statistical methods) to three levels of processing: to symbolic patterns where the underlying pulse is fixed (e.g., a musical score), to symbolic patterns where the underlying pulse may vary (e.g., MIDI data), and to time series data where the pulse may be both unknown and time varying (e.g., feature vectors derived from audio).

All three methods operate well on symbolic patterns where the pulse is fixed. But when the pulse varies, the transform methods fail. Both the oscillators and the statistical methods can follow a changing pulse. Oscillator-based systems entrain to the pulse and the probabilistic methods locate regularities at the beat-level by parsing small chunks of data searching for statistical order. A number of sound examples demonstrate the proper and improper functioning of the methods. The next several chapters discuss how the beat tracking methods can be used in audio signal processing, in musical recomposition, and in musical analysis.

9

Beat-based Signal Processing

There is an old adage in signal processing: if there is information, use it. The ability to detect beat timepoints is information about the naturally occurring points of division within a musical signal and it makes sense to exploit these points when manipulating the sound. Signal processing techniques can be applied on a beat-by-beat basis or the beat can be used to control the parameters of a continuous process. Applications include beat-synchronized special effects, spectral mappings with harmonic and/or inharmonic destinations, and a variety of sound manipulations that exploit the beat structure. A series of sound examples demonstrate.

The ability to automatically detect the beat allows signal processing techniques to be applied on a beat-by-beat basis. There are two ways to exploit beat information. First, each beat interval may be manipulated individually and then the processed sounds may be rejoined. To the extent that the waveform between two beat locations represents a complete unit of sound, this is an ideal application for the Fourier transform since the beat interval is analogous to a single "period" of a repetitious wave. The processing may be any kind of filtering, modulation, or signal manipulation in either the time or frequency domain. For example, Fig. 9.1 shows the waveform of a song partitioned into beat-length segments by a series of envelopes. Each of the segments can be processed separately and then rejoined. Using envelopes that decay to zero at the start and end helps to smooth any discontinuities that may be introduced.

The second method uses beat locations to control a continuous process. For example, a resonant filter might sweep from low to high over each beat interval. The depth of a chorusing (or flanging) effect might change with each beat. The cutoff frequency of a lowpass filter might move at each beat boundary. There are several commercially available software plug-ins (see for example [W: 8] and [W: 46]) that implement such tasks using the tempo specified by the audio sequencer; the performer implicitly implements the beat tracking.

Since certain portions of the beat interval may be more perceptually salient than others, these may be marked for special treatment. For example, time stretching by a large factor often smears the attack transients. Since the beat locations are known, so are the likely positions of these attacks. The stretching can be done nonuniformly: to stretch only a small amount in the vicinity of the

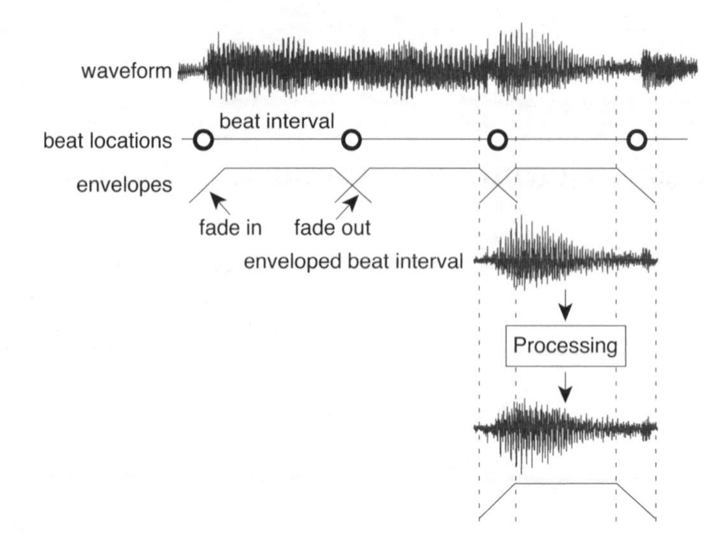

Fig. 9.1. A collection of windows separates the waveform into beat intervals, which can be processed independently. After processing, the intervals are windowed again to help reduce clicks and edge discontinuities. The final step (not shown) is to sum the intervals to create a continuous output.

start of the beat and to stretch a larger amount in the steady state portions between beat locations.

The bulk of this chapter documents a number of experiments with beat-based audio processing. The final section compares the two signal processing techniques used to generate the majority of the sound examples: the phase vocoder and the beat-synchronous FFT.

9.1 Manipulating the Beat

Detected beat information can be used to change durations within a piece in several ways. For example, a recording might have an unsteady beat; an alternative version can be created that equalizes the time span of each beat, as diagrammed in the top part of Fig. 9.2. Sound example [S: 77] equalizes the beat in the *Maple Leaf Rag* so that each beat interval is the same length. A percussion line (from a drum machine) is superimposed to emphasize the metronomic regularity. On the other hand, a piece might suffer from a repetitive metronomic pulse. To create a more expressive performance, the beat intervals may be stretched or compressed at will. Signal processing techniques for changing the length of a passage without changing the pitch are discussed at length in Sect. 9.8.

Such beat manipulations can also be used to change the character of a rhythm. For example, Fig. 9.3 shows several different ways to expand and

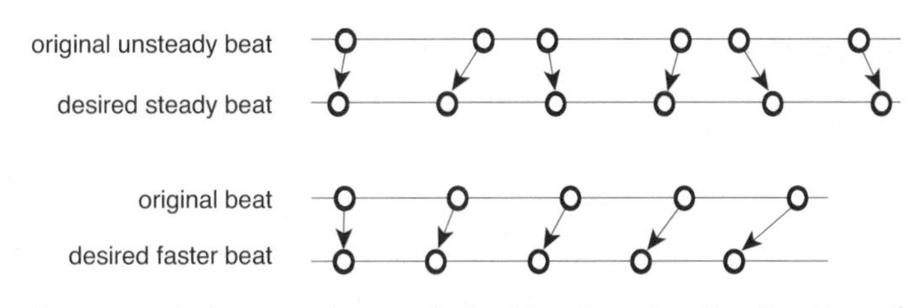

Fig. 9.2. An unsteady beat can be regularized by changing the duration of each beat interval to a desired steady value. A slow tempo can be made faster by shortening the sound in each beat interval.

compress beat intervals. When played together with the original, the sound becomes more complex, often in a rhythmic fashion. The first beat of part (a) is stretched by a factor of $\frac{4}{3}$, the second beat is compressed by a factor of $\frac{2}{3}$, and the third and fourth are left unchanged. Every four beats the stretched/compressed pattern realigns with the original so the overall tempo does not change. This manipulation is applied to the *Maple Leaf Rag* in [S: 78](a). The piano bounces along, having gained some extra flourishes. Part (b) stretches the first and third beats by $\frac{3}{2}$ and leaves the second beat unchanged. Every fourth beat is removed. This pattern is then played simultaneously with the original in [S: 78](b). Again, the piano acquires a rapid rhythmic ornamentation. Similarly, [S: 78](c) manipulates the beat pattern twice and plays all three versions simultaneously. While arbitrarily complex manipulations are possible, at some point the sound will become overly cluttered.

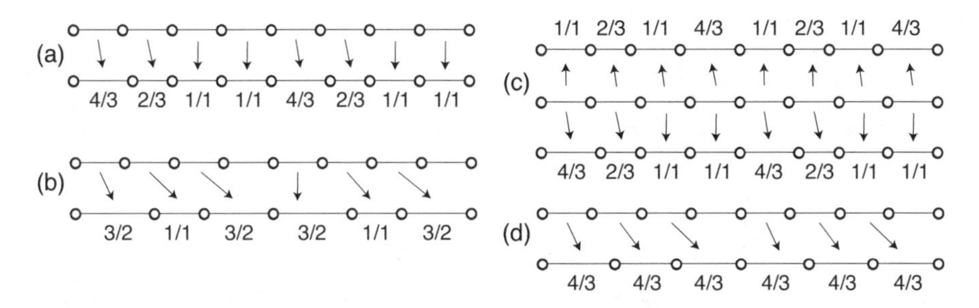

Fig. 9.3. Changing the duration of beat intervals can be used as a kind of beat-synchronized delay processing. Performing different versions simultaneously increases the density, often in a rhythmic way.

Figure 9.3(d) shows a uniform stretching of three out of every four beats by a factor of $\frac{4}{3}$. Thus three of the slower beats occupy the same time interval as four of the faster beats. When played simultaneously, the result is a 4:3 polyrhythm analogous to those of Sect. 3.9 but where individual rhythmic events are replaced by beat intervals. Two versions appear in [S: 79]. In the first version, the last beat of each measure is removed. In the second version, the third beat of each measure is removed. The two lines are panned left and right for the first 16 bars so that it is easier to hear the "four" of the original against the "three" of the stretched. The lines are merged back into stereo for the remainder to emphasize the polyrhythmic percept.

The tempo of the piece can also be changed by increasing or decreasing the length of sound within each beat interval. As discussed in Sect. 5.3.4, one common way to carry out time stretching and compression is to use the phase vocoder. A variable (time dependent) stretching can be used to accomplish a one-to-one time mapping near the start of each beat (to help preserve the attack) and to then speed up (or slow down) to achieve the desired stretching throughout the bulk of the beat interval. Several examples in [S: 80] demonstrate the *Maple Leaf Rag* at a variety of tempos ranging from one-quarter normal speed to sixty-four times normal speed. The *Maple Sleep Rag* [S: 81] develops the half-speed *Maple Leaf Rag* [S: 80](ii) using a variety of subtle (and not-so-subtle) beat-based effects from Sect. 9.2.

Extreme time stretching can be an interesting effect even when it is not tied to beat locations. When sounds become elongated, details that are normally not heard may come to the foreground. In [S: 82](i), a single strike of a gong lasts about four seconds. This is time stretched to over thirty seconds in [S: 82](ii), bringing out details of the evolution of the sound that are lost when heard at normal speed.

A collage of beat intervals from a variety of sources are joined in the *Very Slow* examples of [S: 83](i) and (ii). The uniting aspect of these sounds is that they are stretched (approximately) eight times so that each single beat interval lasts for $\frac{8}{3}$ of a second. Though many of the source beats are percussive, the primary impression of the sounds in the *Very Slow* pieces are of evolving textures, reverberant tonal masses, and transparent sound clouds. Typically, the attack of a sound is a synchronous onset of a complex collection of waves which evolve rapidly. When the sound undergoes extreme time stretching, the synchrony is lost. The "attack" of a drum becomes a cluster of sweeping sounds. Examples [S: 83](iii) and (iv) present the same piece *Inspective Latency* at the original tempo and slowed by a factor of eight.

Kramer's [B: 117] "vertical time" is a regime of perception where a single moment of time appears vastly elongated. In musical composition, vertical time may be evoked by repetition of a set of selected sonic events and by habituation of the listener to the repetition. Extreme time expansion is a signal processing analog of vertical time, where the elongation occurs in the perception of the individual sounds rather than in the perception of sequences of sounds. Beethoven's 9th Symphony, stretched to last 24 hours [W: 1], is a fascinating example of elongated time.

At the other extreme, radical time compression plays with the boundary between composition and timbre. Sound examples [S: 80](iv)–(viii) present a series of increasingly absurd compressions of the *Maple Leaf Rag* by factors of four, eight, sixteen, thirty-two and sixty-four. By the final version, the complete rag is squeezed into less than two seconds! It is no longer a piece of music; it is a complex fluttering timbre.

9.2 Beat-synchronous Filters

Perhaps the best known beat-based effect is the wah-wah pedal. The commercial viability of products such as Dunlop's "crybaby," Ibanez's "weeping demon," and Morley's "Steve Vai" wah-wah attest to its continued popularity. The wah-wah consists of a foot pedal that controls the resonant frequency of a low pass filter (as shown in Fig. 9.4). As the foot presses down, the resonant frequency increases, with the peak moving from (a) to (b) to (c). This imitates (in a crude way) the formants of a voice saying "wah" and gives the device its name.[1] Though this can be used in many ways, one of the most characteristic is a rhythm guitar effect in which the performer rocks the foot up and down in synchrony with the beat while strumming a chord pattern. For example, Isaac Hayes in the theme from *Shaft* [D: 22] and Jimi Hendrix in *Voodoo Child* [D: 23] exploit this style of playing.

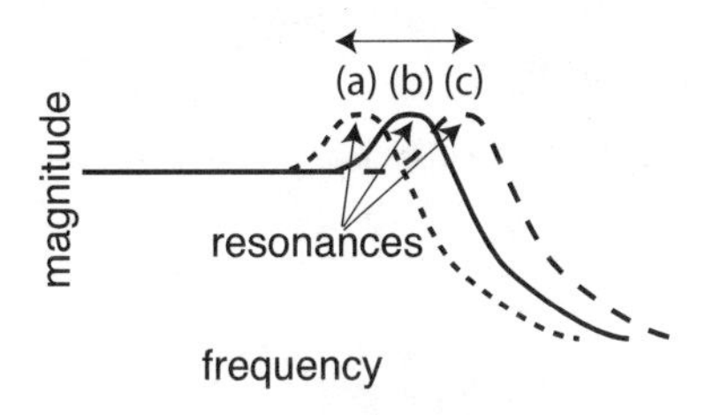

Fig. 9.4. In a wah-wah pedal, the resonant frequency of a lowpass filter is controlled by the position of a footpedal. A common performance technique is to rock the foot back and forth with the beat, sweeping the resonance in time to the music.

With the wah-wah pedal, the performer supplies the beat-synchronization, but the same idea can be applied to the post-processing of sound once the beat locations are known. For example, the resonant frequency of a filter might be controlled by the position within the beat interval: the beat might begin at (a), proceed through (b), reach (c) at the midpoint, and then return smoothly to (a) through the remainder of the beat interval.

The *Beat Filtered Rag* [S: 84] demonstrates some of the possibilities. These include:

(i) linear filters with resonances that change in synchrony with the beat: sometimes on each beat, sometimes twice per beat, sometimes four times per beat
(ii) delays that are synchronized with the beat: portions of the sound are fed back, summed, and delayed using times that are integer subdivisions of the beat

[1] A similar effect is produced when brass players move their hand in the bell of their instruments; for example, Joe "King" Oliver recorded *Wawawa* in the 1920s.

(iii) automatic panning (in the left/right stereo field) in a beat synchronous manner
(iv) flanging (a time variable phase shifting) with delays and time parameters that coincide with the beat

While it may appear as if the *Beat Filtered Rag* has been augmented with synthesizers or other instruments, the only sound source is the solo piano of the *Maple Leaf Rag* [S: 5]. All the sounds are (time varying) linear filters and delays of this single performance.

The parameters of the effects can be automated by a low-frequency oscillation (LFO) that is synchronized to the beat locations. The LFO may be a sinusoid, a ramp up or down, a triangle wave, or some arbitrary shape. It may fluctuate at a rate of one period per beat, or it may oscillate with n periods of the LFO entrained to m beats of the music (where n and m are small integers). Thus the effects may vary more rapidly than the beat or they may evolve more slowly than the beat; the key is that they remain synchronized.

The *Beat Gated Rag* [S: 85] also exploits other techniques:

(v) beat-synchronous gating allows the sound to pass at certain time instants and not at others
(vi) enveloping modulates the sound with predefined amplitude envelopes that synchronize with the beat locations

Again, it may appear at first listening that the *Beat Gated Rag* is supplemented with extra instruments or synthesizers, but it is not. All sounds are derived from the original piano performance by beat-synchronous gating, enveloping, filtering, and delays. How is this possible?

Classical analog synthesizers (such as those by Moog and ARP) begin with a simple waveform such as a sawtooth or a square wave, and then process the waveform using a variety of filters and techniques not dissimilar from the above lists. In the beat-based rags, the source sound is a piano performance (instead of a sawtooth wave), but the kinds of processing tricks, which lend the sound its color and character, are the same. The key is that in order to get intelligible output, it is necessary to synchronize the changes in the parameters of the filters, delays, gates, and envelopes with the changes in the music. In the classical synthesizer used as a musical instrument, the performer orchestrates the timing. In the post processing of a piece such as the *Maple Leaf Rag*, the parameters can be synchronized using beat-based information to control the timing. Additional variations on these processes can be heard in the *Magic Leaf Rag* [S: 141] and the *Make It Brief Rag* [S: 142], which exploit a variety of beat-synchronized gating and filters in addition to beat-synchronized delays such as those of Fig. 9.3. These pieces are discussed further in Sect. 10.3.

There are also many kinds of processing that can be done to a complex sound source that are inappropriate for simpler sounds. The next several sections explore signal processing techniques that require a complex beat synchronized input.

9.3 Beat-based Time Reversal I

A classic effect is to play a recording backwards. The sound of a percussive instrument (like the piano) swells out of nothingness, grows, and then eerily cuts off. Speech becomes unintelligible. For example, *Gar Fael Elpam* [S: 86] plays the first 45 seconds of the *Maple Leaf Rag* backwards. While this kind of reversal can be used as a special effect to generate odd timbres, generally speaking, the process of playing a sound backwards destroys its musical content.

At what level does this loss of meaning arise? Is meaning lost when individual sounds are reversed, or when the composition itself is reversed? A beat-based time reversal can distinguish between these two alternatives. In sound example [S: 87], the audio in each beat interval of the *Beat Reversed Rag* is reversed. The music (the melody, harmony, and rhythm) appears to move forward in the normal manner but the timbre of the sound changes drastically. The familiar timbre of the piano is transformed into an organ or a calliope by the reversal of the time envelope. Thus, time reversal on a small scale (within each beat) changes the timbre of the sound but does not change the composition itself. Time reversal on a large scale destroys the composition.

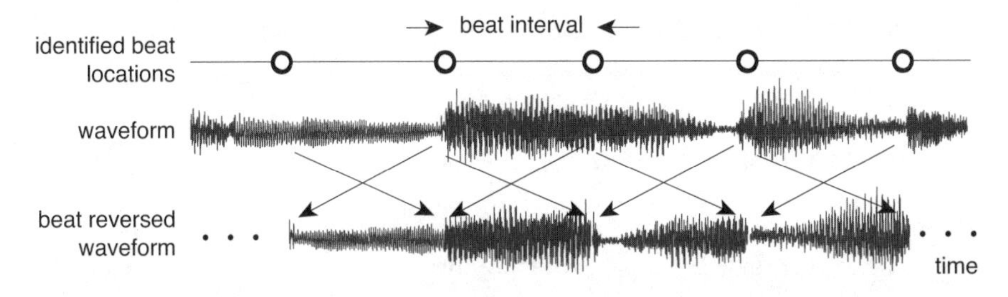

Fig. 9.5. Each beat interval of the *Maple Leaf Rag* is reversed in sound example [S: 87], the *Beat Reversed Rag*. The overall impression is that the music moves forward in the normal manner, but that it is played on a different instrument.

To demonstrate that correctly finding the beat boundaries is crucial in this process, the *Wrongly Reversed Rag* [S: 88] reverses the audio in approximately beat-size chunks (for this example, every 0.4 s). The boundaries of the reversed segments have no sensible relationship to the beat boundaries (which occur about every 0.34 s) and the primary impression is of a confused, or at least confusing, performance.

9.4 Beat-based Averaging

There are many styles of music based on repetitive cycles (any of the styles based on timelines, tala, or clave from Chap. 3) which repeat a rhythmic pattern throughout a significant portion of the piece. Once the beat timepoints have been identified, it is possible to average the signal over successive cycles, creating a kind of variation from the repetition.

Suppose that the individual beat intervals are identified and clustered into cycles so that a_1 represents the first complete cycle, a_2 represents the second cycle, etc. A two-cycle running average defines $b_k = \frac{1}{2}(a_k + a_{k+1})$ and then plays the b_k in succession. More generally, the kth term in an n-cycle average is $c_k = \frac{1}{n}(a_k + a_{k+1} + \ldots + a_{k+n-1})$.

For example, the hip-hop sublime [S: 31] is averaged over $n = 2, 5, 30,$ and 50 cycles in [S: 89]. For $n = 2$, the voice becomes confusing (because Ice Cube is rapping with himself) but the rhythmic feel remains intact. By $n = 5$, the voice overlays itself five times and becomes an incomprehensible chorus. With $n = 30$, the thirty voices have become a swirling sound cloud; the rhythmic material has become phased but is still clearly discernible. By $n = 50$ the voices are no longer recognizable as human and the rhythm has become smeared, but remains recognizable.

This emphasizes the amount of repetition inherent in the music; that which is the same throughout the piece is reinforced at each cycle, that which changes becomes attenuated and blurred.

So far, this chapter has focused on time domain manipulations of the sound within the beat intervals. The next sections turn to frequency domain manipulations of the beat.

9.5 Separating Signal from Noise

Frequency domain methods begin by finding the spectrum using an FFT. Since the spectrum is a sequence of numbers, the numbers can be changed (or "processed" or "mapped") in many possible ways. The result is a new spectrum that can be transformed back into a time signal:

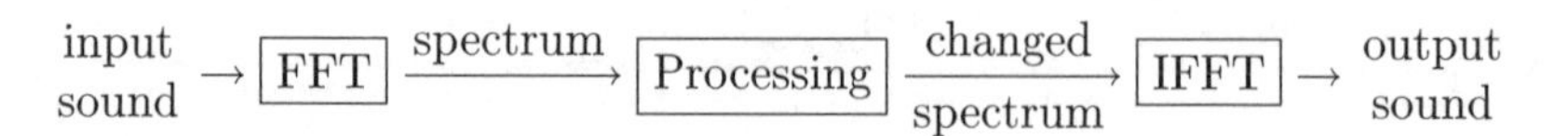

The next sections describe several different kinds of musically sensible processing.

One of the great strengths of a transform-based approach to the processing of musical signals is that the tonal aspects of the sound can be treated differently from the noisy aspects; the periodic components can be treated differently from the aperiodic components. This requires that there be a simple way of separating the signal (loosely, the most salient partials in the sound) from the noise (rapid transients or other components that are distributed over a wide range of frequencies). This separation helps preserve the integrity of the tonal material and helps preserve valuable impulsive information that otherwise may be lost due to smearing [B: 195]. This section shows how to carry out this separation, and then applies the method in a variety of sound examples.

The noise floor can be approximated as the output of a median filter applied to the magnitude spectrum, as shown in Fig. 9.6. Since peaks are rarely more than a handful of frequency bins wide, a median filter with length between $m_L = 21$ and $m_L = 31$ allows good rejection of the highs as well as good rejection of the nulls. For example, the left hand plot in Fig. 9.6 shows the spectrum in a single beat interval of Joplin's *Maple Leaf Rag*. The median filter, of length 25, provides a convincing approximation to the noise floor. If desired for data reduction purposes, this can be approximated using a small number of linear segments without significant loss of detail.

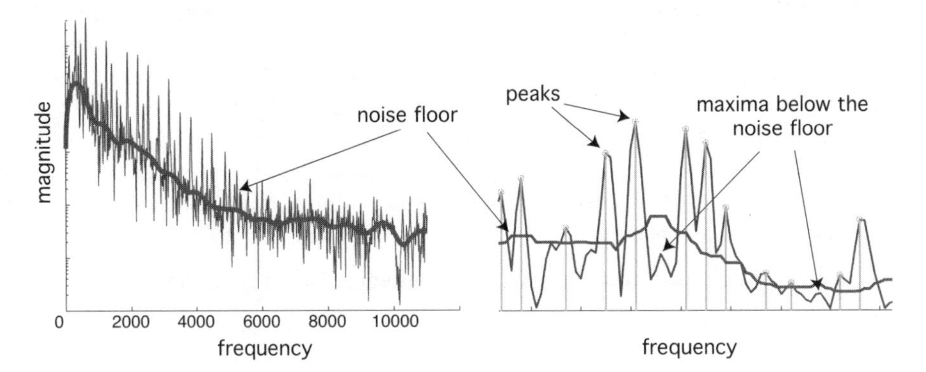

Fig. 9.6. A typical spectrum and the noise floor as calculated by the median filter. The noise floor can be used as a step in the identification of spectral peaks and to help separate the signal from the noise. The plot on the right enlarges a small section of the plot on the left.

To be explicit, the length m_L median filter (with m_L an odd integer) of the sequence $x(k)$ is

$$N(k) = \text{median}\,\{x(j), x(j+1), \ldots, x(j+m_L)\} \tag{9.1}$$

where $j = k - \frac{m_L - 1}{2}$. If the indices refer outside the vector x (as occurs at the start and end), then zero is assumed. In the present application, x represents the magnitude spectrum of the input and $N(k)$ is a magnitude spectrum that approximates the noise floor.

What does the noise floor sound like? The *Noisy Leaf Rag* [S: 90] strips away all information that lies above the noise floor in each beat of the *Maple Leaf Rag* (leaving, for example, only the data below the dark line in Fig. 9.6). After translation back into the time domain, all the significant peaks of the sound are removed and only the noise floor remains. Perceptually, the rhythm of the piece is clearly evoked and the bulk of the pitched material is removed. The *Maple Noise Rag* [S: 91] uses the *Noisy Leaf Rag* as its only sound source, augmenting the rhythmic noise with a variety of beat-based delays and filters as in Sect. 9.2. The *Just Noise Rag* [S: 92] demonstrates another variation. Similarly, *Noisy Souls* [S: 93] removes all but the noise floor from *Soul* and then sculpts a variety of sound textures from this noise.

This technique is particularly interesting when applied to vocal sounds. Sound example [S: 94], the *Noisy StrangeTree*, removes all information above the noise floor in each beat. In the verse, the consonants of the voice are (almost) identifiable. In the chorus, where the voice sings sustained notes, the vocals effectively disappear. This technique is applied more artfully in the second verse of the song *Sixty-Five StrangeTrees* [S: 106], which is discussed more fully in Sect. 9.6.

The opposite manipulation, stripping away all information that lies below the noise floor, has the opposite effect. This is demonstrated in the *Signal Leaf Rag* [S: 96]. Observe that many of the percussive elements (such as the attacks of the piano notes) are removed, leaving the basic tonal material intact. Due to the linearity of the transform process, the sum of the *Noisy Leaf Rag* and the *Signal Leaf Rag* is equal to the original *Maple Leaf Rag* [S: 5].

The noise floor is also important as a step in locating the peaks of the magnitude spectrum. Within each beat, let $X(f)$ be the magnitude spectrum and $N(f)$ the noise floor. A useful algorithm for peak identification is:

(i) Let L be the set of frequencies at which all local maxima of $X(f)$ occur.
(ii) For all $\ell_i \in L$, if $X(\ell_i) < N(\ell_i)$, remove ℓ_i from L.
(iii) Remove all but the M elements of L which have the largest magnitude.

For many purposes, $M = 50$ is a useful maximum number of peaks, though this number might vary depending on the complexity of the sound material and the desired compositional goals. By construction, the algorithm returns up to M of the largest peaks, all of which are guaranteed to be local maxima of the spectrum and all of which are larger than the noise floor. The right hand plot in Fig. 9.6 demonstrates how the peaks (indicated by the small circles) coincide only with maxima that are greater than the noise floor. Observe that this is quite different from a strategy that chooses peaks based on a threshold (which might typically be a function of the magnitude of the largest peak).

What do peaks sound like? Once the M largest peaks are located, it is straightforward to remove the non-peak data and transform the result back into the time domain. Since each peak is actually several samples wide,[2] as can be seen in the right hand plot of Fig. 9.6, it is a good idea to retain a small number of samples surrounding each peak. Removing all but the largest peak results in *Maple One Peak* [S: 97](i). This single sinusoid additive resynthesis[3] of the *Maple Leaf Rag* is barely recognizable. The next several sound examples

[2] due to the windowing and finite resolution of the FFT.

[3] The algorithm is applied separately to the left and right tracks of the original sound file. Because the data is different in each track, the largest peak in the

[S: 97](ii)–(v) increase the number of peaks that are retained: from 3, to 15, to 50, and then to 250. With 3 peaks the piece is recognizable but glassy and synthetic. With 50 and 250 peaks, the character of the piano is clear. In between, with 15 peaks, the timbre is somewhat inconsistent. Certain notes approach the piano timbre while others retain a synthetic feel.

Observe that the process of retaining only the peaks is not the same as the process of removing the noise floor. Nor is retaining the noise floor the same as removing the peaks. For example, the *Atonal Leaf Rags* [S: 98] and [S: 99] remove all data in a band of width ±25 Hz about each of 50 detected peaks. The primary impression is of an atonal rhythmic bed. The atonality occurs because the majority of the (harmonic) peaks are detected and removed, leaving an irregular pattern of peaks unrelated to any harmonic series. The ear interprets this as a random collection of partials that have no tonal structure, yet which retain a rhythmic pulse. The same technique is applied to *Soul* [S: 7] to form the basis of the *Atonal Soul* [S: 100]. The raw atonal output is then beat-gated and beat-filtered as in the previous sections.

Once the tonal components are separated from the noise components of a beat, there are many more interesting ways to process the sound than simple deletion. The next section looks at a technique of mapping the partials from their current location to some desired location.

9.6 Spectral Mappings

The partials of a sound are defined by the peaks of the magnitude spectrum. Since these are represented in the computer as a vector of numbers, the numbers can be manipulated algorithmically. This is called *spectral mapping* [B: 199] because it maps the spectrum of the source into the spectrum of the destination. This changes the frequencies of the peaks (partials) and hence changes the timbre of the sound.

Mathematically, a spectral mapping is a function from $\mathfrak{C}^n \rightarrow \mathfrak{C}^n$, where $\mathfrak{C}^n$ is the n-dimensional space of complex numbers and n is equal to the size of the FFT. In general, the mapping may change from beat to beat because the locations of the source and/or destination partials may change. Thus the mapping is not time invariant.

For example, Fig. 9.7 shows a stylized representation of a spectral mapping from a source spectrum with peaks (partials) at f_i into a destination spectrum with peaks (partials) at g_i. As shown, some of the partials move up in frequency while others move down. If there is a consistent motion (up or down) then the pitch may also change. It can change a harmonic sound into an inharmonic sound. It can change a single pitched note into two or more notes, or it may change a chord into a single tone, depending on the nature of

left may be different from the largest peak in the right. Hence there are times in [S: 97](i) where two sinusoids sound simultaneously.

the source and destination. There are many possible spectral mappings; this section presents some that may give musically interesting results.

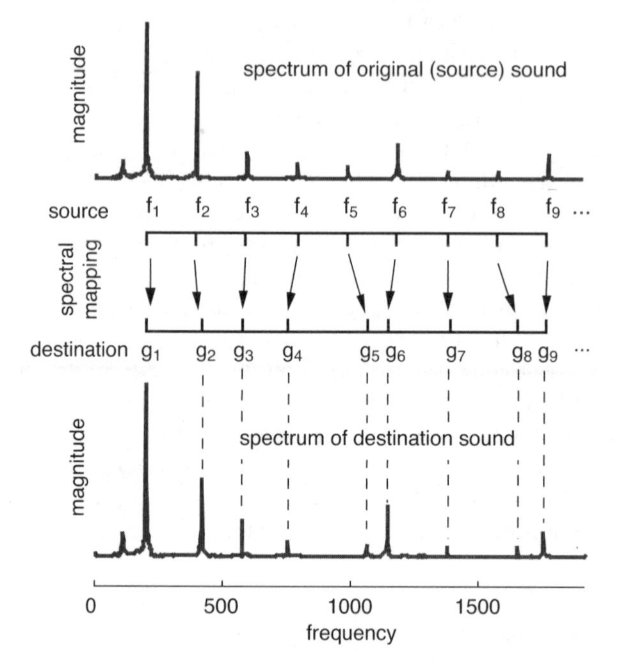

Fig. 9.7. In this schematic representation of a spectral mapping, a source spectrum with peaks at $f_1, f_2, f_3, \ldots$ is mapped into a destination spectrum with peaks specified at $g_1, g_2, g_3, \ldots$. The spectrum of the original sound (the plot is taken from the G string of a guitar with fundamental at 194 Hz) is transformed by the spectral mapping for compatibility with the destination spectrum. The mapping changes the frequencies of the partials while preserving both magnitudes (shown) and phases (not shown).

9.6.1 Mapping to a Harmonic Template

Harmonic sounds play an important role in perception, as discussed in Sect. 4.3.2. What happens when a source (input) sound is spectrally mapped so that all partials are moved to coincide with a single harmonic series? This is a special case of the spectral mapping of Fig. 9.7 where the destination spectrum g_i consists of all integer multiples of a single fundamental frequency. For example, the harmonic series built on 65 Hz is

$$g_1 = 65, \; g_2 = 130, \; g_3 = 195, \; g_4 = 260, \; g_5 = 325, \; \ldots . \tag{9.2}$$

If the source consists of a harmonic series with fundamental f, the spectral mapping is much like a transposition where all partials are multiplied by a constant factor $\frac{g_1}{f}$. But even when the source is inharmonic, the output lies on a single harmonic series. For example, the gong [S: 82](i) is spectrally mapped to the 65 Hz harmonic destination in [S: 101]. The harmonic gong has a definite pitch (the same pitch as a sinusoid with frequency 65 Hz).

Similarly, sound example [S: 102] maps an inharmonic cymbal into a harmonic series. The spectrum of the cymbal (like that of the gong) contains many peaks spread irregularly throughout the audible range. The spectrally mapped version retains some of the noisy character of the cymbal strike, but it inherits the pitch associated with the destination spectrum. The two brief segments of the *Harmonic Cymbal* [S: 102] are:

(i) the original sample contrasted with the spectrally mapped version
(ii) a simple "chord" pattern played by pitch shifting the original sample, and then by pitch shifting the spectrally mapped version

The transformed instrument supports both chord progressions and melodies even though the original cymbal sound does not.

Even greater changes occur when mapping complete musical pieces to harmonic spectra. For example, *Maple in 65 Hz* [S: 103] maps the *Maple Leaf Rag* into the 65 Hz harmonic template (9.2). Chords become static and the harmonic motion is lost. The bass remains rooted in a 65 Hz fundamental even as the pitch moves in the scale defined by the harmonic series. Chord variations in the *Maple Leaf Rag* become variations in timbre in [S: 103]. Throughout, the rhythm remains clear. *Sixty-Five Maples* [S: 104] further develops and elaborates [S: 103]. Other pieces mapped into the same 65 Hz harmonic series are *Sixty-Five Souls* [S: 105] and *Sixty-Five Strangetrees* [S: 106]. Even while remaining tied to a single root, an amazing variety is possible.

Such radical alteration of sound is not without artifacts. The most prominent effect is the 65 Hz "drone" that accompanies each of the pieces. This is unsurprising given the nature of the mapping since all partials between 30 Hz and 97.5 Hz are mapped to 65 Hz. Thus an octave and a half of the bass range is mapped to the fundamental frequency. Similarly, all other partials of the source are mapped to the various harmonics of this drone.

Artifacts associated with time variations can be more subtle. The actual mapping performed within each beat[4] changes as the partials in the source move. For example, at certain times, the partials of the voice in *Sixty-Five Souls* jump around, wiggling up and down. This rapid oscillation between adjacent destination partials is illustrated in Fig. 9.8. Discontinuous outputs can occur even when the input partials vary smoothly. This exhibits an essential tension between the sliding pitch of the original *Soul* [S: 7] and the essentially static pitch of the destination. This artifact can be reduced by incorporating hysteresis in either frequency or in time, though this requires careful tweaking of additional parameters.

[4] or within each frame in a PV implementation.

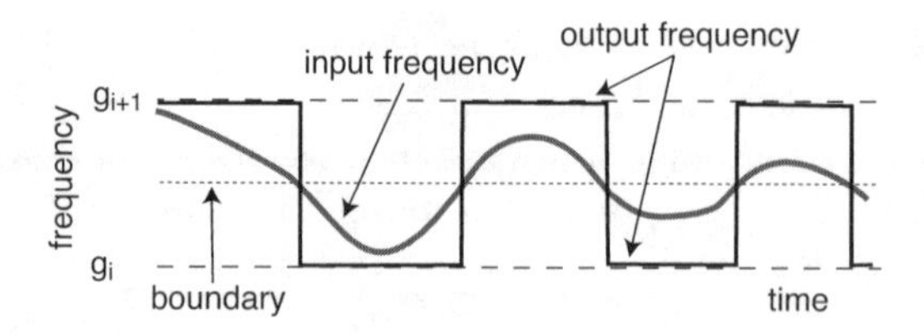

Fig. 9.8. When the partials are mapped to a fixed destination, small changes in the input frequency can cause jumps in the output. In this plot, the smoothly varying input repeatedly crosses the boundary that separates partials mapped to g_i from partials mapped to g_{i+1}.

9.6.2 Mapping to a n-tet Template

Another simple destination spectrum is given by the steps of the n-tone equal tempered scale (abbreviated n-tet). This destination contains the frequencies $F\alpha^i$ for all positive and negative integers i where $\alpha = \sqrt[n]{2}$ and where F is a reference frequency. For example, the familiar 12-tet scale of the Western tradition has $\alpha = \sqrt[12]{2} \approx 1.059463$ and $F = 440$ (though different values of F have been used in various times and places). One motivation for this destination spectrum is that if the partials of a sound are located at intervals corresponding to these scale steps, the sensory dissonance [B: 166] of the sound will be minimized at the intervals of the scale steps [B: 200]. For example, an easy way to achieve consonant chords in (say) 11-tet is to map the partials of the sound to the 11-tet scale steps. A series of examples showing various n-tets and related spectra that minimize dissonance is given in *Tuning, Timbre, Spectrum, Scale* [B: 196].

An n-tet destination spectrum is

$$g_1 = F\alpha^{j_1},\ g_2 = F\alpha^{j_2},\ g_3 = F\alpha^{j_3},\ g_4 = F\alpha^{j_4}, \ldots. \tag{9.3}$$

where the j_i are some subset of the integers. For example, with $n = 11$, it is possible to map individual harmonic sounds into 11-tet sounds using the mapping[5]

$$\begin{array}{ccccccccccc} f & 2f & 3f & 4f & 5f & 6f & 7f & 8f & 9f & 10f & \cdots \\ \downarrow & \downarrow & \downarrow & \downarrow & \downarrow & \downarrow & \downarrow & \downarrow & \downarrow & \downarrow & \\ f & \alpha^{11}f & \alpha^{17}f & \alpha^{22}f & \alpha^{26}f & \alpha^{28}f & \alpha^{31}f & \alpha^{33}f & \alpha^{35}f & \alpha^{37}f & \cdots \end{array}$$

where f is the fundamental of the harmonic tone and $\alpha = \sqrt[11]{2}$. Sound example [S: 107] illustrates this mapping with several instrumental sounds alternating with their 11-tet versions.

(i) harmonic trumpet compared with 11-tet trumpet

[5] These particular destination values (powers of α) are chosen because they are the closest 11-tet scale steps to the harmonic partials of the source. This helps to minimize the perceived changes in the timbre of the sound.

(ii) harmonic bass compared with 11-tet bass
(iii) harmonic guitar compared with 11-tet guitar
(iv) harmonic pan flute compared with 11-tet pan flute
(v) harmonic oboe compared with 11-tet oboe
(vi) harmonic "moog" synth compared with 11-tet "moog" synth
(vii) harmonic "phase" synth compared with 11-tet "phase" synth

The instruments are clearly recognizable and there is little pitch change caused by this spectral mapping. Perhaps the clearest change is that some of the samples have acquired a soft high-pitched inharmonicity: a "whine" or a high "jangle." In others, it is hard to pinpoint any differences.

Isolated sounds do not necessarily paint a clear picture of their behavior in more complex settings. The *Turquoise Dabo Girl* [S: 108] is performed in 11-tet using the spectrally mapped sounds from [S: 107] (along with additional percussion). This demonstrates that some of the kinds of effects normally associated with harmonic tonal music can occur even in such strange settings as 11-tet. For instance, the harmonization of the 11-tet pan flute melody (between 1:33 and 2:00) has the feeling of a kind of (perhaps unfamiliar) "cadence" harmonized by unfamiliar chords.

When mapping complete performances (rather than individual sounds) it is no longer possible to use the pitch (or fundamental frequency) of the source to help define the destination. The simplest approach is to use all possible scale steps $F\alpha^i$ for all integers i. For example, the *Maple Leaf Rag* is mapped into several different n-tet destination spectra in sound examples [S: 109] and [S: 110]:

(i) a 4-tet destination spectrum in *Maple 4-tet*
(ii) a 5-tet destination spectrum in *Maple 5-tet*
(iii) a 10-tet destination spectrum in *Maple 10-tet*
(iv) a 100-tet destination spectrum in *Maple 100-tet*

For small n, there are only a few partials per octave and the source partials must be mapped to distant frequencies. This can have a significant impact on the timbre. For example, the piano timbre in the $n = 4$ and $n = 5$ cases resembles an accordion without a bellow or a cheesy organ with only a single stop. By the time n increases to 100, the destination partials are densely packed and there is little change in the sound. In between, the sound wavers as the source partials switch destinations. In all cases, the rhythmic pulsation is retained.

Each of the *Maple n-tet* examples is generated two ways: using the PV in [S: 109] and using the beat-synchronous FFT in [S: 110]. The FFT method of spectral mapping tends to preserve transients better while the PV method tends to give smoother results. Further discussion and technical details are given in Sect. 9.8. The FFT version of the *Maple 5-tet* is used as the source material for the *Pentatonic Rag* [S: 111]. Applying a series of beat-based filters and gates (as described in Sect. 9.2) allows the development of interesting motifs and rich timbral development even though the basic elements of five equally spaced tones and five equally spaced partials per octave may appear to be a meager resource.

The video [S: 112] displays some of the more subtle points of the spectral mapping procedure by plotting the source spectrum, the output spectrum, and several related features over the first five seconds of *Maple 5-tet* [S: 109](ii). This is annotated in Fig. 9.9, which shows a single frame of the video. The small magenta circles on the horizontal axis are the locations of the 5-tet destination spectrum. There are five of these circles within any octave (between 200 and 400 Hz, or between 400 and 800 Hz, for example). The undulating red spectrum is the input, and the small black circles show the peak detection locating the most prominent of the peaks.

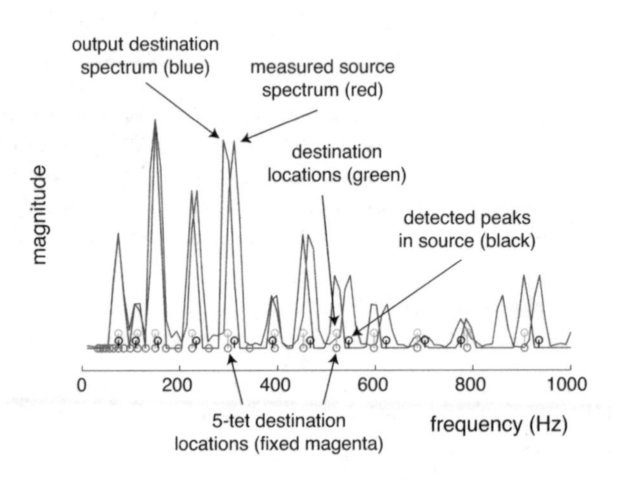

Fig. 9.9. A single frame from the video example [S: 112] is reproduced here and annotated. The video demonstrates the complexity of the spectral mapping process as the partials of the input rise and fall, split and rejoin. The detected peaks wiggle about even when the destination remains fixed. Colors refer to the video.

Each peak in the source must be assigned to one of the destination locations, and these are shown by the nearby green circles in the video. Finally, the output spectrum is shown in blue. This is constructed from the input spectrum through the mapping defined by the black to green circles. The time variation of the mapping is clear in the video, even when the destination remains fixed.

Other examples of n-tet mappings are the *Pentatonic Souls* [S: 113] and *Scarlatti 5-tet* [S: 114] which map *Soul* [S: 7] and Scarlatti's K517 sonata into 5-tet. The *Pentatonic Souls* reduce the song to a single chord. When the band modulates between chords, the 5-tet versions "modulate" from one inversion of the 5-tet chord to another. As in the 65-Hz version, the 5-tet voices jump between discrete pitches where the original varies smoothly (recall Fig. 9.8). In

Scarlatti 5-tet, the piano takes on a metallic character and the piece acquires a gamelan-like flavor. The strong chord progressions of the original are flattened into a continuous recapitulation of the 5-tet mode. The tonal meaning of the piece is warped, perhaps beyond recognition, though the rhythmic activity remains intact.

9.6.3 The *Make Believe Rag*

Spectral mappings to n-tet destinations (for small n) often sound like a single chord. An intriguing idea is to compose using these "notes" and "chords" as basic elements. The *Make Believe Rag* [S: 115] combines spectral mappings of the *Maple Leaf Rag* into 3, 4, 5, and 7-tet sequenced in a beat-synchronous manner so that the sound within each beat interval is transformed to one of the n-tet destinations. The compositional process consists primarily of choosing the order and duration of each mapping and Fig. 9.10 shows two snippets from the musical score.

A																
n-tet destination	3	3	3	4	4	4	4	5	5	5	5	7	7	7	7	3
beat	1	2	3	4	5	6	7	8	9	10	11	12	13	14	15	16

B																
n-tet destination	4	4	4	5	4	4	4	5	5	5	4	5	5	5	4	4
beat	1	2	3	4	5	6	7	8	9	10	11	12	13	14	15	16

Fig. 9.10. Two short segments (labeled A and B) from the score to the *Make Believe Rag*: each beat is transformed to one of the n-tet destination spectra. Various patterns of 3, 4, 5, and 7-tet are used. In sequence A, for example, the first three beat intervals are mapped to 3-tet, the next four beat intervals are mapped to 4-tet, etc. Changes in the tunings (the destination spectra) play a role in the *Make Believe Rag* analogous to the role normally played by changes of harmony in tonal music.

Changes in the tuning (i.e., in the destination spectra) are aligned with rhythmic changes in the piece so that the result "makes sense" even though there are no standard chords or harmonic progressions; changes of tuning play a role analogous to the changing of chords in a tonal context. Observe that while 3-tet and 4-tet mappings are subsets of the familiar 12-tet system, 5-tet and 7-tet are not. Thus the *Make Believe Rag* mixes the familiar with the unfamiliar.

This section has considered only octave-based scales. Non-octave based systems such as stretched pseudo-octaves [B: 141], the tritave-based Bohlen-Pierce scale [B: 142] (and others), proceed along analogous lines. The only requirement is a clear specification of the desired destination spectrum.

9.7 Nonlinear Beat-based Processing

This section discusses four sound transformations that are particularly useful when applied in a beat-synchronous manner: the spectral band filter, spectral freezing, the harmonic sieve, and instantaneous harmonic templates.

9.7.1 Spectral Band Filter

A graphic equalizer separates the frequency into a set of different bands and then amplifies or attenuates the sound in each band individually. A *spectral band filter* separates the magnitude (the vertical axis of the Fourier transform) into a set of different bands and then amplifies or attenuates the sound in each band individually. This tips the standard linear filter on its side. A simple four-band spectral band filter is illustrated in Fig. 9.11. Spectral band filters sound radically different from any linear filter. *Local Variations*, sound example [S: 117], applies a fixed (eight band) spectral band filter to *Local Anomaly* [S: 116]. Within each beat, the relative sizes of the spectral peaks are rearranged, causing drastic timbral changes that nonetheless maintain the rhythmic feel.

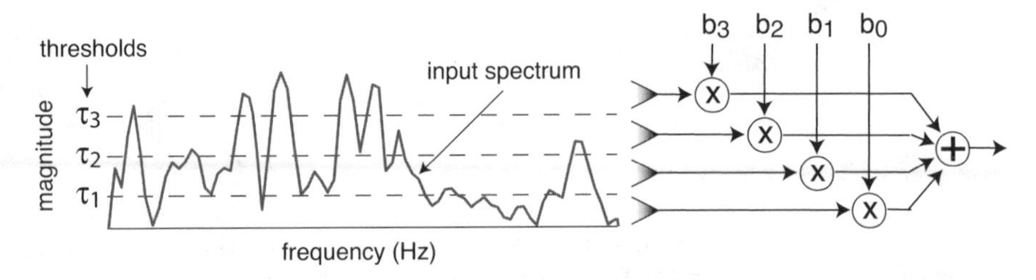

Fig. 9.11. A spectral band filter uses thresholds τ_i to define a collection of magnitude regions. The sound within each region is sent to a different channel and amplified or attenuated (by the b_i) as desired, and the output of all the channels is summed. The FFT that transforms the time signal into the spectrum and the IFFT that transforms the output of the channels back into the time domain are not shown.

9.7.2 Spectral Freeze

One of the remarkable features of algorithms that separate temporal motion from spectral motion is the ability to control the flow of time. Musical time can be sped up, slowed down, run backwards, or stopped completely (frozen) by a simple choice of parameters. If the instants at which time is frozen are synchronized with the beat, this can be an effective way to generate "new" rhythmic material from existing material.

Figure 9.12 shows two rhythmic patterns in the necklace notation of Sect. 2.1.3. The inner and outer circles indicate times when the left and right audio tracks are frozen; each freeze is held until the next one begins so that there is no pause in the sound. Applying these patterns to the *Maple Leaf*

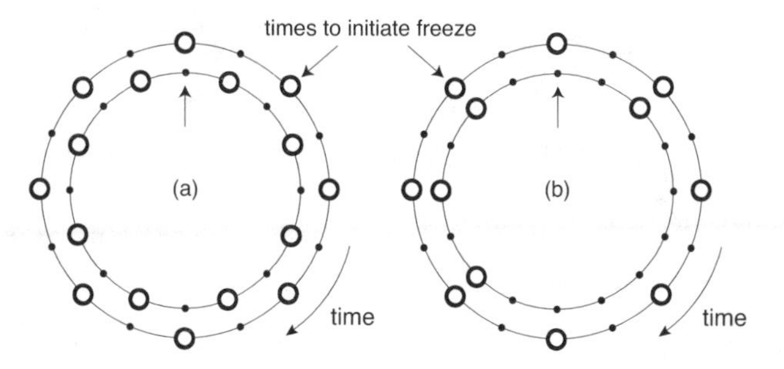

Fig. 9.12. The *Maple Leaf Rag* and *Soul* are frozen in these two patterns in [S: 118] and [S: 119]. The inner and outer necklaces indicate when the left and right tracks are frozen.

Rag results in [S: 118]. In (a), the two tracks are frozen at a steady rate of once per (eighth-note) beat, but the right track freezes at the start of the beat while the left track freezes in the middle. Together, they move at the tatum (sixteenth-note) rate. In (b) the right track again moves at the steady beat rate while the left track executes a simple syncopated pattern. New rhythms result from the interaction between the rhythm of the piece and the rhythm of the freezing pattern. The *Soul Freezes* [S: 119] apply the same patterns to *Soul* [S: 7], and *Frozen Souls* [S: 120] elaborates pattern (b) using the standard complement of beat-synchronous tricks.

9.7.3 The Harmonic Sieve

A harmonic filter $H_g(f)$ centered at g Hz passes only frequencies in the neighborhood of the harmonic series $g,\ 2g,\ 3g, \ldots$, as illustrated in Fig. 9.13. The output of the harmonic filter is the product of (a) the spectrum $S(f)$ of the input and (b) the transfer function $H_g(f)$ of the harmonic filter. Thus the output spectrum (c) is $R_g(f) = H_g(f)S(f)$. Let $||R_g(f)||$ be the energy of the output in each beat interval. The *harmonic sieve* chooses the $R_g(f)$ with the greatest energy in each beat. Formally, the output of the sieve in a given beat is $R(f) = R_{g^*}(f)$ where

$$g^* = \underset{g}{\operatorname{argmax}} \, ||R_g(f)|| = \underset{g}{\operatorname{argmax}} \, ||H_g(f)S(f)||$$

is the fundamental frequency of the harmonic filter with the greatest output energy. The harmonic sieve can be thought of as a collection of harmonic comb

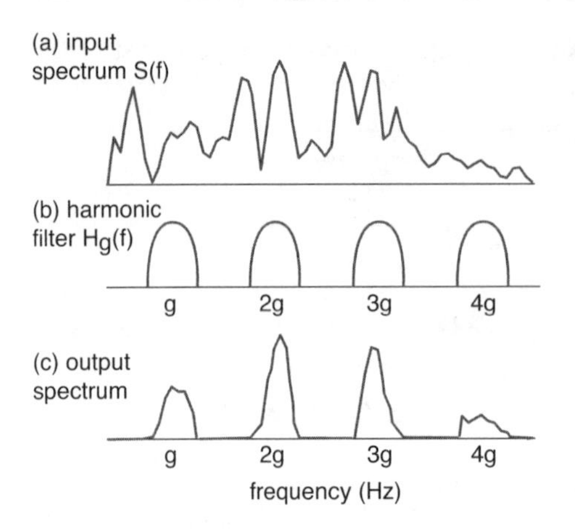

Fig. 9.13. The spectrum of the input $S(f)$ is multiplied by a set of harmonic filters $H_g(f)$, each of which passes frequencies around the partials of the harmonic series $g,\ 2g,\ 3g, \ldots$. The output of each harmonic filter is $R_g(f)$. The $R_g(f)$ which has the greatest energy is the output of the harmonic sieve in that beat interval.

filters with fundamental frequencies g_i along with a switch that chooses the comb with the maximum output in each beat interval.

The harmonic sieve forms the basis of three sound examples. The first applies the sieve to *Three Ears* [S: 121]; the result is *Mirror Go Round* [S: 122]. The many odd sounds produced by the harmonic sieve in *Mirror Go Round* are brought to the foreground in the mix; these include a large collection of gentle bubbling, sweeping, and tinkling artifacts. These could be annoying, but because they synchronize with the underlying sound, they ornament the timbre. Similarly, the sieve transforms the bulk of the percussion into accents that merge into the flow of the timbre. Perceptually, one of the primary features of the sieve is that it transforms transients such as the attacks of notes into timbral ornaments and accents.

A collection of harmonic sieves are applied to *StrangeTree*, to give *Sievetree* [S: 123]. The various sieves are sometimes synchronized with the beat, sometimes with the half-beat, and sometimes with every second beat. Each sieve shifts the voice differently in time; the effect is especially striking in the verses (for instance, near 0:22 and near 1:10) where a single voice echoes itself in a ghostly manner. It is also interesting to compare with *Sixty-Five Strangetrees* [S: 106] which applies a spectral mapping to a fixed 65 Hz harmonic template. This induces a fixed drone at the destination frequency; *Sievetree* changes harmonic center as the piece progresses.

The third application of the harmonic sieve is to *Phase Space* [S: 124], resulting in *Reflective Phase* [S: 125]. Here, a large number of sieves were used, each synchronized to a different multiple or submultiple of the beat. The outputs of all these sieves were then rearranged, mixed and edited; no other kinds of processing were used (other than some reverberation added to the final mix). Yet the piece is almost unrecognizable.

One reason to use different time scales for different sieves is that this allows different gs to occur simultaneously in the same beat interval. For example, each segment in the top line of Fig. 9.14 represents a beat interval and g_1

through g_5 are the frequencies of the harmonic filter with the greatest energy. The second and third lines show the two possible double-width beats, and these are also labeled with g values representing the harmonic filters with the greatest energy. Even though (for instance) the time interval g_6 contains the same sound as the time intervals g_1 and g_2, they may (or may not) be equal. Similarly, there is no necessary relationship between the gs in the second and third lines; thus g_6 and g_8 may or may not be equal. By layering over several time scales, complex timbres can occur even though each individual sieve passes only the harmonics of a single fundamental. In terms of Fig. 9.14, the g_is may all be different.

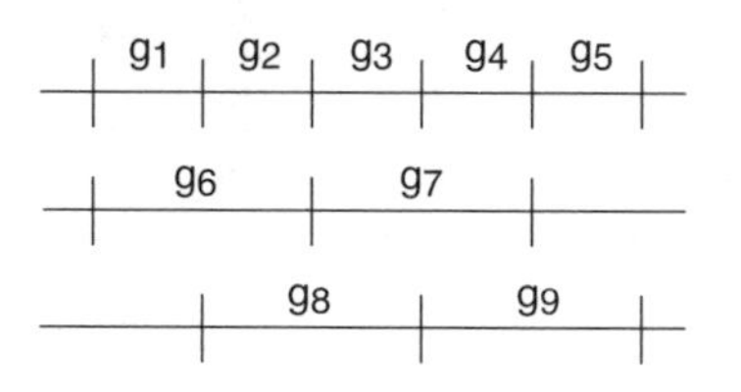

Fig. 9.14. The g_i represent the fundamental frequencies of the harmonic filter with the greatest energy in each beat interval. There is no necessary relationship between the gs at the various levels. Layering several such harmonic sieves allows complex timbral variations.

9.7.4 Instantaneous Harmonic Templates

The harmonic sieve is a way of choosing a harmonic destination spectrum that changes with each beat interval. An analogous idea can be applied using spectral mappings; all that is required is a way of specifying the destination spectrum. One approach is to attempt a pitch identification and to use the fundamental frequency of the pitch to define the harmonic template. Another approach is to use a periodicity detector such as the Periodicity Transform to locate the desired fundamental(s). These tend to work well when the pitch and/or periodicity identification succeed.

The technology for extracting the pitch of a single (monophonic) sound source is well advanced, but complex sound sources remain problematic. Sometimes the pitch extraction technique of [B: 138] and [W: 56] works well and sometimes it fails spectacularly. In particular, it will fail when the sound has no pitch (for example, the rustling of leaves or the crunching of snow underfoot) or when the sound has many pitches. Indeed, what is "the" pitch of an *A*♭ major chord?

Nonetheless, it is possible to apply the pitch extraction method to the sound in each beat interval and to use the detected pitch as a fundamental frequency to define a harmonic destination spectrum. The *Instant Leaf Rag* [S: 126] applies this to the *Maple Leaf Rag*. In places, the song is completely recognizable. In others, the mapping has completely changed the melodic and chordal progression, sometimes in peculiar and unexpected ways.

Using the periodicity transform does not produce results that are significantly more reliable, but it does allow the simultaneous detection of multiple periodicities. Each periodicity corresponds to a different fundamental frequency; together they define the destination spectrum. Like the *Instant Leaf Rag*, this can result in unexpected and even amusing juxtapositions of tonalities. For example, the *Instant Nightmare* [S: 127] uses three simultaneous periodicities in each beat interval. Percussion has been added and segments have been rearranged to emphasize the complete atonality of the output; the original piece has been changed (or perhaps mutilated) beyond recognition. As expected, the rhythmic motion is preserved.

9.8 The Phase Vocoder vs. the Beat-synchronous FFT

The bulk of the sound examples of this chapter have been generated using either a (modified) phase vocoder or a beat-synchronous FFT. This section compares and contrasts these two approaches in terms of the details of implementation and then in terms of the kinds of sounds that result.

9.8.1 Implementations

The PV and the beat-synchronous FFT are alike in many ways, both

(i) use a collection of windows to parse the signal into segments
(ii) take the FFT of each segment
(iii) apply a mapping to the transformed data
(iv) apply the inverse FFT, and finally
(v) piece together the output from the modified segments

The fundamental difference between the two methods is in the choice of segments: the PV uses a series of overlapping fixed-size frames while the beat-synchronous FFT uses adjustable segments that coincide with the beat intervals. Many of the details of implementation are driven by the differences in the width of the FFTs: the windowing, frequency resolution, methods of locating spectral peaks, and the kinds of spectral mappings. These differences are summarized in Table 9.1 and then discussed in detail throughout the bulk of this section.

Windows: The two windowing strategies are shown pictorially in Fig. 9.15. Typically, the windows (frames) of the PV are between 1K and 4K samples wide,[6] and overlap by a factor of about four. The windows in a beat-synchronous FFT are the same size as a beat: somewhere between 200 ms and 1.5 s. At the standard CD sampling rate, this corresponds to 8K to 60K samples. As shown in Fig. 9.15, only a small overlap at the beat boundaries (typically 128 or 256 samples) is needed to help remove clicks that might occur between successive beats.

[6] Smaller windows give better time localization and better reproduction of transients; larger windows allow more faithful reproduction of low frequencies.

Table 9.1. Implementations of the phase vocoder and the beat-synchronous FFT

	Phase Vocoder	Beat-synchronized FFT
windows	small frames from 1K–4K with 2 to 8 times overlap	large beat-sized windows $\frac{1}{5}$–$\frac{1}{2}$ s, zero padded to a power of two
FFT resolution	40 Hz –10 Hz (improved by phase adjustment)	3 Hz – 1.5 Hz (phase adjustment possible)
peak finding	all local max above median or threshold (see Sect. 9.5)	plus distance parameter (forbidding peaks too close together)
spectral map	direct resynthesis: output frequencies placed in FFT vector with phase adjustment	resampling with identity window [B: 199], no phase adjustment
alignment	See Fig. 9.18	See Fig. 9.18
beat detection	optional	required

Resolution of FFT: The small FFTs of the PV imply a poor frequency resolution, especially in the bass. Fortunately, the accuracy of the frequency estimation can be increased using the phase values as discussed in Sect. 5.3.4. This strategy can also be applied to the beat-synchronous FFT by using two shorter overlapping FFTs within the same beat interval but it is probably not necessary in most cases due to the increased accuracy of the frequency estimates of the longer FFTs.

Peak Finding: When searching for peaks in the spectra, the PV can use the median-based method of Sect. 9.5 directly. With longer magnitude vectors, however, more false peaks may be detected. For example, Fig. 9.16 shows an ambiguous collection of "peaks" centered around 490 Hz. Using the median-based method directly, up to five peaks are detected. Yet, because these are all so tightly packed in frequency, it is likely that they are all the result of some single physical action, and hence should be mapped together as a group. One way to help insure that only a single peak is detected in situations like

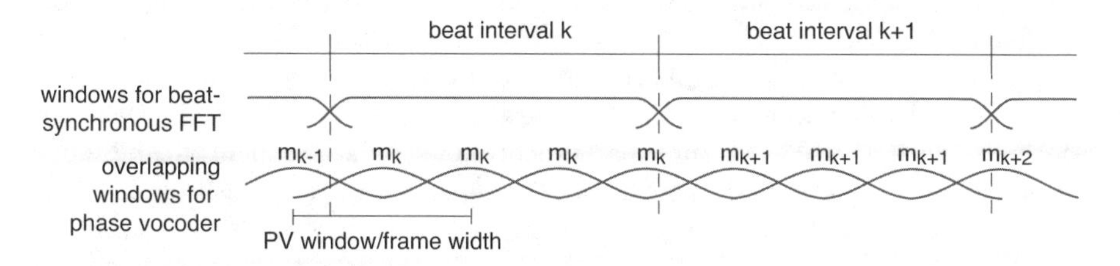

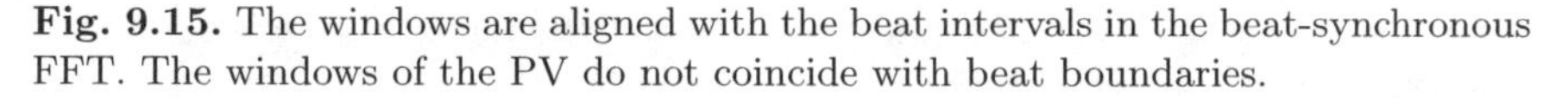

Fig. 9.15. The windows are aligned with the beat intervals in the beat-synchronous FFT. The windows of the PV do not coincide with beat boundaries.

this is to incorporate a parameter that requires adjacent peaks to be at least a minimum distance apart.

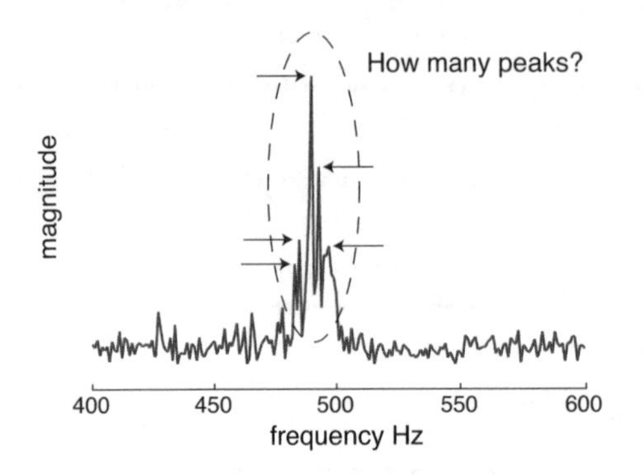

Fig. 9.16. This shows a small portion of the spectrum from a single beat in the *Maple Leaf Rag*. How many peaks lie within the dotted region? The median method finds up to five, depending on the median length. A more sophisticated peak finding algorithm that includes a parameter specifying the minimum allowable distance between adjacent peaks might detect only the single largest peak.

Spectral Mapping: Different processing techniques are also needed for the spectral mappings. For the PV, it is possible to directly construct a frequency-domain magnitude vector that contains the desired frequencies. The phase adjustment technique of Sect. 5.3.4 ensures that the output partials align correctly across the frame boundaries. The beat-synchronous FFT uses the resampling with identity window (RIW) technique for spectral mapping, which is shown in Fig. 9.17. One assumption underlying spectral mappings is that the most important information (the partials that define the sound) is located at or near the spectral peaks. RIW relocates these peaks to the appropriate destinations. Thus the spectrum is divided into regions associated with the peaks (these are copied verbatim from the source to the destination) and the relatively empty regions between the peaks (these are stretched or compressed via resampling). In order to help preserve the temporal envelope within each beat interval, the phases accompany the magnitudes (in both the identity and the resampling portions of the mapping). A more complete discussion of this technique can be found in [B: 196] and [B: 199].

Data Alignment: Consider a set of source locations $s_1,\ s_2, \ldots, s_n$ and a set of destination locations $d_1,\ d_2, \ldots, d_m$ where n may be different from m. The problem of how to assign the s_i to the d_j is an example of the data alignment problem, and there is no simple (unique) solution. The "nearest neighbor" and the "sequential alignment" approaches are shown in Fig. 9.18, and either may be used with the PV or the beat-synchronous FFT. Sequential alignment is ideal for simple sounds which may undergo some transposition (observe that the sequential alignment between two harmonic sounds with different fundamentals is just a transposition). But aligning the partials sequentially may also map partials to distant destinations, which can have a large impact on the timbre.

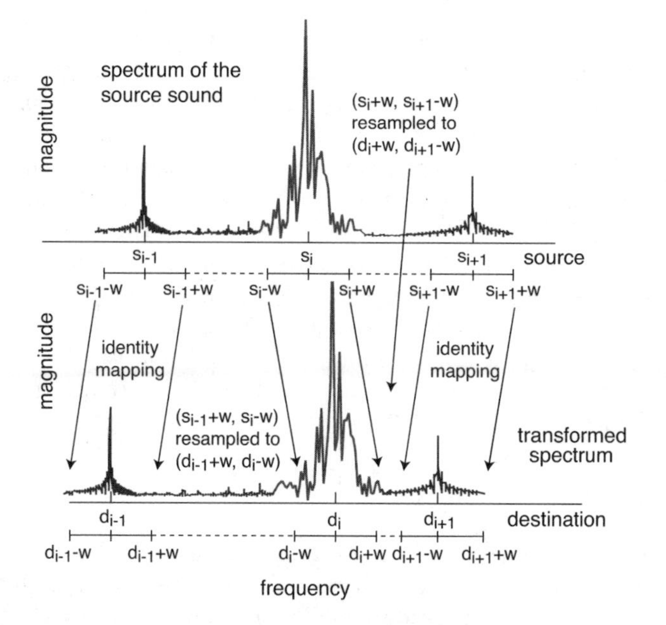

Fig. 9.17. Resampling with identity windows (RIW) preserves the information in a region about each peak by copying the data from the source spectrum to the destination spectrum. In between, the data is stretched or compressed via resampling. The phases (mapped similarly) are not shown.

The nearest neighbor method ensures that source partials are mapped to nearby destination partials, but allows two source partials to map to the same destination. (The parallel problem that some destination locations may not be assigned to a source is less important.) This conflict may be resolved by either discarding the smaller of the two, or by summing the partials. Both resolutions can themselves be problematic in certain situations since both irretrievably lose information. The simulations and sound examples of this chapter have taken a pragmatic approach: try the various alignment methods and use whichever sounds best in a given situation.

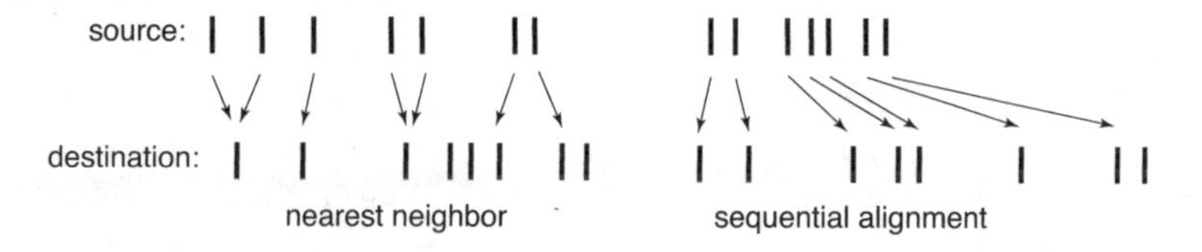

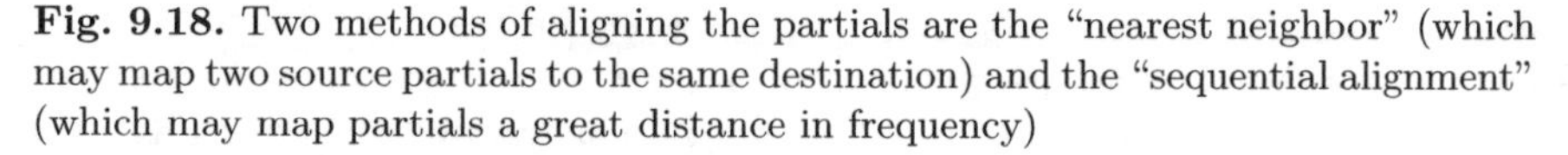

Fig. 9.18. Two methods of aligning the partials are the "nearest neighbor" (which may map two source partials to the same destination) and the "sequential alignment" (which may map partials a great distance in frequency)

Beat Detection: Since the beat-synchronous FFT partitions the audio into segments that coincide with the beat intervals, accurate beat tracking is essential. In contrast, the segmentation of the audio in the PV is specified without reference to the beat locations. For some spectral mappings (such as *Sixty-Five Maples* [S: 104] which uses a single fixed destination spectrum), the PV can operate without beat information. But if the destination locations change, (such as in the *Make Believe Rag* [S: 115]) it becomes crucial that the changes in the mapping align with the rhythmic structure of the piece. Figure 9.15 shows that the PV can exploit beat information by changing the mappings (indicated in the figure by the m_k, the mapping at the kth beat interval) whenever the center of a window crosses a beat boundary.

More important than the mechanics of how the methods operate is the question of how they influence the sound.

9.8.2 Perceptual Comparison

In any spectral mapping (other than the identity mapping) there is an inherent ambiguity: are the perceived changes due to the nature of the mapping (the given source and destination pair) or are they caused by the particular algorithms used to carry out the mapping procedure? The similarities and differences between the two different algorithms for spectral mapping help untangle this ambiguity.

The *Maple n-tet* sound examples (discussed on p. 237) conduct a series of spectral mappings for $n = 4$, 5, 10, and 100. These are computed using the PV strategy in [S: 109] and using the beat-synchronous FFT in [S: 110]. Aspects that appear similar between the corresponding examples are likely due to inherent properties of the mapping while differences are due to details of the implementations.

For example, in the $n = 4$ and $n = 5$ cases, the primary perception is that of a single inharmonic chord. Even as the input modulates from *A*♭ major to *E*♭ major, the output remains centered on the same 4-tet (or 5-tet) tonal cluster. Thus, this is likely to be an inherent feature of the mapping. On the other hand, there is a soft "underwater" phasiness to the FFT versions and a smearing and smoothing of the attacks in the PV versions. These are likely to be artifacts of the methods themselves. Indeed, both of these can be influenced and ameliorated by careful control of the parameters of the algorithms.

Both the PV and the FFT versions of the *Maple* 10*-tet* are similar, suggesting that the sound is not dominated by implementation artifacts. The spectral mapping of 12-tet performances into 10-tet destinations may be intrinsically subject to such out-of-tune (or more properly, out-of-timbre [B: 196]) effects. Finally, both $n = 100$ versions reliably reproduce the original chordal motion (for example, the arpeggiation of the *A*♭ and *E*♭ chords) and recreate recognizable piano timbres. But there are also significant differences: the PV tends to smear attacks, while the FFT tends to overemphasize (and sometimes misplace) transients.

The mapping of *Soul* [S: 7] into the 5-tet *Pentatonic Souls* is also conducted both ways in [S: 113]. Again, the PV is smoother while the beat-synchronous FFT retains greater rhythmic articulation. The vocal artifacts are conspicuous in both versions though they are very different from each other. The PV is prone to oscillate between adjacent destination partials as in Fig. 9.8 while the voice in the beat-synchronous FFT has more of a "chipmunk" effect. Similarly, when mapping *Soul* into the 65 Hz harmonic template of [S: 105], the PV tends to smear attacks while the beat-synchronous FFT tends to induce a kind of phasiness (especially in the vocals). Thus, while both the PV and the beat-synchronous FFT can carry out the spectral mappings, each has its own idiosyncrasies; each has its own strengths and weaknesses.

10

Musical Composition and Recomposition

The beats of a single piece may be rearranged and reorganized to create new structures and rhythmic patterns including the creation of beat-based "variations on a theme." Musical uses are discussed, and new forms of rhythmic transformation and modulation are introduced. Two pieces may be merged in a time-synchronous manner to create hybrid rhythmic textures that inherit tonal qualities from both. A series of sound examples demonstrate.

There are many ways to create music. Chapter 2 showed an interplay between various kinds of notation and the kinds of strategies that composers adopt. Many traditional methods rely on elaboration and repetition; others rely on conceptual strategies involving random actions or algorithmic processes. Schoenberg writes:

> Smaller forms may be expanded by means of external repetitions, sequences, extensions, liquidations and broadening of connectives... derivatives of the basic motive are formulated into new thematic units. [B: 191]

Compositional strategies typically involve sequences of elements that are combined, arranged, and organized according to some artistic, aesthetic, or logical principles. Wishart [B: 242] distinguishes the *field*, the sound elements used in the composition, from the *order*, the arrangement of the sound elements. Both the field and the order may operate at a variety of time scales.

For a traditional composer who works by placing individual notes on a score, the field consists of the note events and the order is given by their placement on the score. Individual notes have no inherent rhythm, and only achieve a timbre when they are realized in a musical performance. This parallels a MIDI-based compositional paradigm that orders MIDI note events into a sequence. The individual MIDI notes contain no inherent rhythm and have a timbre only when assigned to a sound module for output. At a higher level, the MIDI sequence itself can be considered a compositional element.[1]

[1] Commercial libraries of short MIDI files contain musical extracts, drum patterns, bass, guitar, piano, and percussion lines intended to be "cut and pasted into your own compositions" [W: 21]. They are available in a variety of musical styles (such as Latin, funk, jazz, Brazilian, and country) and are created by a variety of well known musicians.

The field consists of MIDI sequences that can be (re)arranged into a composition; these elements contain inherent rhythmic and melodic structures but are devoid of timbral information.

For a hip-hop composer using audio loop-based "construction kits," the field is the collection of sound loops and the order is defined by the way the loops are arranged and superimposed [B: 119]. Each individual loop has an intrinsic rhythm and an intrinsic texture; the art lies in creating a montage of elements that is greater than the individual elements themselves.[2]

Between the note-based time scale and the sequence or loop-based time scale lies the realm of the beat. Beat intervals are typically longer than individual notes but shorter than the four or eight-bar phrases typical of sequence or loop-based composition.[3] Like the loop, a beat interval (in an audio source) has a recognizable timbre and sound texture. Like a note, the beat interval does not have an intrinsic rhythm. Table 10.1 summarizes the properties of the elements of the sound field at various (approximate) time scales and levels of organization.

Table 10.1. Intrinsic properties of events at various time scales

Field	Time Scale	Intrinsic Rhythm?	Intrinsic Timbre?
note	$\frac{1}{8}$-4 s	N	N
beat	$\frac{1}{5}$-2 s	N	Y
sequence	2-20 s	Y	N
loop	2-20 s	Y	Y

This chapter explores the use of beats as basic compositional elements. This is composing at a higher level than when composing with notes or individual sounds since beat intervals taken directly from an audio source have an internal timbre, structure and consistency. On the other hand, composing with beat-elements is at a lower level than the kinds of building blocks commonly associated with construction kits, drum loop libraries, and MIDI sequences. These often contain two or four measures that express an internal rhythm and are essentially recordings of miniature performances. Composing in this middle realm allows the (re)use of timbres and expressions from musical pieces without being tied to the rhythms of the original.

[2] Commercial software for such "beat splicing" is available in Sony's `Acid` [W: 50], Native Instrument's `Traktor` [W: 38], and Propellerheads `ReCycle` [W: 41], the "ultimate tool for sampled grooves." These tools rely on extensive human intervention in the location of beat timepoints.

[3] While some notes may last as long as several beats, typical beat intervals involve many simultaneous notes.

10.1 *Friend of the Devil of the Friend*

Before proceeding, it is worthwhile to give a simple example of a beat-based "composition." Consider the beginning of the song *Friend of the Devil* [D: 19] by the Grateful Dead, which is beat tracked in [S: 71](11). An acoustic guitar plays a four note descending pattern in the first measure and is then joined by a second guitar. The bass enters at the end of measure five, and the three instruments continue until the ninth measure, when the voice enters. Thus there are eight measures (thirty-two beats) of introduction. Suppose that the introduction is played forwards and then backwards, where backwards means on a per beat basis. Symbolically, this is

$$1,\ 2,\ 3,\ \ldots,\ 30,\ 31,\ 32,\ 32,\ 31,\ 30,\ \ldots,\ 3,\ 2,\ 1$$

where the ith number represents the ith beat interval of the song. In this extended version, the three instruments continue together until the bass drops out at the start of the thirteenth measure. The two guitars continue through the fifteenth, and the single guitar plays a four note ascending pattern in the final measure.

Listen to *Friend of the Devil of the Friend* [S: 128]. Can you hear where the audio reverses direction? Yes. Does it sound greatly different backwards from forwards? No. Consider how different this is from *Gar Fael Elpam* and from the *Beat Reversed Rag* ([S: 86] and [S: 87]). In both of these, the audio was played backwards, causing a large change in the timbre of the piano. In *Friend of the Devil of the Friend*, however, the audio is played forwards; it is the composition that is played backwards. As in all such manipulations, a small window (128 or 256 samples) is used on both sides of the beat interval to help remove clicks and to cover up any (small) inaccuracies in the beat tracking.

10.2 "New" Pieces from Old

The *Friend of the Devil of the Friend* would not be possible without the ability to accurately track beat timepoints and the bulk of this chapter develops a variety of compositional strategies based on beat interval manipulations. These descend from tape splicing techniques used on tape recorders since the *musique concrète* of the 1950s, though rapid-cutting montage techniques were used in the 1920s by filmmakers such as Lev Kuleshov to manipulate the apparent flow of time. With the modern use of sampling and computer-based editing, the splicing operation became easy and practical, and has been credited with inspiring much of the sound and form of hip-hop [B: 119]. Breaking apart a piece at the beat boundaries opens another level of this kind of processing.

There are also important theoretical implications to beat-based processing. For example, the *Friend of the Devil of the Friend* is clearly derived from the original *Friend of the Devil.* But how far can the manipulations be taken before the piece is transformed into a "new" piece? This brings up questions about the fundamental identity of a musical work, what it means to "be" a piece of music. This is discussed further in Sect. 12.4.

Another implication can be most easily stated in terms of Kramer's [B: 117] dichotomy between linear and nonlinear musics. "Linearity" in this context is "the determination of some aspects of music in accordance with implications that arise from earlier events in the piece." Thus familiar chord progressions and tonal systems are examples of structures that allow the creation of linear music. "Nonlinear" music is memoryless, without directed temporal implications: timeless, atmospheric, stationary, repeating with variation but without progression or climax. Kramer cites gamelan music with its rhythmic cycles, and modern "trance" and "techno" styles as examples of primarily nonlinear musics. Moments of time in a nonlinear piece are intended to be moments of pure sound, of harmonic stasis, and are not intended to be part of a linear, goal-oriented progression.

Of course, no music is completely linear or completely nonlinear, but rather different genres, styles (and musical cultures), lie somewhere between. The kinds of beat-based rearrangements of pieces conducted in this chapter can be used to conduct a concrete test of the linearity of a piece: if the beat intervals of a piece can be rearranged without changing its essence, then it has strong nonlinear component. If the meaning of the piece is destroyed by rearrangement, then the piece has a strong linear component. The *Friend of the Devil of the Friend* shows that this Grateful Dead song has a significant nonlinear component. On the other hand, similar rearrangements of the verse and chorus are more disjoint, as *Devil of a Friend* [S: 129] shows. Thus the linearity of the song is also strong.

10.3 The *Maple Leaf Waltz*, *Julie's March*, and *Take Four*

Perhaps the simplest beat-based manipulation is to delete beat intervals. For example, if a piece is in a $\frac{4}{8}$ time signature and one beat in every four is removed, it is transformed into $\frac{3}{8}$. Thus the *Maple Leaf Rag*[4] is transformed into the *Maple Leaf Waltz.* Ragtime, which literally means "time in tatters," has been shredded even further. The procedure is shown schematically in Fig. 10.1, and the results can be heard in the *Maple Waltzes #1* and #2, sound examples [S: 130] and in the more fully developed *Maple Leaf Waltz* [S: 131].

Since there are four (eighth note) beats per measure, there are four possibilities, only two of which are shown in Fig. 10.1. The top diagram omits the

[4] As shown in Fig. 2.3 on p. 27, the notated time signature is $\frac{2}{4}$, with the quarter note receiving one beat. The beat tracking of [S: 6] locates the beat at the eighth note, which corresponds to an identified time signature of $\frac{4}{8}$.

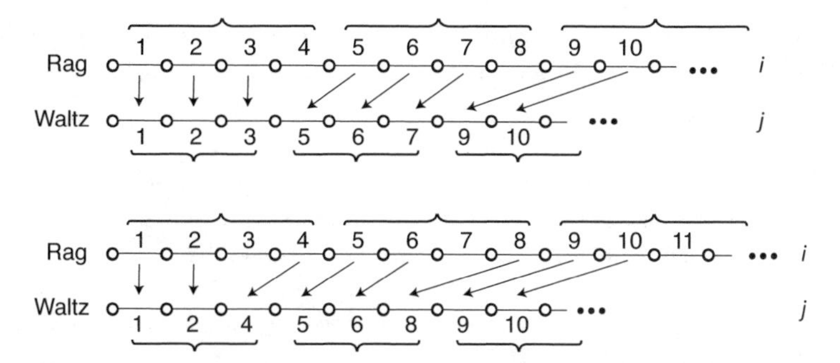

Fig. 10.1. From the *Maple Leaf Rag* to the *Maple Leaf Waltz*: the black circles represent the beat boundaries and the numbers represent the sound in the corresponding beat interval. In both figures, every fourth beat interval is deleted. In the top, the fourth beat of each measure is removed while in the bottom, the third beat of each measure is removed. Listen to these examples in [S: 130] and [S: 131].

fourth beat interval of each measure. This can be heard in the *Maple Waltz #1*. The bottom diagram deletes the third beat in each measure, resulting in the *Maple Waltz #2*. The other two possibilities do not sound as smooth. These two are combined and elaborated into the *Maple Leaf Waltz* in [S: 131].

More generally, it is possible to remove one out of every nth beat. If the first deleted beat is $n-k$ (with $0 \le k < n$) then the jth beat in the output can be written directly in terms of the ith beat of the input as

$$j = i + \lfloor \frac{i+k-1}{n-1} \rfloor$$

where the floor function $\lfloor \cdot \rfloor$ rounds down to the nearest integer. The $n=4$, $k=0$ case is shown in the top of Fig. 10.1 while the $n=4$, $k=1$ case is shown in the bottom.

This provides a simple recipe for manipulating the meter of a piece, and several other examples are presented. The *Soul Waltzes* [S: 132] reshape the $\frac{4}{4}$ hard rock rhythm of [S: 7] into a hard rock $\frac{3}{4}$. The truncation of the lyrics can be a bit disconcerting, but the rhythmic motion is as clear in $\frac{3}{4}$ as in $\frac{4}{4}$. Observe that each of the four possible versions has a "different" rhythm. Similarly, *Bond's Waltz* [S: 133] drops one beat in each four. Somewhat more absurdly, *Take Four* [S: 134] removes one out of every five beats from Grover Washington's [D: 45] version of Dave Brubeck's classic *Take Five*. Both the melody and rhythm work well in four, though perhaps something is lost from an aesthetic perspective.

There are many ways to reorganize the metrical structure. For example, to change from a triple to duple meter it is possible to duplicate one beat in each measure (for instance, to perform beats 1 2 3 3 or 1 2 2 3) or to omit one beat from each measure (hence to change from $\frac{3}{4}$ to $\frac{2}{4}$). Deletion tends to sound smoother because duplicated beats make it appear that the recording is skipping. *Julie's March* [S: 135] removes one out of every three beats from *Julie's Waltz* [S: 8], leaving a clear $\frac{2}{4}$ feel. Similarly, *Howell in* $\frac{2}{4}$ [S: 136] appears to progress more rapidly in $\frac{2}{4}$. No time compression (other than the removal of every third beat) is done.

Similarly, it is possible to remove even more than one beat per measure. The *Half Leaf Rags* [S: 137] delete two out of every four beats from the *Maple Leaf Rag*. While there are occasional glitches (especially at section boundaries), the overall flow is smooth and the essence of the piece is preserved. These are further developed as the *Magic Leaf Rag* [S: 141] using the techniques of Sects. 9.1 and 9.2. *Half a Soul* [S: 138] removes two out of every four beats from *Soul* and the rhythmic drive is clear although the lyrics are shredded. Even more extreme is to remove all but one beat from each measure. *Quarter Soul* [S: 140], which "fits" the complete two and a half minute song into forty seconds, is somewhat more successful than the *Quarter Leaf Rag* [S: 139]. When the *Quarter Leaf Rag* is processed using the beat-gating and beat-filtering strategies of Sect. 9.2, the result is the inimitable *Make It Brief Rag* [S: 142].

10.4 Beat-based Time Reversal II

The transformation of a piece in $\frac{5}{4}$ into $\frac{4}{4}$ (or $\frac{4}{4}$ into $\frac{3}{4}$) preserves the basic harmonic motion of the piece. Continuity is interrupted at only a single level, the measure. Phrases still point in the same direction and tonal structures are preserved. Thus, while these may be interesting metrical manipulations of the compositions, they do not address the question of the linearity (or nonlinearity) of the music.

Kramer [B: 117] writes: "When a composition blurs the distinction between past, present, and future, forwards and backwards become in some sense the same." By making time move backwards, the normal tonal flow is disrupted. For example, the finalizing cadence *C Am Dm F G C* would become *C G F Dm Am C*, which is no longer a finalizing cadence. This toys with tonal progressions, turning them backwards. If a piece were truly nonlinear then this reversal should leave the overall effect of the piece unchanged. Highly linear pieces should be destroyed beyond all recognition.

How linear is the *Maple Leaf Rag*? Using beat-based processing, it is easy to reverse the temporal flow without unduly changing the timbre. Instead of performing the beat intervals in numerical order, they are performed in reverse numerical order in the *Backwards Leaf Rag* [S: 143]. Moment by moment, the *Backwards Leaf Rag* is plausible. The rhythm bounces along much like normal despite occasional glitches. In the large, however, the *Backwards Leaf Rag* wanders aimlessly. There are moments that are clearly recognizable (such as the prominent rhythmic motifs) and there are moments where the reversal is surprisingly fresh, but overall the tonal progression is upset.

Backwards Soul [S: 144] provides another example. The driving rhythm is as compelling backwards as forwards. Since there is little harmonic motion in the original, the time reversal does not greatly perturb the chord patterns. Perhaps the most interesting aspect of the *Backwards Soul* is Beil's voice. Many individual words fit within a single beat interval, and these are preserved. Phrases are invariably sliced into incomprehensible bits. The scat singing and the screaming are rearranged, but are not particularly disturbed by the reversal. Similarly, *Devil of a Friend* [S: 129] and a beat reversal of the *Theme from James Bond* [S: 145] play havoc with the semantic flow. Yet the rhythmic motion persists.

10.5 Beat Randomization

Nonlinear composition occurs when a piece exists in "moment time," when the order of succession appears arbitrary. Rearranging the order of beats provides a concrete way to test the nonlinearity of a composition: if a rearranged version appears qualitatively similar to the original, the order of succession is unimportant. If the rearrangement fundamentally disturbs the piece, the order of succession is crucial to the meaning of the piece. This section discusses several ways that such rearrangements may be conducted.

Perhaps the most extreme way to reorder the beats of a piece is to choose them randomly. To create the *Random Leaf Rag #1* [S: 146], all the beat intervals of the *Maple Leaf Rag* were numbered from 1 to N. Each beat interval of the *Random Leaf Rag #1* was chosen randomly (uniformly) from the list. All sense of tonal progression is gone. The rhythm is intermittent and the flow is broken. Only the timbre of the piano remains. Similarly, *Random Soul #1* [S: 147] haphazardly rearranges the beats of *Soul*. Again, the rhythm of the piece is fundamentally disturbed, though some of the aggressive feel remains. The timbre of the voice and guitars is preserved.

Randomization is perhaps best applied judiciously and with constraints. Cognitive studies suggest that items that occur at the start and the end of a grouping are particularly important. Indeed, both the start and the end are quite clearly important since they lie on boundaries. Consider the following text:[5]

> Aoccdrnig to a rscheearch at Cmabrigde Uinervtisy, it deosn't mttaer in waht oredr the ltteers in a wrod are, the olny iprmoetnt tihng is taht the frist and lsat ltteer be at the rghit pclae. The rset can be a total mses and you can sitll raed it wouthit probelm.

It is not much more difficult to read than if the words spelled correctly. A musical analog of this might equate words with a measure or short phrase

[5] I have been unable to document the original source of this text; it may not be from "rscheearch at Cmabrigde Uinervtisy."

and letters with the beat. The idea would be that an arbitrary scrambling might not make much sense but that if the most salient of the locations were maintained, then the others could be safely randomized. Figure 10.2 shows five of the 24 possible rearrangements of the four beats in a measure, and these are applied to the *Maple Leaf Rag* in the *Permutations of the Rag* [S: 148].

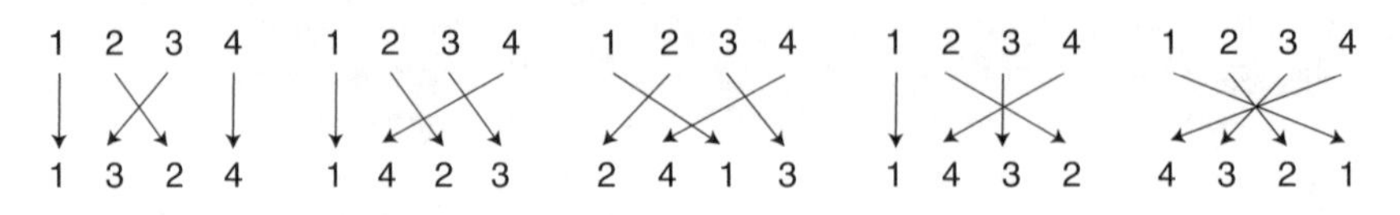

Fig. 10.2. Several different local permutations of the beat intervals can be heard in *Permutations of the Rag* [S: 148]

Somewhat more generally, the *Permutation Leaf Rag* [S: 149] applies a different randomly chosen permutation to each measure. Though these versions have a somewhat wandering character, they are considerably more comprehensible than the *Random Leaf Rag #1*. *Permutations of Soul* [S: 150] applies the same technique to *Soul.*

Another way to constrain the randomization is to choose beats randomly from among those that occupy the same relative location in the measure. Formally, let the beat intervals of a piece be labeled from 1 to N. Create n subsets indexed by $j = 1, 2, \ldots, n$

$$B_j = \{j + ni\}, \text{ for } i = 0, 1, \ldots, \lfloor \frac{N}{n} \rfloor.$$

For example, if n is the number of beats in each measure, then B_1 contains all of the beat intervals that start each measure and B_n contains all the beat intervals that end each measure. A piece may be recomposed from the B_j by choosing the first beat randomly from B_1, the second beat randomly from B_2, and the kth beat randomly from $B_{mod(k-1,n)+1}$. For example, $n = 4$ for both the *Maple Leaf Rag* and *Soul,* and there are four such subsets. The resulting pieces are *Random Leaf Rag #2* and *Random Souls #2* ([S: 151] and [S: 152]). These are slightly more coherent than with complete randomization of the beats. There are places in *Random Souls #2* where the rhythm is quite solid, though there are also odd glitches and unexpected halts.

Such random rearrangements of beats and measures sometimes makes musical sense and sometimes does not. When the piece has little goal-oriented motion, randomization is unlikely to cause substantial changes. Kramer [B: 117] says: "Listening to a vertical music composition can be like looking at a piece of sculpture... we determine for ourselves the pacing of our experience: we are free to walk around the piece, view it from many angles, concentrate on some details... leave the room when we wish, and return for further viewings." Randomization does not disturb such a leisurely viewing experience.

On the other hand, if the piece is primarily linear (or horizontal), randomization of the beat structure destroys the sense of directedness and strips the sound of its tonal implications. This may freeze the piece into a nonlinear sound collage where harmony becomes a static element akin to timbre. Without tonality as a driving force, the sense of time moving may be suspended.

Randomization can also be carried out at levels above and below that of the beat. Randomizing larger segments is a time honored tradition in algorithmic musical composition and stems back at least to Mozart's "dice-game" in which the performer rolls dice to determine the order of the measures to be played [W: 36]. In the present setting, it is straightforward to paste the beats into small clusters (i.e., into measures) and to then randomize the clusters.

Using snippets of sound much smaller than the beat tends to destroy the integrity of the sound. For example, [S: 153] subdivides each beat of the *Maple Leaf Rag* into N segments that can be rearranged at will. The subdivision into $N = 2$ pieces has a bouncy rhythmic feel,[6] but as N increases, the fluttering sound of the fragmentation dominates. Since such granularization (recall Sect. 2.2.3) tends to destroy the integrity of the beat (to change the timbre in a fundamental way), this is typically thought of as a technique of sound synthesis (or resynthesis) rather than a method of composition. Nonetheless, Roads [B: 181] demonstrates that careful choice of materials and algorithms can blur the distinction between sound synthesis and musical composition.

10.6 Beat-synchronous Sound Collages

Several musical performances can be simultaneously rearranged, reorganized, combined, and recombined. Such sound collages juxtapose contrasting elements across time whereas collages in the visual arts juxtapose elements across space. The *Maple Leaf Rag* is ideal for these experiments because it has been performed in a large variety of styles over the years as indicated in Table A.2 on p. 290.

In preparation, the 27 versions of the *Maple Leaf Rag* were beat regularized[7] so that the beat interval in each performance was a constant 0.33 s. When necessary, the performances were transposed so that all were in the same key. The beat intervals were numbered and aligned so that the first beat in each version occurred at the pick-up note indicated in the score (this was necessary because some of the versions have introductions, which vary in length).

Picking and choosing the beats from among the various renditions creates a sound collage. The simplest procedure selects beats in their normal sequence from versions chosen at random. For example,

---

[6] For the *Maple Leaf Rag*, $N = 2$ represents the tatum level, which is the smallest subdivision that does not display the characteristic fluttery boundaries of the rearrangement.

[7] See Sect. 11.2 for details.

Beat 1 might be the first beat of Blumberg's piano [S: 76](1)
Beat 2 could be the second beat of Dorsey's big band [S: 76](14)
Beat 3 may be the third beat of Glennie's marimba [S: 76](20)
Beat 4 might be the fourth beat of Van Niel's guitar [S: 76](22)

And this is only the first measure! The resulting *Maple Leaf Collage* [S: 154] sounds disjoint and discontinuous. Nonetheless, the motion through the rag is clear. The timbral variations are extreme, but also somewhat fascinating. The fragmentation in the *Maple Leaf Collage* is caused by several factors: the lack of continuity of any single timbre, the sudden jumps in volume from beat to beat, and the unevenness of the left/right balance. A smoother variation addresses these issues by:

(i) normalizing the power in each beat
(ii) using four simultaneous versions of each beat (rather than one)
(iii) ensuring that there is always at least one piano version
(iv) balancing the four versions in the left/right mix
(v) encouraging versions to persist for more than one beat[8]

The results appear in *Rag Bag #1* and *Rag Bag #2* [S: 155] which presents two runs of the collaging algorithm.[9] Not only is the rhythmic and harmonic motion of the *Maple Leaf Rag* clear in these examples, but the collage creates a new kind of sound texture. While the two *Rag Bags* contain no sections that are actually the same (except perhaps by accident of the random numbers) they achieve an overall unity defined by the rapid changes in timbre.

A typical visual collage juxtaposes collections of photographs, fabric, and papers by arranging them side by side in space to create a larger work. The *Rag Bags* superimpose the beats of the various versions of the *Maple Leaf Rag* in time and juxtapose them across time to create a beat-synchronous sound collage.

10.7 Beat-synchronous Cross-performance

The superposition in the previous section was the simplest possible kind: summation. There are many other interesting ways to combine simultaneous sounds. Merging two (or more) individual sounds is often called cross-synthesis; merging two (or more) different musical performances is a kind of cross-performance. This section discusses combining pairs of compositions in a beat-synchronous manner. As in the collages of Sect. 10.6, the segments of the performances are aligned with the beats and the temporal motion is linear throughout the piece. Thus the underlying composition is (more-or-less)

[8] This is implemented in the following way: if version i is chosen to sound in beat j, then i will also be chosen to sound in beat $j+1$ with probability $p = 0.7$.

[9] The algorithm was run twelve times; these two were my favorites.

maintained while the timbres of the two performances interact to generate a "new" performance.

To be concrete, let t_1 and t_2 be the ith beat interval of the two performances and let M_1, θ_1 and M_2, θ_2 be the corresponding magnitude and phase spectra. The merged performance t and the corresponding spectra M and θ can be calculated in a variety of ways:

(i) summation: $t = t_1 + t_2$
(ii) convolution: $M = M_1 M_2$, $\theta = \theta_1 + \theta_2$
(iii) cross-modulation: $M = M_1$, $\theta = \theta_2$
(iv) square root convolution: $M = \sqrt{M_1 M_2}$, $\theta = \frac{1}{2}(\theta_1 + \theta_2)$
(v) cross-product #1: $M = M_1 M_2$, $\theta = \theta_1$
(vi) cross-product #2: $M = M_1 M_2$, $\theta = \theta_2$
(vii) cross-product #3: $M = M_1 M_2$, $\theta = \theta_1 + \theta_2$

Table 10.2 shows 22 cross-performances of the *Maple Leaf Rag* from [S: 156]. Each begins with two renditions (from Table A.2) and combines them in a beat-synchronous manner using one of the methods (i)–(vii).

Table 10.2. Examples of beat-synchronous cross-performance in [S: 156]. Performances refer to Table A.2 on p. 290 and methods refer to the ways (i)–(vii) of combining the merged spectra.

	Performances	Method	Sound File	Comment
1	13, 23	(i)	`Bechet-GGate`	demonstrates synchronization
2	13, 20	(i)	`Bechet-Glennie`	
3	1, 18	(i)	`Blumberg-Bygon`	
4	1, 20	(i)	`Blumberg-Glennie`	
5	1, 23	(i)	`Blumberg-GGate`	
6	8, 1	(ii)	`Joplin-Blumberg`	method simplifies the sound
7	1, 20	(ii)	`Blumberg-Glennie`	marimba becomes a chime
8	8, 1	(iii)	`Joplin-Blumberg`	cross two pianos
9	1, 8	(iii)	`Blumberg-Joplin`	cross two pianos
10	18, 13	(iii)	`Bechet-Bygon`	instruments with vocal color
11	8, 14	(iii)	`Dorsey-Joplin`	horns take rhythm from piano
12	23, 18	(iii)	`GoldenGate-Bygon`	banjo strum applied to voices
13	26, 13	(iii)	`Kukuru-Bechet`	strumming horns
14	1, 26	(iii)	`Blumberg-Kukuru`	piano-banjo hybrid
15	18, 27	(iii)	`Bygon-MamaSue`	bluegrass with a wah-wah pedal
16	1, 22	(iii)	`Blumberg-VanNiel`	piano-guitar mixture
17	8, 1	(iv)	`Joplin-Blumberg`	beat boundaries perceptible
18	26	(iv)	`Kukuru-Kukuru`	rhythmic simplification
19	13, 1	(v)	`Bechet-Blumberg`	
20	13, 26	(vi)	`Bechet-Kukuru`	
21	1, 20	(vii)	`Blumberg-Glennie`	
22	18, 1	(vii)	`Bygon-Blumberg`	

The first method, intended primarily for comparison, demonstrates the amazing synchrony that is possible between independent performers. The mixture of Bechet and Golden Gate (1) works because the banjos comp a relaxed rhythm while the horns provide a frenetic lead. The mixture of Bechet and Glennie (2) is surprising because the marimba plays in an apparently free manner, that just happens to coincide with the horns. In many places in (1)–(5), it is hard to believe that the performers are not listening carefully to each other.

Method (ii) multiplies the magnitude spectra and adds the phase spectra. This is a linear filter that convolves the beat of one song with the corresponding beat of the second. Frequencies that are common to both performances are emphasized; frequencies that are different are attenuated. This tends to "simplify" the sound, to make it less complex. For example, cross-performances (6) and (7) are typical; the pianos of Joplin and Blumberg combine to form a calliope, the marimba merges with the piano to form a chime. Both follow the basic harmonic progression of the rag, though with a somewhat simplified rhythm.

The cross-modulation method (iii) combines the magnitude of one sound with the phase of another. Thus cross-performance (8) differs from (9) in which piano donates the magnitude and which donates the phase. Because both are piano performances of the same piece, they retain their piano-like quality, although (8) has gained some percussiveness and (9) has gained some phasing artifacts. Crosses between different timbres are more interesting. For example, (14) and (16) merge instrumental timbres to create piano-banjo and piano-guitar hybrids. In (11), Dorsey's horns assume some of the rhythmic comping of Joplin's piano while the horns in (13) bounce to the strumming of the banjo. Bechet's horn takes on a vocal character in (10) and Mama Sue's banjo band (15) acquires a "wah" sound.[10]

Method (iv) takes the square root of the product of the magnitudes. This nonlinear filtering tends to equalize the power in all common frequencies and to attenuate frequencies that are unique to one performance. The sound becomes richer, though noisier. As in cross-performances (17) and (18),[11] the rhythm is simplified to one strong stroke per beat, though again the harmonic motion remains clear.

The cross-product methods (v)–(vii) do not sound radically different: all tend to emphasize the beat boundaries. All tend to have a calliope-like timbre and a straight rhythmic feel, though the mixture with Glennie in (21) has a chime-like quality and the vocal accents of (22) are engaging.

At the risk of crossing into absurdity, it is also possible to apply the sound collaging technique of Sect. 10.6 to these cross-performances. Three collages of cross-performances appear in the *Grab Bag Rags* of [S: 157].

---

[10] Recall the discussion of the wah-wah pedal surrounding Fig. 9.4 on p. 227.

[11] This piece is not really a *cross*-performance except in the sense that it crosses the same piece with itself.

11

Musical Analysis via Feature Scores

Traditional musical analysis often focuses on the use of note-based musical scores. Since scores only exist for a small subset of the world's music, it is helpful to be able to analyze performances directly, to probe both the literal and the symbolic levels. This chapter uses the beat tracking of Chap. 8 to provide a skeletal tempo score *that captures some of the salient aspects of the rhythm. By conducting analyses in a beat-synchronous manner, it is possible to track changes in a number of psychoacoustically significant musical variables. This allows the automatic extraction of new kinds of symbolic* feature scores *directly from the performances.*

While there is a long history of speculation about the meaning and importance of rhythms, the first scientific investigations of the perception of rhythm began in the last few years of the 19th century. Much of our current knowledge builds on the foundations laid by researchers such as Bolton, Woodrow, Meumann and their colleagues. For example, Bolton [B: 17] documented the subjective rhythm that arises when listening to a perfectly uniform sequence (recall [S: 33] and the discussion in Sect. 4.3.1). Meumann [B: 147] discussed the role of accents in rhythm and Woodrow [B: 245] considered the interactions between pitch and rhythmic perception. Investigators such as Hall and Jastrow [B: 86] began to try to understand the limits of human abilities: how well do people estimate short durations and what is the just noticeable difference for duration? Observing that people sometimes overestimate and sometimes underestimate durations, Woodrow [B: 244] investigated the "indifference interval," that is neither over- nor under-estimated.

Over the years, there are clear shifts in the methods of investigation. As the questions became more difficult, the experiments became increasingly rigorous. While early authors might be content to write based on reflection and to run tests based on common musical pieces, later authors needed to conduct controlled experiments. In order to achieve unambiguous results that are easily reproducible at other laboratories, researchers needed to simplify the experimental methods: "probe tones" and "sine waves" replaced instrumental sounds, simplified patterns of "pure duration" replaced rhythmic performances. While early papers such as [B: 215] have an appendix listing the musical pieces used in the experiments, later papers have tables with lists of frequencies and durations. Thus, as the results become more concrete and less arguable, they also become narrower and less applicable. This tension haunts us today.

The beat-finding techniques of *Rhythm and Transforms* provide a tool for the investigation of rhythmic and temporal features of musical performances directly from a sound recording. This allows a return to the analysis of music in place of the analysis of artificially generated patterns of duration. The methods are necessarily an analysis of musical performance, and only incidentally of the underlying musical work. It provides a tool for those interested in comparing rhythmic features of different performances of the same piece.

The central idea of this chapter is to use the output of the beat tracking algorithms of Chap. 8 (the beat timepoints and the beat intervals) as a starting point for further inquiry. In one approach, the beat timepoints are used directly to examine the temporal structure of a piece: how the tempo changes over time and how long sections endure. This provides an ideal tool for the comparison of different performances of the same piece. In another approach, the beat intervals are used to define a psychoacoustically sensible segmentation over which various features of the piece can be measured. For example, it is possible to quantify how the "brightness" of a performance changes over time by comparing the number of high harmonics to the number of low harmonics within each beat interval, to quantify the "noisiness" of a performance by measuring the noise-to-signal ratio within beat intervals, and to calculate the "sensory roughness" in a beat-synchronous manner.

11.1 Accuracy of Performance

A recurrent theme in rhythm analysis is to try to determine how accurately performers play. No one can perform a sequence with exact regularity, play from a musical score with clockwork precision, or repeat any performance exactly. The deviations shed light on the motor skills of the performer, on the expectations of the listener, and on the idea of expressive deviations.

Sears' [B: 192] preliminary experiments looked at the regularity of a mechanical music box. By attaching a mechanical counter to a rotating fan within the box, it was possible to calculate the time interval between adjacent notes and measures very accurately. The average length of the measures in one piece was 578 ms (with a mean variation (MV) of 16.8 ms), quarter notes averaged 190 ms with MV 10.7 ms, eighth notes averaged 105 ms with MV 16.7 ms. Sears observes some trends in the deviations: eighth notes tend to be longer than half the length of quarter notes. These provide an indicator of the kinds of errors (deviations from the score) that would normally go unobserved.

Sears then constructed a special keyboard that tracked the depressions of the keys of a keyboard as performers played. The output was a trace on a scrolling piece of paper that could be read to an accuracy of a few hundredths of a second. Of course, electronic musical instruments now provide more accurate methods of measuring the timing accuracies of performers, but Sears basic conclusion still holds: that the relative lengths of the tones were variable and "do not follow exactly the ratios expressed by the written notes. They were sometimes too great and sometimes too small." In studying the performance practices of violin players, Small [B: 215] came to essentially the same conclusion: "...the violinists use of pitch and intensity involves deviations from the score, interpolations of unindicated factors, and a rather constant but orderly flux around certain variable levels." More recently, Noorden and Moelants investigate the most natural rates at which people tap [B: 156] while Snyder and Krumhansl explore the cues needed to invoke the sensation of a pulse [B: 220].

How steady is a steady beat? It should be no surprise that the length of time occupied by a beat is not completely fixed throughout a performance (except perhaps in some modern dance styles where the rhythm is dictated by a drum machine). For example, Fig. 11.1 shows a plot of the beat interval (the difference between successive beat timepoints) vs. the beat number for the 27 performances of the *Maple Leaf Rag* in [S: 76] (as described in Table A.2 on p. 290). Such *tempo scores* show how the tempo evolves over time. Each trace

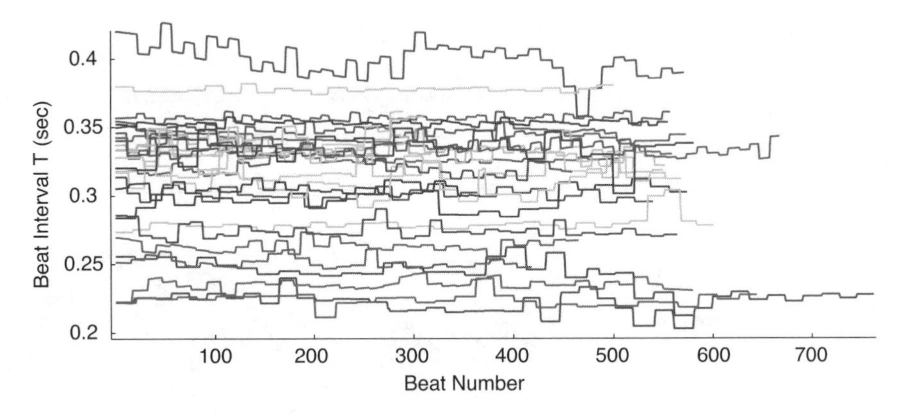

Fig. 11.1. A *tempo score* is a plot of the duration of each beat vs. the beat number; it shows how the tempo changes over time. In this plot, 27 performances of the *Maple Leaf Rag* are played in a variety of tempos ranging from $\mathcal{T} = 0.22$ to $\mathcal{T} = 0.4$ s per beat. The plot shows how the tempo of each performance varies over time.

represents a single performance. The rapid big band renditions by Sidney Bechet [S: 76](13) and Tommy Dorsey [S: 76](14) at the bottom are longer than the others because they repeat sections and include solos by individual band members. The two slowest renditions (at the top) are played on guitar by Van Ronk [S: 76](21) and Van Niel [S: 76](22). Using the data in the plot, it is easy to gather statistics on the performances and the mean and standard deviation of the tempo are tabulated in Table A.2. Despite the similarity in tempo, the two guitar performances are among the most varied in terms of expressiveness: the Van Niel[1] performance has the lowest standard deviation while Van Ronk's fluid solo guitar is among the most variable. The piano versions (the instrument for which the rag was composed) tend to cluster with a tempo around $\mathcal{T} \approx 0.33$ s, though there are large differences in the steadiness of the tempos among the different renditions.

Accuracy of beat tracking

The discussion surrounding Fig. 11.1 should give an idea of some of the uses of tempo scores, but before it is possible to use the technique to seriously investigate musical performances, it is necessary to "calibrate" the method, to determine the accuracy to which the beat locations can be found. There are two parts to this: how repeatable is the output from one simulation to another,[2] and how closely does the output of the algorithm correspond to the beat points identified by a typical listener.

To investigate the repeatability, *Julie's Waltz* [S: 8] was beat-tracked eight times using eight different frame sizes between 4 and 13 s. The output of the algorithm is plotted as a tempo score in Fig. 11.2. The beat interval begins at

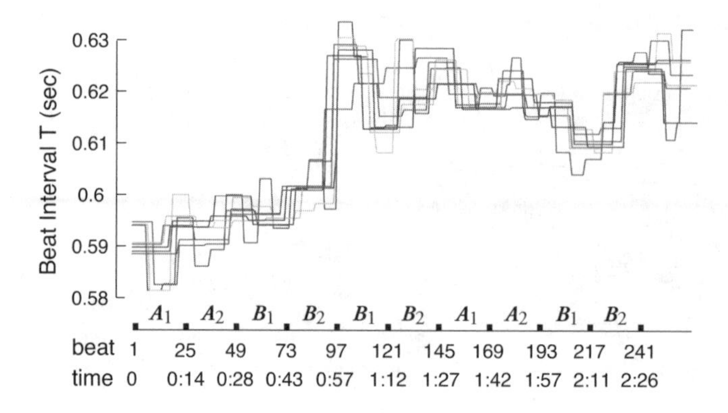

Fig. 11.2. Eight runs of the statistical method of beat tracking (from Chap. 8) differ by the frame size, which is varied between 4 and 13 s. The estimates follow similar trajectories throughout the analysis of *Julie's Waltz* [S: 73].

$\mathcal{T} \approx 0.59$ s and increases slowly to $\mathcal{T} \approx 0.60$. At beat 97, there is a jump to $\mathcal{T} \approx 0.62$ and then a long slow decline to $\mathcal{T} \approx 0.61$. The last few beats slow again. (The musical significance of these tempo variations is explored further in Sect. 11.4.) These $\mathcal{T}$ values represent the mean of the eight tempo estimates

[1] Since the recording consists of two acoustic guitars both played by Van Niel, it may have been recorded using a metronome or click track. This would be consistent with the remarkable steadiness.

[2] Recall that the algorithm relies on random "particles" and hence does not return exactly the same results each time it is run even when all parameters are fixed.

as the piece progresses. The standard deviation about this mean (averaged over all beats) is 0.017, which quantifies the repeatability. Each of the tracks is piecewise linear because the state is only estimated once per frame. Thus the short plateaus show the widths of the frames which vary between the different runs. The narrowest and the widest frames have the highest variance, suggesting that intermediate values might generally be preferred.

To compare the output of the algorithm to listeners perceptions, eight amateur musicians agreed to participate in a beat tracking exercise. The subjects (adults of both sexes with an average of about 12 years practice on a musical instrument) listened to *Julie's Waltz* through headphones and "tapped along" by hitting keys on a laptop computer. The program `TapTimes` [B: 209] instructed participants to "Strike any key(s) on the keyboard to indicate the foot-tapping beat (i.e., tap on the beat as you hear it)." When the subject reached the end of the piece, they were given the option to save the tapping or to repeat. Most repeated the procedure two or three times before saving.

Figure 11.3 compares the means and standard deviations (the error bars) at each beat for the manual tapping experiment (on the left) and for the eight runs of the algorithm (on the right). Since the vertical scale is the same, it is clear that the algorithm centers around the same mean(s) but has much smaller deviation(s) than the subjects. The eight listeners did not all tap at

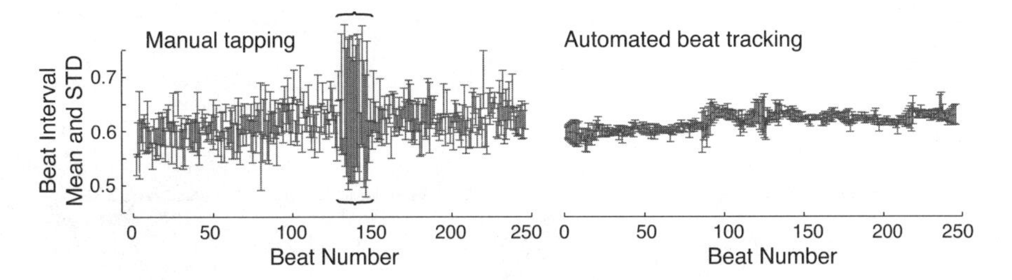

Fig. 11.3. *Julie's Waltz* [S: 73] is used to compare experimental data (listener's tapping on the left) to the output of the algorithm (on the right). The mean tempos are comparable, but the algorithm has significantly smaller standard deviation, as shown by the error bars.

the same metrical level. Six tapped at the beat level (i.e., near $\mathcal{T} = 0.6$) which can be compared directly to the output of the algorithm. One chose to tap at the measure (near $\mathcal{T} = 1.8$ s). This subject's taps were amalgamated with the others by aligning the beats in time (thus this subject had only $\frac{1}{3}$ as many data points as the others). The final subject chose to tap primarily at the beat level but also tapped intermittently at the tatum level between beats 121 and 145. This is the source of the large variance indicated between the two brackets. With this data included, the average standard deviation for the experiment was 0.042. With this data removed, the average deviation was 0.036. Both are much greater than the analogous term (0.017) for the automated beat tracking.

Of course, this is only a preliminary experiment designed to give some confidence in the use of the output of the beat tracking algorithm. Additional investigation is needed to untangle more subtle effects such as the difference between the accuracy of the perception of a passage and the motor ability of listeners to synchronize their tapping with that perception.

11.2 Beat Regularization

When the beat is fixed throughout a piece, the Periodicity Transform (PT) can often reveal important relationships between various levels of the metrical structure. For example, *Jit Jive*, *Take Five*, and the *Baris War Dance* were all analyzed successfully in Sect. 8.3.1 (recall Figs. 8.21–8.23 on pp. 213–215). But when the underlying tempo changes, the periodicity transform may fail to provide a meaningful analysis. For example, the PT of *Julie's Waltz* [D: 40] is shown in the left hand plot of Fig. 11.4; it shows no recognizable features. This should not be unexpected because the tempo of the waltz changes slowly over time as revealed in the tempo score Fig. 11.2.

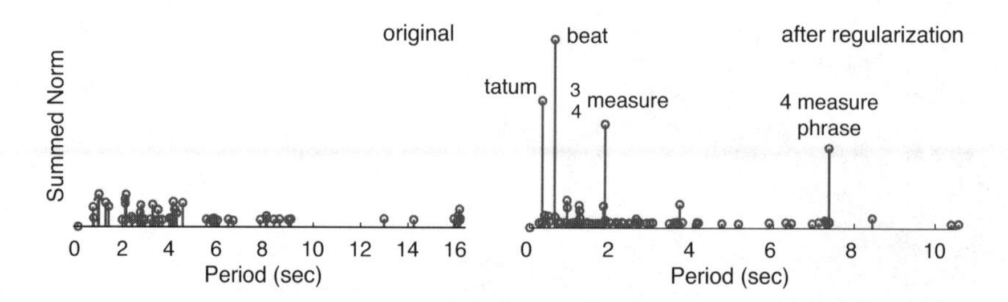

Fig. 11.4. Applying the Periodicity Transform to *Julie's Waltz.* Without regularization of the beat, the transform is unable to provide a meaningful analysis. After regularization, the PT locates several levels of the metrical hierarchy: the eighth-note tatum, the quarter-note beat, the $\frac{3}{4}$ measure, and the 4-bar phrase.

Using the output of the beat tracking algorithm, an unsteady tempo can be regularized by changing the duration of each (varying) beat interval to a desired steady value.[3] Thus, when the tempo is slow the beat intervals can be shortened. When the tempo is fast, the beat intervals can be lengthened, creating a performance with a completely regular pulse. (The procedure is diagrammed in Fig. 9.2 on p. 225 and applied to the *Maple Leaf Rag*

[3] Alternatively, the beat tracking information can be used to resample (and regularize) the feature vectors rather than the audio. This offers only a slight improvement in computation since the bulk of the calculations occur in the PT itself and not in the regularization.

in [S: 77].) The PT can then be applied to the processed (regularized) performance. *Julie's Waltz* was regularized so that each beat occupies exactly $\overline{T} = 0.61$ s, which is the mean value of the period parameter over the complete performance. The PT of the regularized version is shown in the right hand plot of Fig. 11.2. It clearly displays regularities at several levels: the tatum, the beat, the $\frac{3}{4}$ measure and the 4-bar (12 beat) phrase.

11.3 Beat-synchronous Feature Scores

Traditional musical scores focus on notes, each with their own pitch and (approximate) timing. Tempo scores such as Figs. 11.1 and 11.2 focus on the performed timing of the flow of a musical passage. Missing from these is any notion of the quality, timbre, or color of the sound. Bañuelos [B: 8] comments:

> Ironically, sound color is one of the elements of music for which our vocabulary is most limited and least precise, perhaps because the property itself is one of the most elusive and hardest to define and pinpoint. Thus far it has been difficult to talk objectively about sound color due to the complexity of factors that affect its structure. And yet, its impact on the listener is usually immediate and powerful. A single texture played for as little as one bar before any melodic material in the proper sense is introduced can serve to set up the tone or mood of an entire movement.

A significant amount of effort (for example, [B: 83, B: 85, B: 167, B: 196]) has been devoted to the attempt to define what timbre is, and what it is not. Timbre involves spectral characteristics of a sound, but it is also heavily influenced by temporal features. Timbre involves the number and distribution of the peaks of the spectrum as well as the way the peaks evolve over time. Spectrograms (such as Fig. 4.5) provide one tool for the visualization of sound color, but (like all literal notations) they contain too much information that is either irrelevant or redundant. Accordingly, Bañuelos suggests focusing "on specific qualitative properties of sound and tracing their development through time within a musical context."

Feature vectors (such as those of Fig. 4.19 used to accomplish the beat tracking) are an example of "tracing a particular property through time" and the four feature vectors of Sect. 4.4.3 were designed specifically to extract auditory boundaries. By choosing to measure aspects of the signal that are relevant to the perception of timbre (rather than to the detection of auditory boundaries), it is possible to create "scores" that demonstrate concrete changes over time. Bañuelos provides an example:

> For instance, we can determine how much energy is contained in the higher frequencies compared to that in the lower frequencies. This

> gives a reading that could be called the "brightness" or the "acuteness" of a sound. If we analyze in this way an entire beat... we will get a single number assessing its "brightness." Naturally, this number itself is irrelevant as far as its magnitude is concerned, but we can perform the same operation on a number of consecutive beats within a musical context and plot them over time. With the new graph – a kind of "brightness score" – we can gather invaluable insight on the way this aspect of sound develops over the course of a passage.

Conducting the analysis over a time span comparable to a beat is important because the beat interval is among the largest units of time over which a musical signal is likely to remain uniform. Since the beat forms a coherent perceptual unit, there is a greater chance that the sound remains consistent within a single beat than across beat boundaries. For example, Fig. 11.5 shows a small snippet of audio with the beat timepoints identified. Segmenting the audio into frames j that are aligned with the beats i allows the analysis to reflect only the contents of a single beat. In contrast, if the segments cross beat boundaries like the frames labeled k, the analysis will amalgamate and confuse the contents of the two beats. This is particularly important when trying to measure timbrally relevant parameters because the timbre is more likely to change at the beat boundaries than within a single beat.

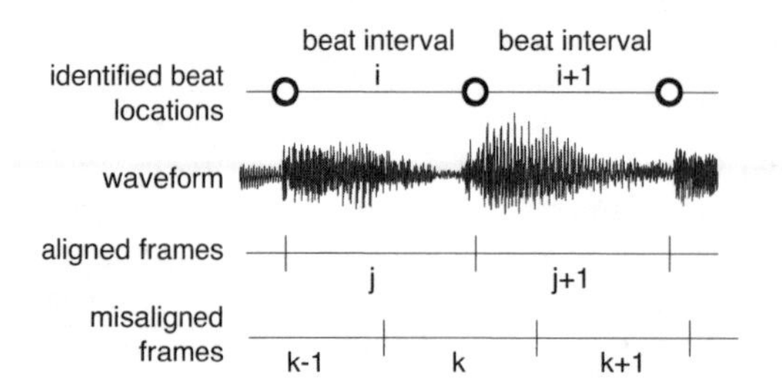

Fig. 11.5. When the frames are aligned with the beat intervals, each frame represents a single sound entity. When misaligned, the frames blend and confuse information from adjacent beats.

Bañuelos [B: 8] details several psychoacoustically motivated features that are particularly useful in an analysis of Alban Berg's *Violin Concerto*, subtitled *In Memory of an Angel*, that merges standard analytical techniques with new feature scores in an elegant and insightful way. Bañuelos' *Beyond the Spectrum* appears on the CD and adds considerable depth to the use of beat-synchronous feature scores in musical analysis. He comments on the goals of such a study:

> A number of graphs and alternate "scores" ... enable me to discuss comprehensively the ways in which sound color is used in the concerto to dramatize its program and to provide unity and form... In the end, the point is to incorporate the concrete information acquired about the particular sounds in a piece of music into a comprehensive and living-breathing interpretation...

The remainder of this section describes four measures useful in the creation of feature scores.

Brightness Score: Grey and Gordon [B: 84] observe that the balance in spectral energy between the low and high registers is a significant indicator of the perception of the steady-state character of a sound. Suppose that the spectral peaks $\mathcal{P}$ in a given beat are at frequencies $f_1 < f_2 < \ldots < f_n$ with amplitudes $a_1, a_2, \ldots, a_n$, which may be calculated using the peak picking technique of Sect. 9.5. Define the function

$$B(\mathcal{P}) = \frac{\log(p_n) - \log(c)}{\log(c) - \log(p_1)}$$

where c is the centroid of the frequencies. This quantifies the closeness of the centroid to both the upper and lower limits of the sound and corresponds roughly to the perception of the brightness of a sound.[4] A plot of $B(\mathcal{P})$ over time (one value for each beat interval) is called a "brightness score." The brightness score for *Julie's Waltz* is shown in Fig. 11.6 along with several other feature scores. These are interpreted in Sect. 11.4.

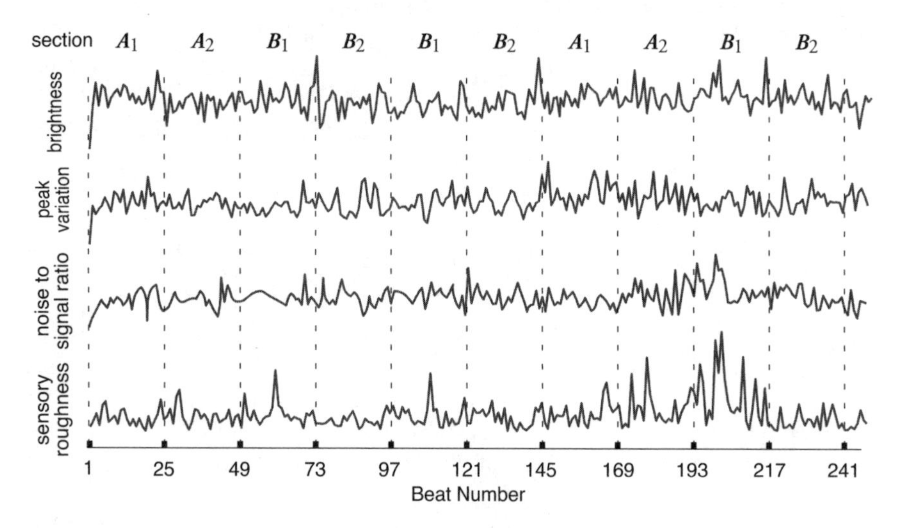

Fig. 11.6. Four feature scores for *Julie's Waltz* demonstrate how simple beat-synchronous measurements can reflect various timbral aspects of a musical performance

Peak Variation Score: The "peak variation" measures fluctuations in the height of adjacent spectral peaks on a decibel scale

[4] Bañuelos suggests two variations called registral and timbral brightness.

$$PV(\mathcal{P}) = \frac{1}{n} \sum_{i=1}^{n-1} |\log_{10}(\frac{a_{i+1}}{a_i})|.$$

This relates to the perception of poignancy or saturation of the sound within the beat. A plot of $PV(\mathcal{P})$ over time (one value for each beat interval) is called a "peak variation score." Figure 11.6 demonstrates.

Noise-to-Signal Score: The "noise-to-signal ratio" calculates the ratio of the amount of energy not in the partials (i.e., the energy in the noise) to the total energy. Using the median-based strategy of Sect. 9.5 to calculate the noise spectrum[5] $N(k)$, define

$$NSR(\mathcal{P}) = \frac{\sum_k N^2(k)}{\mathcal{E}}$$

where $\mathcal{E}$ is the total energy of sound in the beat interval. A plot of $NSR(\mathcal{P})$ over time (one value for each beat interval) is called a "noise-to-signal score," as in Fig. 11.6.

Sensory Roughness Score: The final feature score exploits recent psychoacoustical work on "sensory roughness," though the idea that the beating of sine wave partials is related to the tone quality of a sound was introduced by Helmholtz more than a century ago [B: 94]. Plomp and Levelt [B: 166] quantified this observation experimentally and related the range of frequencies over which the roughness occurs to the width of the critical band, thus providing a physically plausible mechanism. The roughness between two sinusoids with frequencies f_1 and f_2 (for $f_1 < f_2$) can be conveniently parameterized [B: 200] by an equation of the form

$$d(f_1, f_2, \ell_1, \ell_2) = \min(\ell_1, \ell_2)[\mathrm{e}^{-b_1 s(f_2-f_1)} - \mathrm{e}^{-b_2 s(f_2-f_1)}]$$

where $s = \frac{0.24}{s_1 f_1 + s_2}$, ℓ_1 and ℓ_2 represent the loudnesses of the two sinusoids, and where the exponents $b_1 = 3.5$ and $b_2 = 5.75$ specify the rates at which the function rises and falls. A typical plot of this function is shown in Fig. 1 of [B: 200], which is included on the CD. When a sound contains many partials, the sensory roughness is the sum of all the $d()$s over all pairs of partials

$$SR(\mathcal{P}) = \frac{1}{2} \sum_{i=1}^{n} \sum_{j=1}^{n} d(f_i, f_j, \ell_i, \ell_j),$$

If the sound happens to be harmonic with fundamental g, then the ith partial is ig. For such a sound, a typical curve (a plot of $SR(\mathcal{P})$ over all intervals of interest) looks like Fig. 3 of [B: 200]. Minima of the curve occur at or near musically sensible locations: the fifths, thirds, and sixths of familiar usage and the general contour of the curve mimics common musical intuitions regarding consonance and dissonance. This model is discussed at length in *Tuning, Timbre,*

[5] This can be done as in Eqn. (9.1), and the results depicted in Fig. 9.6 on p. 231.

Spectrum, Scale [B: 196], where it is used to draw "dissonance scores" which trace the sensory roughness of a musical performance through time. This chapter refines this idea by requiring that the segmentation of sound used to carry out the calculations be defined by the beat intervals. This allows "sensory roughness scores" to be much smoother and more accurate than the earlier "dissonance scores."

11.4 *Julie's Waltz*: An Analysis

The previous sections have shown how the tempo score and the feature scores display a variety of surface features of a musical performance. These are quite different from what is normally available in a musicological analysis, and this section compares the output of the algorithms with a more conventional view.

Tempo scores display temporal aspects of a performance and the stately variation in the tempo of *Julie's Waltz* is readily observable in Fig. 11.2. Though the changes are subtle and might go unnoticed in a casual listening, they add to the relaxed and informal feeling.[6] When the slide guitar begins its solo (at beat 97, the start of section $\mathcal{B}$), the tempo slows by about 5%, but it is heard as the arrival of a new voice, or perhaps the return of an old friend (the coming of the slide is foreshadowed by its background appearance in the first presentation of $\mathcal{B}$). This new voice slowly gathers momentum and eventually passes the theme to the violin when section $\mathcal{A}$ returns. The melody continues to accelerate until the final presentation of $\mathcal{B}$ and the tag. Despite the objective slowing at beat 97, the motion does not flounder because the instrumentation becomes more dense and the melody becomes more active. This increased activity is reflected in the brightness and noise-to-signal feature scores, which exhibit increased variance as time progresses.

Several levels of the metrical hierarchy appear in the periodicity transform of the beat-regularized version in Fig. 11.4. The beat at $\mathcal{T} \approx 0.61$ s is the strongest periodicity, and it is divided into two equal parts by the tatum. Three beats (or, equivalently, six tatum timepoints) are clustered together into the next highest level. If the beat is notated as a quarter note, then the time signature[7] of the measure would be $\frac{3}{4}$. The highest level is the four-bar phrase at $\mathcal{T} \approx 7.3$ s, which contains twelve beats.

The tempo score in Fig. 11.2 and the feature score in Fig. 11.6 have been annotated with section numbers; these are formal sections of the piece, each eight measures long. The harmony changes steadily each measure, and the

[6] The conversation-like interplay between the instruments emphasizes the easy-going tone.

[7] The time signature $\frac{3}{8}$, where the beat would be represented as an eighth note, is also consistent with the data. The important clustering of the beats into groups of three (or tatum timepoints into sixes) remains consistent through any such representation.

chord pattern within each section is clearly divided into two phrases (separated by ||)

$$\begin{array}{ll} \mathcal{A}_1: & \text{I} \;\; \text{IIIm IV V} \;||\; \text{IV I II7sus4 V7} \\ \mathcal{A}_2: & \text{I} \;\; \text{IIIm I IV} \;||\; \text{VI}\flat\ \text{I IIm/V} \;\; \text{I} \\ \mathcal{B}_1: & \text{VIm VIm V} \;\; \text{I} \;||\; \text{IV I II7sus4 V7} \\ \mathcal{B}_2: & \text{I} \;\; \text{IIIm I IV} \;||\; \text{VI}\flat\ \text{I IV/V} \;\; \text{I} \end{array}$$

where I represents the tonic, V the dominant, etc. Given the tight two-phrase structure of the sections, it is not surprising that the rhythmic analysis detects groupings of four bars (twelve beats). The internal structure of the sections emphasizes the phrases by repetition: the last four measures of $\mathcal{A}_1$ and the last four measures of $\mathcal{B}_1$ are identical; sections $\mathcal{A}_2$ and $\mathcal{B}_2$ are the same but for the penultimate chord. The first phrase of $\mathcal{A}_1$ and the first phrase of $\mathcal{A}_2$ make analogous statements: the first two chords are the same while the last two move by a fifth (IV to V vs. I to IV). The most unusual harmony is the VI♭ that begins the final cadences in $\mathcal{A}_2$ and $\mathcal{B}_2$. The use of the II7sus4 to lead back to the V is also notable: the suspension anticipates the root of the V chord and the 7th suspends (and resolves to) the third from above.

The sections are often distinguished by instrumentation, which is listed in Table 11.1. The plucked mandolin and finger-picked guitar begin together, and continue throughout the piece. They are joined by slide guitar (in $\mathcal{B}_1$) and by bass and violin in $\mathcal{B}_2$. The lead instruments alternate melody and harmony throughout the remainder of the piece. The feature scores of Fig. 11.6 show some of the timbral changes due to variations in instrumentation. For example, all four features become more active and exhibit greater variance as the piece progresses, reflecting the increased complexity[8] and density of the sound. The brightness score, for instance, often has small peaks at the ends of the sections. Such peaks can occur because there is more energy in the treble or because there is less energy in the bass (since it shows the ratio of the location of the centroid in terms of the highest and lowest peaks). The noise-to-signal score clearly shows the increase in complexity (increase in variance or wiggliness of the curve) as the piece proceeds.

The roughness curve is interesting because of what it shows and because of what it fails to show. In many circumstances, roughness can be interpreted as a crude measure of the dissonance (it is also called "sensory dissonance" [B: 196]) since intervals such as octaves, thirds and fifths tend to have small values while seconds and sevenths tend to have larger values. Since the VI♭ chord that starts the final cadence is outside the key of I, it would appear to be a point of tension. Yet the roughness score does not show large values at these points. This is because the calculations are done moment-by-moment (beat-by-beat, to be exact) and so the roughness calculation during the VI♭ chord has no way of "knowing" that it is imbedded inside a piece in the

[8] This is not the same as loudness, which is removed from these curves. Loudness could be used to form another, independent feature score.

Table 11.1. Sections in *Julie's Waltz* are differentiated by instrumentation as well as harmonic motion

Section	Beat	Time	Instrumentation	Comment
$\mathcal{A}_1$	1	0:00	mandolin & guitar	statement
$\mathcal{A}_2$	25	0:14		answer
$\mathcal{B}_1$	49	0:28	slide in background	alternate (minor) theme
$\mathcal{B}_2$	73	0:43	add bass & violin	answer
$\mathcal{B}_1$	97	0:57	slide melody	alternate
$\mathcal{B}_2$	121	1:12		answer
$\mathcal{A}_1$	145	1:27	violin melody	statement
$\mathcal{A}_2$	169	1:42		answer
$\mathcal{B}_1$	193	1:57	mandolin & violin & slide	alternate
$\mathcal{B}_2$	217	2:11	mandolin & violin in unison	final answer
	241	2:26	mandolin & guitar	tag (last phrase of $\mathcal{A}_2$)

key of I. Hence its roughness is comparable to that of other major chords throughout the piece. Also interesting are the peaks that occur as the I moves to the IV in the middle of each $\mathcal{C}$ section. This is hardly a chord pattern where roughness would be expected. However, in each case, there are a number of passing tones (on mandolin and/or on slide) that are outside the I – –IV chords. These passing tones increase the tension and create the roughness peaks. Perhaps most striking is the overall trend of the roughness: it begins low, builds to a climax (in the final $\mathcal{C}$ section) and then relaxes. This is one of the characteristic trademarks of Western music, and it is clearly displayed (along with a few subsidiary peaks and valleys) in the roughness score.

Feature and tempo scores depict surface features of a performance that are quite different from the deeper analyses possible from a close reading of the score. The tempo score quantifies the temporal motion by depicting the time interval of each beat. The transform methods can, with suitable care, sensibly cluster the beats into groups that correspond to measures and phrases. The feature scores present a variety of properties that often correlate with subjective impressions such as changes in timbre, density, complexity, and/or dissonance.

This analysis of *Julie's Waltz* is intended only as a "teaser" to show the kinds of insights that are possible when using tempo scores and beat-synchronous feature scores. The interested reader will find a more complete (and, from a musical point of view, far more compelling), analysis of Berg's *Violin Concerto* that uses many of the same techniques in Bañuelos [B: 8]. This is included on the CD.

12

Speculations, Interpretations, Conclusions

Rhythm and Transforms *begins with a review of basic psychoacoustic principles and a brief survey of musical rhythms throughout the world. The three different pattern finding techniques of Chaps. 5–7 lead to a series of algorithms that show promise in solving the automated beat tracking problem. Chapters 9–11 explore the uses of a beat-based viewpoint in audio signal processing, composition, and musical analysis. Most of the book stays fairly close to "the facts," without undue speculation. This final chapter ventures further.*

The beat tracking methods of *Rhythm and Transforms* identify regular successions directly from feature vectors and not through the intermediary of a list of note events. Section 12.1 wrestles with the question of whether notes are a perceptual experience or a conceptual phenomenon. Do "notes" correspond to primary sensory experiences, or are they cognitive events assembled by a listening mind? Section 12.2 suggests how beat intervals may represent one of the primary constituents of musical meaning. The resulting hierarchy of musical structures has clear perceptual correlates, and the beat-level analysis is not bound to a single musical culture. The distinction between a musical composition and a performance of that composition in Sect. 12.3 sets the stage for a discussion of the implications of the beat manipulation techniques. These are not only technologies of processing audio signals but also of creating performances and even compositions. The many ways to combine existing beats into new patterns and new structures raise questions about the integrity of a musical piece: at what point does the piece cease to be the "the same" as the original(s) from which it is derived? Some practical suggestions are made in Sect. 12.4. There is an obvious relationship between musical rhythms and the nature of time. Section 12.5 looks at some of the implications of *Rhythm and Transforms* in terms of how time is perceived. Finally, the three technologies for locating temporal patterns are summarized. Each has its particular area of applicability and its own weaknesses.

12.1 Which Comes First, the Notes or the Beat?

One of the peculiarities of the methods in *Rhythm and Transforms* is the avoidance of individual note events. Instead, the methods parse feature vectors to find templates that fit well, to find regular successions hidden within the feature vectors. In terms of perception, this presents a chicken-or-egg question. Do we perceive a collection of individual note events and then observe that these events happen to form a regular succession? Or do we perceive a regular succession of auditory boundaries which is then resolved by the auditory system into a collection of aligned note events? Said another way: are note events primary elements of perception, or do we first perceive a rhythmic flux that is later resolved into notes?

At first glance, the question seems nonsensical since "notes" feature so prominently in the human conception of music. Notes are used to generate music by striking keys, plucking strings, or blowing through air columns; in each case producing a recognizable note event. Notes are used in musical "notation" to record collections of such events. Notes are enshrined in language as the primary syntactic level of musical meaning. But, as a practical matter, it has turned out to be a very difficult task to reliably detect note events from the audio as the recent MIREX note detection contest shows[1] [W: 32].

The experience of *Rhythm and Transforms* suggests that a psychological theory of the perception of meter need not operate at the level of notes. People directly perceive sound waves, and the detection of notes and/or regular successions are acts of cognition. To the extent that computational methods are capable of tracking the beat of musical performances without recourse to note-like structures, the "note" and "interonset" levels of interpretation may not be a necessary component of rhythmic detection. This reinforces the argument in Scheirer [B: 188] where metric perceptions are retained by a signal containing only noisy pulses derived from the amplitude envelope of audio passed through a collection of comb filters. Similarly, the "listening to feature vectors" sound examples [S: 57]–[S: 59] are able to evoke rhythmic percepts using only noisy versions of feature vectors.

Nevertheless, most beat tracking and rhythm finding algorithms operate on interonset intervals (for instance [B: 43, B: 48, B: 76]), which presuppose either a priori knowledge of note onsets (such as are provided by MIDI) or their accurate detection. The methods of *Rhythm and Transforms*, by ignoring the "notes" of the piece, bypass (or ignore) this level of rhythmic interpretation. This is both a strength and a weakness. Without a score, the detection of "notes" is a nontrivial task, and errors such as missing notes (or falsely detecting notes that are not actually present) can bias the detected beats. Since *Rhythm and Transforms* does not detect notes it cannot make such mistakes. The price is that the explanatory power of a note-based approach to musical understanding remains unexploited. Thus the beat tracking technologies of

[1] When the performance has clearly delineated amplitude envelopes such as percussion, the algorithms work well, with the best methods detecting (for example) 321 notes with only three false positives and three false negatives in the solo-bars-and-bells category. When the input is more complex, however, even the best of the algorithms reported (for example) 143 detected notes with 286 false positives and 86 false negatives in the solo-singing-voice category.

Rhythm and Transforms are more methods of signal processing at the level of sound waveforms than of symbol manipulation at the note level. Though the methods do not attempt to decode pitches (which are closely tied to a note level representation), they are not insensitive to frequency information. Indeed, the feature vectors of Sect. 4.4.1 process each critical band separately, which allows the timbre (or spectrum) of the sounds to directly influence the search for appropriate regularities in a way that is lost if only note events and/or interonset interval encoding is used.

Beat tracking is a form of audio segmentation, and may also have uses in the processing of language. One of the major problems confronting automated speech recognition systems is the difficulty of detecting the boundaries between words. As the rhythm finding methods locate regular successions in music without detecting notes, perhaps it is possible to use some kind of rhythmic parsing to help segment speech without (or before) recognizing words: to parse the audio stream searching for rhythmic patterns that can later be resolved into their constituent phonemes and word events.

12.2 Name That Tune

The fundamental unit of analysis in *Rhythm and Transforms* is the individual "beat interval" (a short duration of time delineated by adjacent metrical positions on the timeline) rather than the "note." "Notes" are relegated to their proper role in the notation of common practice music. Indeed, much of the writing about musical rhythm is not about the experience of sound but about its symbolic representation. An emphasis on beat intervals as the primary rhythmic experience is one way to counterbalance this tendency. Of course, notes remain a convenient way to talk about the kind of auditory events initiated by the pressing of a key on a piano or by the striking of a stick against the side of a bell. The following two anecdotes highlight some of the differences between the note-level conception of music and a beat-level conception.

Contestants in the television game show *Name That Tune* [W: 37] compete to name a song in the fewest notes. As the announcer presents clues to the identity of the song, a contestant might bid "I can name that tune in five notes," and bidding continues until one contestant challenges the other to "Name that tune." The first five notes of the melody are then performed and the contestant must correctly identify the song. Even with the hints, this is an extraordinary ability, and it obviously relies heavily on a shared background of melodies: chosen by the producers, performed by the pianist, and recognized (rapidly and with few clues) by the contestants.

In my family, a favorite pastime on car trips is to compete to name each new song as it comes on the radio. When listening to a familiar station playing a well known musical genre, it is often possible to identify a piece within only a beat or two, well before the "melody" has even begun. This is identification of the timbre or tone color of a familiar recording. Similarly, if you have been listening along to the sound examples, you are by now very familiar with the *Maple Leaf Rag*. The four short sounds in [S: 158] each contain a single beat interval. Most likely, you can recognize the rag, and moreover, where in the rag the beat occurs. Somewhat more challenging are the 39 short sounds in [S: 159]. Again, each is one beat long. They are randomly selected from among thirteen of the pieces in Tables A.1 and A.2. Can you tell which sound belongs to which performance? Good luck![2]

Thus it is often possible to recognize a performance of a song from a single beat of sound. On the other hand, the brief sound snippets of [S: 153] (recall the discussion on p. 259) show how the timbre is changed when the duration is significantly shorter than a beat. Together, these suggest that the beat (or perhaps the tatum) is the smallest atom of musical "meaning," the shortest duration over which the character of a piece is maintained. A beat interval contains information about the timbre, the instrumentation, the instantaneous harmonization, and the sound density. It also contains information about the recording process itself: room reverberation, compression, noises, etc. The beat is capable of reflecting the piece on a larger scale. All of these are lacking from a symbolic note-oriented analysis. This extra information may explain why the "name that tune" contestants need several notes and extensive verbal hints while the teenager in the car needs to hear only the downbeat[3] before exclaiming "it's my favorite tune..."

In this conception, the hierarchy of musical structures is:

(i) Audio rate *samples* are nearly inaudible
(ii) *Grains* are individually devoid of meaning, but can be easily combined in masses, clouds (and other higher level structures) to convey density and timbre
(iii) The *beat* (or *tatum*) intervals convey information about timbre and instrumentation as well as the recording process
(iv) *Rhythms* are built from sequences of beat (or tatum) intervals

This also reflects a perceptual hierarchy. The continuous ebb and flow of neural messages that lies beneath awareness parallels audio rate samples. Elementary sense impressions, which have no intrinsic meaning until they are bound into larger structures, are analogous to grains. For example, the puff of air that initiates a flute sound and the following sinusoidal tone are bound together into the perception of "a flute playing middle C." The beat corresponds to the lowest level of musical meaning, and lies fully within the perceptual present.

---

[2] Perhaps because I have become hypersensitized to the music in Tables A.1 and A.2 from working with them for so long, I can tell instantly (i.e., within one beat), which piece a random beat is from (excluding the many similar-sounding piano renditions).

[3] An analogous feat is the ability to recognize a friend on the telephone after hearing only a brief "hello."

To be understood, rhythms typically involve interaction with long term memory.[4] Rhythms evoke expectations (anticipation of future events) and require attention (e.g., the will of the observer must focus on the rhythmic pattern).

Finally, by focusing on the beat (or tatum) level, it is possible for the results in *Rhythm and Transforms* to aim for a kind of cultural independence that is rare in musical studies. While not all of the world's music is beat-based (recall Sect. 3.1), a large portion is. The Western conception of meter, the African conception of the timeline, the Latin use of the clave, the Indian *tala*, and the rhythmic cycles of the gamelan, are all built on a repetitive beat structure. All depend on a regular succession of timepoints that may be sounded directly or may be implicit in the "mind's ear" of the performer and listener. While the basic rhythm of a timeline or clave is provided by an irregular pattern of sounded events, the events synchronize with the underlying lattice of tatum timepoints.

12.3 Variations on a Theme

While the distinction may sometimes be blurred in common usage, it is important to distinguish the performance of a piece from the piece itself. The hierarchy of musical structures continues:

(v) A musical *performance* is the act of making music and, by extension, the results of this action
(vi) A musical *composition* is an abstract form from which a variety of performances may be realized

Thus the *Maple Leaf Rag*, a composition by Scott Joplin, provides a set of instructions (in this case, a musical score) that musicians can follow in order to make a variety of different performances[5] of the *Maple Leaf Rag*. The 27 versions in Table A.2 demonstrate the wide variety of possible realizations of this same composition.

There are many ways to compose. The classic image of a tortured master sitting in a silent room with staff paper and quill pen is but one compositional paradigm. An important alternative is improvisation, where the composer plays an instrument or sings during the compositional process, often in a setting with other musicians. Improvisational composition necessarily involves performance, though improvisational performance need not be composition.

There is a long tradition of composition using algorithmic processes. In Mozart's musical dice game [W: 36], chance determines which measures will be played. In recent years, the use of randomness and deterministic chaos

[4] Recall Fig. 4.1 on p. 77.

[5] It is sometimes useful to also distinguish the act of making music from the record of that act (for example, to distinguish the concert hall experience from a CD recording of the concert). The present discussion blurs these two levels, referring to both as the same "performance."

have been extended to other levels of the compositional process. Xenakis [B: 247] creates the form of the composition *Analogique B* using Markov chains. Roads [B: 181] creates compositions from stochastically generated collections of sound grains. In these techniques, the audible result is dictated by the formal specifications of the composer; the composition is embodied in an algorithm, not in a score.

Many modern compositional trends emphasize the creation and selection of sound materials such as the collage aesthetic in musique concrète [B: 210] and the multi-layering of rhythmic loops and samples in hip-hop [B: 119]. These methods transform existing sound into rhythmic, tonal, and timbral variations that form the raw material for new pieces. The chapters on beat-based signal processing and beat-based composition arrange, cross, mutate, and rearrange pre-existing beats to generate a surprising array of sounds that are variations of both the performance and the composition. These may remain faithful to the original composition (as in the *Rag Bags* [S: 155]) or may appear unrelated (as in the *Pentatonic Rag* [S: 111] or the *Atonal Leaf Rags* [S: 98]). The continuum of novelty in recomposition extends from pieces that are clearly related to the original to those that disguise the source completely.

Typically, the form of a composition allows for many possible performances, each a variation on the original. The beat-level techniques extend the idea of "variations on a theme" in two ways. First, new performances of a composition can be created directly from existing performances without explicit reference to the underlying composition. Thus the *Rag Bags* and the *Grab Bag Rags* generate new performances automatically by rearranging beats, without using the composition as a guide. Second, new compositions can arise from the manipulation and rearrangements of a performance. Thus the *Make Believe Rag* [S: 115] has little apparent connection to the *Maple Leaf Rag* [S: 5], *Local Variations* [S: 117] has little perceptual affinity to *Local Anomaly* [S: 116] and *Mirror Go Round* [S: 122] is not obviously related to *Three Ears* [S: 121]. Yet, in all three cases, one is derived from the other via an algorithmic process.

12.4 What is a Song?

Compositions only exist in symbolic form, while performances are literal renderings. Notations help musicians learn new music, help performers realize a composers vision, and help clarify the internal structure of music. Each kind of notation has strengths and weaknesses in terms of its ability to represent high level features in ways that are easy for composers and musicians to comprehend and manipulate. Each notation has its own aesthetics and ways of composing: with pencil and paper for musical scores, with sequencers for MIDI notation, by algorithm for granular synthesis, by image editing software for spectrograms.

Notations also reflect attitudes and values about music, and raise questions of musical ownership. Aristotle might have laughed at the idea that someone could claim ownership of a song. The US copyright office does not find it funny at all. Western music has adopted the musical score as a primary form for the dissemination and preservation of compositions. In order to copyright a musical piece in the US under the copyright act of 1909, it was necessary to write the music in standard notation and to send it (along with a registration fee) to the copyright office. Thus a musical composition was defined, legally, to require a symbolic representation. The law was changed in 1976 to allow literal notations (sound recordings), and now recognizes two kinds of copyright: the copyright of a performance and the copyright of a composition (the underlying work) [W: 11]. The two kinds of copyright are very different. For example, while the underlying composition of the *Maple Leaf Rag* has passed into the public domain (is now out of copyright), most of the versions in Table A.2 are under performance copyright by their creators.

Thus, prior to 1976, it was only possible to legally protect music that was represented in symbolic form. Since then, it has become possible to legally protect music represented in either symbolic or literal form. While the copyright office tries to be clear, there are many gray areas. What criteria distinguish two performances of the same underlying work? For example, there is very little to distinguish my version of the *Maple Leaf Rag* [S: 5] from the versions by Blumberg A.2(1), Copeland A.2(3), and Joplin A.2(8). The tone of the piano varies somewhat, the detailed timing is subtly different, but all four are played in a straightforward manner from the sheet music. Yet "clearly," these four are separately eligible for performance copyright since they are performed by different artists.

In contrast, consider the *Beat Filtered Rag* [S: 84] and the *Beat Gated Rag* [S: 85]; both are derived algorithmically from [S: 5]. In a logical sense they are "clearly" derivative works; hence they would infringe on the performance copyright.[6] Yet the *Beat Filtered Rag* and the *Beat Gated Rag* sound very different from the original and from each other; far more different than the four straightforward piano versions! Hence the performance copyright rewards (with legal protections) versions that sound nearly identical and refuses such protection to versions that sound as different as any two renditions in Table A.2.

Even more difficult is the question of how to distinguish compositions. Is the *Beat Gated Rag* a performance of the same work as the *Maple Leaf Rag*? Yes, since the chord pattern and occasionally the rhythm are maintained. But neither the *Make Believe Rag* nor the *Atonal Leaf Rag #2* inherit the harmony and rhythm from the *Maple Leaf Rag*. A naive listener would be unlikely to judge them to be "the same piece." Yet the algorithmic relationship

[6] This is not actually a problem in this particular case, because I am the creator (and hence automatically the owner of the copyright) of the performance [S: 5].

argues that they are derived from the *Maple Leaf Rag* and hence that they are "derivative" works.

An old conundrum asks how many trees can be removed from a forest and still remain a forest. Similarly, it is tempting to ask how many beats can be removed from a song and still remain the same song. Are the *Maple Leaf Waltz's* [S: 130, S: 131] the "same" as the *Maple Leaf Rag*? How about the *Half Leaf Rag* [S: 137]? The *Quarter Leaf Rag* [S: 139]? By their timbre, these are clearly derived from the original performances. Logically, they are derived from the original composition, but this is not at all clear perceptually. Because the *Quarter Leaf Rag* moves four times as fast as the *Maple Leaf Rag*, the basic feel of the rhythm changes. Removal of three quarters of the melody renders it unrecognizable. The *Make It Brief Rag* [S: 142] is constructed from the *Quarter Leaf Rag* using the beat-level delay, filter, and gating techniques of Sect. 9.2, to remove the timbral similarity. If the melody is different and the rhythm is different, and the timbre is different, in what way is the *Make It Brief Rag* the same composition as the *Maple Leaf Rag*? The issue here is the fundamental "identity" of a song; what it means to be a composition.

As Kivy [B: 112] points out, in the art of painting there is an original physical object that represents the work. In music, where there is no single original, it is not even clear what it means to "be" a piece of music. Here are some possibilities:

(i) a CD on which music is recorded
(ii) a pattern of sound waves
(iii) a mind perceiving the sound waves
(iv) a mind (silently) rehearsing and remembering the performance
(v) a musical score, MIDI file, or other high level symbolic representation
(vi) an idea (or algorithm) specifying some set of meaningful musical actions or parameters

Possibilities (i) and (ii) represent performances while (v) and (vi) lie at the compositional level. Yet (iii) (and possibly (iv)) are the levels at which decisions about the "sameness" and "difference" of musical works must necessarily be made. Undoubtedly, the legal situation will continue to evolve, and it is not clear how the kinds of sound manipulations represented by beat-oriented processing will fit into the legal framework, either at the performance or the compositional levels.

Some modern artists are offering listeners the chance to participate more in the formation of the final sound. For example, Todd Rundgren's interactive CD *New World Order* [D: 38] allows users to manipulate various components of the music as they listen. David Bowie has sponsored a series of "mash-up" competitions [W: 6] where fans remix Bowie's songs. What part of the final sound is Rundgren or Bowie, and what part is the listener-participant? Again, this raises the question of how a musical work is defined.

12.5 Time and Perceived Time

> What is time? As long as no one asks me, I know what it is; but if I wish to explain it to someone, then I do not know. – St. Augustine, *Confessions XIV*.

St. Augustine is not alone. It is almost impossible to define *time* in a way that is not essentially circular. Definitions made in terms of motion (a year defined as one revolution of the Earth around the sun, a day as one rotation of the Earth) are problematic because motion is itself defined in terms of time. The *Oxford English Dictionary* defines time as a "continuous duration regarded as that in which a sequence of events takes place." But can one really know what a "duration" is, unless one already understands time? The modern scientific definition of one second as

> the duration of 9,192,631,770 periods of the radiation corresponding to the transition between the two hyperfine levels of the ground state of the cesium 133 atom at 0K.

is no more enlightening, though it is very precise.

For Aristotle, "time does not exist without change." For Kant, the changes are in the mind (not in the world), implying that the understanding of time is not a matter of knowledge but of psychology. Time, according to Clifton [B: 32] is a relationship between people and the events they perceive; time is that which orders our experiences. Bergson [B: 12] writes that rhythm is a "pure impression of succession, the way that we directly apprehend the flow of time." Even more recent is the idea that time arises through perception. But how exactly do we perceive the passage of time? Unlike all the other senses, there is no (known) sensory apparatus that perceives time. Gibson [B: 73] argues that "time is an intellectual achievement, not a perceptual category."

Much of the early philosophical interest in rhythm arises from its obvious relationship with the passage of time. Music reflects the structures of time. Music, like time, unfolds linearly. As in time, there is repetition. Time is punctuated by the eternal cycles of day and night; music is punctuated by the recurrence of rhythmic structures such as meter, timecycles and timelines.

Music is temporal: abstract sounds "move" through time, interpenetrating each other in ways that objects in space cannot. Music takes place in time, but also helps to shape our perception of time. An enjoyable piece can flow by in an instant while a loud song overheard from a neighbor appears to last an eternity. Thus time as experienced may differ from "clock time," and we may experience both kinds of time simultaneously. Observation of this close connection between rhythm and time is not limited to the Western philosophical tradition. Blacking [B: 16] writes that "Venda music is distinguished from nonmusic by the creation of a special world of time." Here, the ability of music to shape the perception of the flow of time is presented as the defining characteristic of music.

There is little that *Rhythm and Transforms* can say about time (whatever it may be), but there are some tantalizing possibilities concerning the way time is perceived. Models of time perception are either based on the idea of an internal clock that marks the passage of time [B: 120], or on the idea that perceived duration is related to the amount of information processed or stored [B: 65]. This division parallels the event/boundary dichotomy of Fig. 4.12 (on p. 94) and the two models seem to explain different aspects of time perception. The clock models explain the ability of performers to maintain steady tempos, while the information models are consistent with studies where the apparent duration of intervals changes depending on the content of the interval. Clark and Krumhansl [B: 31] suggest that both mechanisms may be important, with the clock models addressing short durations (and performance behaviors that involve motor skills), and the information models addressing phenomena with longer durations. The ability to process and rearrange sound may provide interesting ways to process the sound for further testing such hypotheses. For example, consider stretched and compressed versions of the same piece. Presumably these contain (approximately) the same amount of information, and so they might be used to test models of perceived duration.

The duration of the perceived present depends on the richness of its contents and the possibilities for its organization into groups. In general, the more changes perceived in a stimulus, the longer it tends to appear. The order is important in terms of the ability to organize and cluster the elements (recall Figs. 3.1 and 3.2). The ability to reorder segments of a piece in a simple fashion could be used to investigate the perceptions of duration as a function of the ease of clustering. Presumably sensible orderings should be easier to remember and reproduce, while nonsense orderings should be more difficult. Similarly, it may be possible to use the beat-splicing techniques as a way to concretely test the linearity (goal directedness) or the nonlinearity of a particular piece or of a musical genre.

Cross-cultural misunderstandings are common when talking about rhythm. The Westerner listens for the steady beat and the regular metrical grouping. The Indian searches for the *tala*, the underlying circular pattern with its characteristic downbeats (*sam*) and unaccented beats (*kali*). The Indonesian searches for the regularly punctuated gong cycle, and without it cannot find the inner melody. The African searches for the timeline, the uneven repetitive pattern against which all other musical lines play and are heard.

What is the organizing principle underlying rhythmic structures? *Rhythm and Transforms* takes an approach one level below the grouping of the meter/ tala/ gong/ timeline, at the level of the beat. This is one of the major features that distinguishes the present work from most other studies of rhythmic structure: music is conceived as sound waves (that can be searched for beats) rather than as notes. This has disadvantages: all the power of a note-based representation are lost. But there are also advantages. From a perceptual perspective, it bypasses a very difficult stage of processing, that of detecting notes from a continuous sound. From a philosophical perspective it is interesting to see what can be done without the concept of notes. The short answer is: a lot more than is at first apparent. The focus is on direct interpretation of the sound, and not of a symbolic representation of the sound.

12.6 The Technologies for Finding Patterns

The location of beats in a musical passage presents a challenging technological problem and *Rhythm and Transforms* has approached it from three quite different points of view. Transforms represent the classical method of locating periodicities in data as a linear combination of basis (or frame) elements. Dynamical systems consisting of clusters of adaptive oscillators mimic the biologically inspired idea that a listener follows a musical rhythm by synchronizing an internal clock to the pulsations of the music. Probabilistic methods are based on parameterized models of repetitive phenomena and statistical techniques help locate optimal values for those parameters. Applying all three methods to the same set of problems highlights the strengths and weaknesses of each approach.

Transforms appear to be good at finding large scale patterns and at locating multiple patterns simultaneously. They fail, however, when confronted with repetitive phenomenon that are not really periodic, as when the underlying pulse of the music changes over time. Oscillators operate very rapidly, both in terms of computation time and in terms of the speed with which they can respond. But they tend to fail when there are certain kinds of (common) noises that may contaminate the data. The statistical methods seem to provide workable solutions for the location of the desired parameters, but they employ oversimplified models whose assumptions are clearly violated in practice (for example, that successive data points in a feature vector are statistically independent of each other).

In terms of the musical problem, this suggests that transforms should be used whenever possible, though with the realization that they may often be inapplicable. When the data is noise free (as in a MIDI rendition or when following a musical score), adaptive oscillators may be the method of choice, especially if the pulse is likely to change rapidly. The more difficult problem of locating beats from low level audio features may require the more sophisticated statistical methods.

The field of beat tracking is just beginning to approach maturity. The yearly MIREX competitions [W: 32] provide a challenging playground on which various methods can be tested. A special session of the IEEE conference on acoustics, speech, and signal processing, organized by Mark Plumbley and Matthew Davies (to occur in mid 2007), will bring researchers together to plan for future directions. Cemgil and coauthors will be talking about generalizing from tracking beats to following complete rhythms using a "bar pointer" [B: 239]. Dixon and Gouyon, winners of the 2005 MIREX beat tracking competition, will be talking about their recent work. Eck will be discussing innovative kinds of feature vectors.

Finding temporal patterns in nature and extrapolating into the future is one of the essences of intelligent behavior; it is the kind of thing that people master intuitively but that often stymies machines. Surely there are other problem areas where the three technologies for locating temporal patterns can be put to good use. Hopefully this case study will be of some use to those trying to automate solutions to these other problems.

Beat Tracked Musical Performances

The statistical method of audio beat tracking, detailed in Sects. 7.5 and 7.6 and discussed in Sect. 8.3.2, has been successfully applied to a large number of musical pieces (about 300). Table A.1 shows 16 pieces that are representative of the various musical genres that have been attempted. Excerpts (approximately thirty seconds each) can be heard in [S: 71] with the detected beat locations indicated by short bursts of noise.

Table A.1. Audio beat tracking of a variety of musical excerpts is demonstrated by superimposing a burst of white noise at each detected beat [S: 71]

	Style	Title	Artist	File	Ref.
1	Pop	*Michelle*	The Beatles	`Tap-Michelle`	[D: 2]
2		*Tambourine Man*	The Byrds	`Tap-Tambman`	[D: 7]
3	Jazz	*Take Five*	Dave Brubeck	`Tap-Take5Bru`	[D: 9]
4		*Take Five*	Grover Washington	`Tap-Take5Wash`	[D: 45]
5		*Maple Leaf Rag*	Scott Joplin	`Tap-Maple1916`	[D: 41]
6	Classical	*K517*	D. Scarlatti	`Tap-ScarK517`	[D: 39]
7		*Water Music*	G. Handel	`Tap-Water`	[D: 21]
8		*Pieces of Africa*	Kronos Quartet	`Tap-Kronos`	[D: 28]
9	Film	*James Bond Theme*	Soundtrack	`Tap-Bond`	[D: 6]
10	Rock	*Soul*	Blip	`Tap-Soul`	[S: 7]
11	Folk	*Friend of the Devil*	The Grateful Dead	`Tap-Friend`	[D: 19]
12		*Howell's Delight*	Baltimore Consort	`Tap-Howell`	[D: 1]
13	Country	*Angry Young Man*	Steve Earl	`Tap-Angry`	[D: 14]
14	Dance	*Lion Says*	Prince Buster	`Tap-Ska`	[D: 10]
15	Bluegrass	*Julie's Waltz*	M. Schatz	`Tap-Julie`	[D: 40]
16	Gamelan	*Hu Djan*	Gong Kebyar	`Tap-HuDjan`	[D: 17]

The *Maple Leaf Rag* has been performed many times by many artists in many styles over the years. In addition to CDs in shops and over the web, many versions are available using the gnutella file sharing network [W: 15]. Table A.2

details 27 different performances, showing the artist, instrumentation and style. Excerpts (approximately thirty seconds of each) can be heard in [S: 76] with the detected beat locations indicated by short bursts of noise.

The mean value of the length of the beat intervals is given (in seconds) for each version and the column labeled "STD" gives the standard deviation about this mean (the number is multiplied by 1000 for easier comparison). See Fig. 11.1 on p. 265 (and the surrounding discussion) for a more detailed analysis.

Table A.2. Audio beat tracking of excerpts from a variety of performances of the *Maple Leaf Rag* is demonstrated by superimposing a burst of white noise at each detected beat [S: 76]

	Instrument	Artist	File	Mean	STD	Ref.
1	piano	D. Blumberg	Tap-MapleBlumberg	0.32	5.36	[W: 5]
2		C. Bolling	Tap-MapleBolling	0.30	4.42	[D: 5]
3		Copeland	Tap-MapleCopeland	0.32	5.75	[W: 10]
4		Cramer	Tap-MapleCramer	0.35	3.32	[D: 12]
5		Entertainer	Tap-MapleEnter	0.33	12.54	[D: 18]
6		G. Gershwin	Tap-MapleGershwin	0.30	8.23	
7		D. Hyman	Tap-MapleHyman	0.24	9.95	[D: 26]
8		S. Joplin	Tap-Maple1916	0.34	4.49	[D: 41]
9		Jelly Roll Morton	Tap-MapleMorton	0.33	10.50	[D: 32]
10		Motta Junior	Tap-MapleMotta	0.31	10.93	[W: 34]
11		M. Reichle	Tap-MapleReichle	0.36	2.16	[D: 34]
12		J. Rifkin	Tap-MapleRifkin	0.33	6.80	[D: 35]
13	big band	S. Bechet	Tap-MapleBechet	0.22	7.31	[D: 3]
14		Dorsey	Tap-MapleDorsey	0.22	4.48	[D: 13]
15		Butch Thompson	Tap-MapleButch	0.34	4.84	[D: 44]
16	symphonic	Unknown	Tap-MapleOrch	0.34	9.33	
17		Unknown	Tap-MapleQLK	0.25	3.92	
18	a capella	Unknown	Tap-MapleBygon	0.33	9.43	
19	brass	Canadian Brass	Tap-MapleBrass	0.33	10.14	[D: 11]
20	marimba	Glennie	Tap-MapleGlennie	0.26	3.89	[D: 16]
21	guitar	D. Van Ronk	Tap-MapleRonk	0.40	12.51	[D: 37]
22		Van Neil	Tap-MapleVanNiel	0.38	1.79	[W: 55]
23	banjo	Golden Gate	Tap-MapleGGate	0.33	9.97	
24		Heftone Banjo Orch.	Tap-MapleHeftone	0.28	5.72	[W: 17]
25		Klezmer	Tap-MapleKlez	0.30	3.63	
26		Kukuruza	Tap-MapleKukuruza	0.27	5.01	[D: 30]
27		Big Mama Sue	Tap-MapleMamaSue	0.23	3.85	[W: 4]

I have been unable to locate proper references for several of the versions in this table. I apologize to these artists and would be happy to complete the references in a future edition.

Glossary

The usage of many terms in the literature of rhythm and meter is not completely consistent. This glossary provides definitions of terms as used in *Rhythm and Transforms.* Sources for the definitions are noted where they have been taken verbatim.

absolute time: the time that is shared by most people in a given society and by physical processes [B: 117]

accent: a rhythmic event that stands out clearly from surrounding events; regularly recurring patterns of accented beats are the basis of meter [B: 219], see also p. 54

agogic accent: accents caused by changes in duration

attack point: the perceived location in time of the beginning of an event [B: 219]

auditory boundary: a timepoint at which a perceptible change occurs in the acoustic environment, see Sect. 4.2

auditory event: a way of talking about what happens at an auditory boundary; a perceptible change in the acoustic environment, see p. 85

"a" beat^1: a single timepoint that is musically significant because of its importance in a metric hierarchy [B: 117], beats mark off equal durational time units [B: 21]. Informally, the times at which a listener taps the foot to the music, see p. 87 and Fig. 1.9 on p. 15

"the" beat^2: a succession of beat timepoints (see beat^1) that are approximately equally spaced in time, commonly used in phrases such as "the beat of the song" or "keeping the beat"

beating^3: the slow undulation of the envelope of a sound caused by repeated cancellation and reinforcement, see p. 32

beat interval: a short duration of time delineated by adjacent metrical positions on the timeline, the duration between adjacent beat^1 timepoints

beat location: same as a beat^1 timepoint

boundary: see auditory boundary

bps: number of beats^1 per second

bpm: number of beats[1] per minute
cent: there are 100 cents in a musical semitone, 1200 cents in an octave; hence one cent represents an interval of $\sqrt[1200]{2} \approx 1.0005779$ to 1
chunking: mental grouping of stimuli (or events) in a manner conducive to understanding and remembering them [B: 117]
clock time: absolute time as measured by a clock [B: 117]
composition: an abstract (symbolic) form from which a variety of performances may be realized, see Sect. 12.3
duration: the perceptual correlate of the passage of time
entrainment: the mutual synchronization of two or more oscillators
event: see auditory event
Fast Fourier Transform (FFT): an algorithm that transforms signals from the time domain into the frequency domain
feature vector: a method of highlighting relevant properties of a signal and de-emphasizing irrelevant aspects, see Sect. 4.4
feature score: a feature vector calculated in a beat[2]-synchronous manner as discussed in Sect. 11.3
frequency: measured in cycles per second (Hz), frequency is the physical correlate of pitch
Gaussian: the familiar bell-shaped noise distribution
Gestalt: a structure or pattern of phenomena that constitute a complete unit with properties that are not derivable from the sum of its parts (from German, literally, shape or form)
harmonic sound: a sound is harmonic if its spectrum consists of a fundamental frequency f and partials (overtones) at integer multiples of f
Inverse Fast Fourier Transform (IFFT): an algorithm that transforms signals from the frequency domain into the time domain
interonset interval (IOI): the time interval between adjacent note events
layering: the process of building elaborate musical textures by overlapping multiple looped tracks [B: 119]
linear music: principle of musical composition and of listening under which events are understood as outgrowths or consequences of earlier events [B: 117]
literal notation: a representation of music from which the performance can be recovered completely, for example, a `.wav` file or a spectrogram, see Sect. 2.2
loudness: or volume, the perceptual correlate of sound pressure level, see p. 80
median: in a list of numbers, there are exactly as many numbers larger than the median as there are numbers that are smaller; the middle value
meter: the grouping of beats into recurrent patterns formed by a regular temporal hierarchy of subdivisions, beats and bars; meter is maintained by performers and inferred by listeners and functions as a dynamic temporal framework for the production and comprehension of musical durations in the Western tradition

noise[1]: a sound without pitch; the spectrum will typically be continuous, with significant energy spread across a range of frequencies

noise[2]: random changes to a sequence, signal, or sound

nonlinear music: principle of musical composition and of listening under which events are understood as outgrowths of general principles that govern the entire piece [B: 117] (contrast with linear music)

normal: the Gaussian distribution

overtone: synonym for partial

partials (of a sound): the sinusoidal constituents of the sound, often derived from the peaks of the magnitude spectrum

perception: the process whereby sensory impressions are translated into organized experience, see Chap. 4

perceptual present: the time span over which short term memory organizes perceptions, see p. 77

performance[1]: the act of making music, see footnote on p. 281

performance[2]: a recording (or other literal notation) resulting from a musical performance[1], see Sect. 12.3

pitch: that aspect of a tone whereby it may be ordered from low to high, the perceptual correlate of frequency

pulse: an event in music that is directly sensed by the listener and typically reflected in the physical signal, see Fig. 1.9 on p. 15

pulse train: a sequence of pulses that may give rise to the beat and/or the tatum

phase vocoder (PV): a variation on the short-time Fourier transform that extracts accurate frequency estimates from the phase differences of successive windowed FFTs

regular succession: any grid of equal time durations such as the tatum, the beat, or the measure, see pp. 54 and 86

rhythm[1]: a temporally extended pattern of durational and accentual relationships [B: 21]

rhythm[2]: a musical rhythm is a sound that evokes the sensation of a regular succession of beats [B: 158], see Sect. 4.3.6

short time Fourier transform (STFT): takes the FFT of widowed segments of a longer signal, see Fig. 5.8 on p. 124

sound pressure level (SPL): roughly equal to signal power, SPL is the physical correlate of loudness, see p. 80

spike train: see pulse train

symbolic notation: a representation of music which depicts salient features in pictorial, numeric, or geometric form, for example, a note-based musical score, a MIDI file, or a functional notation, see Sect. 2.1

synchronization: an oscillator synchronizes with a repetitive phenomenon when it locks its frequency (and/or phase) to the period (and/or phase) of the phenomenon

tactus: synonym for beat

tatum: the regular repetition of the smallest temporal unit in a performance, the "temporal atom" [B: 15] or most rapid beat[2], see Fig. 1.9 on p. 15
tempo: the rate at which the beats occur, see p. 87
timecycle: a circle that represents the passage of time; each circuit of the circle represents one repetition
timeline: an axis or dimension representing the passage of absolute time
timepoint: an instant, a point on a timeline or timecycle analogous to a geometric point in space, see Fig. 1.9 on p. 15
vertical time: temporal continuum of the unchanging, in which everything seems part of an eternal present [B: 117]

Sound Examples on the CD-ROM

The sound examples may be accessed using a web browser by opening the file `html/soundex.html` *on the CD-ROM and navigating using the html interface. Alternatively, the sound files, which are saved in the* `.mp3` *format, are playable using* `Windows Media Player`*,* `Quicktime`*, or* `iTunes`*. Navigate to* `Sounds/Chapter/` *and launch the* `*.mp3` *file by double clicking, or by opening the file from within the player. References in the body of the text to sound examples are coded with* [S:] *to distinguish them from references to the bibliography, discography, and web links.*

Sound Examples for Chapter 1

[S: 1] *A Heartbeat* (`Heartbeat.mp3 0:32`) The well known rhythmic "lub-dub" of a beating heart. See Fig. 1.1 on p. 2.

[S: 2] *Where does it start?* (`TickPhaseN.mp3 0:20`) This series of sound examples `N=0,1,...,8` performs the rhythmic pattern shown in Fig. 1.3. When `N=0`, all notes are struck equally. Which note does the pattern appear to begin on? For `N=1,...,7`, the `N`th note is emphasized. How many different starting positions are there? For `N=8`, All notes are struck equally, but the timing between notes is not exact. Does timing change the perceived starting position?

[S: 3] *Regular Interval 750* (`RegInt750.mp3 0:15`) This series of identical clicks with exactly 750 ms between each click is diagrammed in Fig. 1.4(a) on p. 6. Does the sequence *sound* identical? Observe the natural tendency of the perceptual system to collect the sounds into groups of 2, 3, or 4.

[S: 4] *Ever-Ascending Sound* (`EverRise.mp3 5:19`) The pitch of this organ-like sounds rises continuously, yet paradoxically returns to the place where is started. See p. 12.

[S: 5] *Maple Leaf Rag* (`MapleLeafRag.mp3 1:52`) Scott Joplin's (1868–1917) most famous ragtime piano piece, the *Maple Leaf Rag* was one of the first instrumental sheet music hits in America, selling over a million copies. For other versions, see Table A.2 on p. 290.

[S: 6] *Maple Tap Rag* (`MapleTapRag.mp3 1:51`) Audio beat tracking of the *Maple Leaf Rag* is demonstrated by superimposing a burst of white noise at each detected beat. In terms of the musical score, Fig. 2.3, the detected beat coincides with the eighth note pulse. See Sect. 8.3.2 for more detail.

[S: 7] *Soul* (`Soul.mp3 2:47`) *Soul* is used to illustrate various sound manipulations throughout *Rhythm and Transforms* in a "hard rock" context. *Soul* is written by Ruby Beil and Phil Schniter, who are joined by Ami Ben-Yaacov (on bass) and

Bill Huston (on drums) in this energetic performance by the band *Blip*. The song is used (and abused) with permission of the authors.

[S: 8] *Julie's Waltz* (`JuliesWaltz.mp3 2:39`) *Julie's Waltz* is used in Chap. 11 as a case study showing how beat-based feature scores can display detailed information about the timing and timbre of a performance. *Julie's Waltz*, written by Mark Schatz, appears in *Brand New Old Tyme Way* [D: 40]. The song is used with permission of the author. See also [W: 47].

Sound Examples for Chapter 2

[S: 9] *Variations of King's "Standard Pattern"* (`StanPat(a)G.mp3 0:32`), (`StanPat(a)N.mp3 0:32`), and (`StanPat(a)CA.mp3 0:32`) These demonstrate part (a) of Fig. 2.5 on p. 29; `G` refers to the starting point of the Ghana pattern using the high bell, `N` refers to the Nigerian starting point and `CA` to the Central African. (`StanPat(b).mp3 0:32`) and (`StanPat(c).mp3 0:32`) perform parts (b) and (c) using a bell for the outer ring and a drum for the inner ring.

[S: 10] *Patterns in and of Time* (`TimePat.mp3 0:25`) Demonstrates the various numerical notations of Fig. 2.6 on p. 30.

[S: 11] *Example of Drum Tablature* (`DrumTab.mp3 0:38`) This rhythmic pattern is described in Fig. 2.8 on p. 32.

[S: 12] *First two measures of Bach's Invention No. 8* (`Invention8.mp3 0:11`) Used in Fig. 2.9 on p. 32 to demonstrate Schillinger's notation.

[S: 13] $\frac{4}{4}$ *Bell* (`Bell44.mp3 0:09`) This bell pattern in $\frac{4}{4}$ is shown in timeline notation in Fig. 2.11. It is given as a MIDI event list in Table 2.2, and is shown in piano-roll notation in the first measure of Fig. 2.12. See p. 35.

[S: 14] *Midi Drums* (`MidiDrums.mp3 0:11`) MIDI piano roll notation, such as shown in Fig. 2.12 can be used to specify percussive pieces by assigning each MIDI "note" to a different drum sound. This sound example repeats measures 2, 3, and 4 of Fig. 2.12. See p. 35.

[S: 15] *Sonification of an Image* (`StretchedGirlSound.mp3 0:24`) *Metasynth* turns any picture into sound. This is the sound associated with Fig. 2.21 on p. 45.

[S: 16] *Filtering with an Image* (`MapleFiltN.mp3 0:06`) for `N=1,2`. *Metasynth* uses an image to filter a sound. The first few seconds of the *Maple Leaf Rag* [S: 5] are filtered by the images (a) (and (c)) in Fig. 2.22. The resulting spectrogram is shown in (b) (and (d)). See p. 46.

[S: 17] *Listening to Gabor Grains* (`GaborGrains.mp3 0:53`) and (`Grains2.mp3 0:36`) Individual grains as well as small sound-cloud clusters appear in these examples. See p. 47.

[S: 18] *Grains Synchronized to a Rhythmic Pattern* (`GrainRhythm.mp3 5:06`) A variety of different grain shapes are chosen at random and then synchronized to a shuffle pattern, creating an ever changing (but ever the same) rhythmic motif.

Sound Examples for Chapter 3

[S: 19] *Clustering of a Melody* (`ClusterMel.mp3 0:24`) One possible realization of the melody in Fig. 3.1 on p. 55.

[S: 20] *Playing Divisions* (`PlayDiv.mp3 0:39`) A melody is divided into simple and compound rhythm, melody, and time as described in Fig. 3.3 on p. 58.

[S: 21] *Agbekor Timelines* (`Agbekor(a).mp3 0:29`) and (`Agbekor(b).mp3 0:29`) Renditions of the timelines of Fig. 3.4 on p. 59.

[S: 22] *Drum Gahu* (`DrumGahu.mp3 0:32`) The *gankogui* bell, the *axatse* rattle, and a drum simulate this rhythmic pattern from Ghana called drum Gahu. See Fig. 3.5 on p. 60.

[S: 23] *Three Clave Patterns* (`Clave(a).mp3` 0:19), (`Clave(b).mp3` 0:32), and (`Clave(c).mp3` 0:19) from Fig. 3.7 on p. 62.

[S: 24] *The Samba* (`Samba(a).mp3` 0:32) and (`Samba(b).mp3` 0:32) simulate the two rhythmic patterns of Fig. 3.8 on p. 64.

[S: 25] *Vodou Drumming* (`Vodou(a).mp3` 0:32) and (`Vodou(b).mp3` 0:32) demonstrate the two rhythmic patterns of Fig. 3.9 on p. 65.

[S: 26] *Tala* (`TalaN.mp3` 0:30) for `N=(a),(b),(c),(d),(e)` simulate the five tala of Fig. 3.10 on p. 66.

[S: 27] *Polyrhythms* See p. 68.

(i) (`Poly32-1200.mp3 0:10`) The three vs. two polyrhythm of Fig. 3.11(a) is performed at a rate of 1.2 s per cycle. Typically, the cycle is heard as a single perceptual unit rather than as two independent time cycles.

(ii) (`Poly32-120.mp3 0:05`) The three vs. two polyrhythm is performed at a rate of 120 ms per cycle. The rhythm has become a rapid jangling.

(iii) (`Poly32-12.mp3 0:05`) The three vs. two polyrhythm is performed at a rate of 12 ms per cycle. The rhythm has become pitched, and the two rates are individually perceptible as two notes a fifth apart.

(iv) (`Poly32-12000.mp3 0:30`) The three vs. two polyrhythm is performed at a rate of 12 s per cycle. All rhythmic feel is lost.

(v) (`Poly32b.mp3 0:15`) The three vs. two polyrhythm of Fig. 3.11(b) is performed at a rate of 1.0 s per cycle. Typically, the cycle is heard as a single perceptual unit rather than as two independent time cycles.

(vi) (`Poly43.mp3 0:20`) The four vs. three polyrhythm of Fig. 3.11(c) is performed at a rate of 2.0 s per cycle.

(vii) (`Poly52.mp3 0:20`) The five vs. two polyrhythm.

(viii) (`Poly53.mp3 0:20`) The five vs. three polyrhythm.

(ix) (`Poly54.mp3 0:20`) The five vs. four polyrhythm.

(x) (`Poly65.mp3 0:20`) The six vs. five polyrhythm.

[S: 28] *Persistence of Time* (`PersistenceofTime.mp3 4:55`) This composition exploits the three against two polyrhythm as its basic rhythmic element.

[S: 29] *Gamelan Cycle* (`GamelanCycle.mp3` 0:36) A four-part realization of a *balungan* melody. See Fig. 3.12 on p. 70.

[S: 30] *Funk* (`Funk(a).mp3` 0:15) and (`Funk(b).mp3` 0:15) excerpt single cycles from James Brown's *Out of Sight* and *Papa's Got a Brand New Bag* [D: 8] to highlight the relationship between the funk groove and the African timeline. See Fig. 3.13 on p. 71.

[S: 31] *The Hip-hop Sublime* (`Sublime.mp3` 0:15) The rhythmic patterns of the sublime parallel the interlocking structures of traditional African timelines. Further examples appear in [S: 89]. See Fig. 3.14 on p. 73.

Sound Examples for Chapter 4

[S: 32] *Equal Power Noises* (`EqualPower.mp3 0:16`) The sound alternates between a Gaussian noise and a Uniform noise every second. The noises have been equalized to have the same power. Though the waveform clearly shows the distinction (see Fig. 4.3 on p. 81), it sounds steady and undifferentiated to the ear.

[S: 33] *Regular Interval T* (`RegIntT.mp3 0:20`) This series of sound examples `T=2,5,10,20,33,50,100,333,500,750,1000,3000,5000` performs a regular sequence of identical clicks with exactly `T` ms between each click, as diagrammed in Fig. 4.12(a) on p. 94. Perceptions vary by tempo. Tones are heard with small `T`, rhythms for medium `T`, and unconnected isolated clicks for large `T`.

[S: 34] *Speed-Up Events* (`SpeedUpEvent.mp3 2:33`) As the tempo increases, the events become closer together, eventually passing through the various regimes of perception. Condenses the examples of [S: 33] into one sound file.

[S: 35] *Randomly Spaced Ticks* (`GeoTick.mp3 0:20`) (`NormTick.mp3 0:20`) The interval between successive clicks in these examples is random, defined by either the geometric distribution or the normal distribution.

[S: 36] *Irregular Successions* (`IrregT.mp3 1:00`) for `T=40,400,4000,40000`. The normal (Gaussian) distribution is used to specify the times for the events. The `T` values specify the average number of events per second that occur at the peak of the distribution.

[S: 37] *Sweeping Sinusoids* (`SineSweep.mp3 0:15`) Three sinusoids begin at frequencies 150, 500, and 550 and move smoothly to 200, 400, and 600 Hz, respectively. Many complex interactions can be heard as the sinusoids change frequency. When they come to rest on the harmonic series, they merge into one perceptual entity: a note with pitch at 200 Hz. See Fig. 4.9 on p. 89.

[S: 38] *Sweeping Rhythms* (`SweepRhyN.mp3 1:18`) Three steady beats begin with periods 0.38, 0.44, and 0.9 per second and move smoothly to 0.33, 0.5, and 1.0 per second, respectively. Many complex interactions can be heard as the successions change period. When they come to rest on the periodic sequence (about 2/3 of the way through) they merge into one perceptual entity: a single rhythmic pattern with period 1 s. The example is repeated `N=4` times. For `N=1`, the phases of the three beats are aligned when they synchronize. For `N=2,3`, the phases take on arbitrary values (though the periods remain synchronized). For `N=4`, the three sequences are performed on different sounds (`stick.wav`, `clave.wav`, and `tube.wav`). See Fig. 4.10 on p. 90 for a visual representation.

[S: 39] *Windchime* (`WindChime1.mp3 1:33`) The “ding” separates from the “hum” in these wind chimes as the sound initially fuses and then separates. See p. 92.

[S: 40] *A Single Chime* (`WindChime2.mp3 1:20`) Created using a normal distribution to specify the times for the events as in [S: 36]. Each strike is allowed to ring for its full length. As in [S: 39], the “hum” separates from the “strike” though both are clearly present throughout. See p. 92.

[S: 41] *Streaming Demo* (`Streaming.mp3 1:57`) The three notes defining a major chord are repeated. The notes alternate in timbre, a synthetic trumpet followed by a synthetic flute. When played slowly, the outline of the major chord is prominent. When played more rapidly, the instruments break into two perceptual streams. See Fig. 4.11 on p. 93.

[S: 42] *Regular Durations T* (`RegDurT.mp3 0:20`) This series of sound examples `T=2,5,10,20,33,50,100,333,500,1000,3000,5000` performs a regular sequence of durations, each of length `T` ms, as diagrammed in Fig. 4.12(b) on p. 94. The sounds are a guitar pluck and a synth chord. Perceptions vary by tempo. Tones are heard with small `T`, rhythms for medium `T`, and unconnected isolated clicks for large `T`. Thus the perceptions of rhythms may be elicited equally from empty time sequences (like the clicks of [S: 33]) or from filled-time durations.

[S: 43] *Speed-Up Durations* (`SpeedUpDuration.mp3 2:33`) As the tempo increases, the durations occur closer together, eventually passing through the various regimes of perception. Condenses the examples of [S: 42] into one file.

[S: 44] *Irregular Successions of Durations* (`IrregDurT.mp3 1:00`) for `T=40, 400, 4000`. The normal (Gaussian) distribution is used to specify the durations. The `T` values specify the average number of durations per second that occur at the peak of the distribution. See Sect. 4.3.5 on p. 93.

[S: 45] *Best Temporal Grid* (`TempGrid(N).mp3 0:20`) for `N=b,c,d`. These perform the sequences shown in Fig. 4.13 parts (b), (c), and (d) on p. 95.

[S: 46] *Changing Only Pitch* (`ChangePitch.mp3 0:32`) The loudness of each note is equalized; notes are defined by changes in pitch. See p. 97.

[S: 47] *Changing Only Bandwidth* (`ChangeBWN.mp3 0:32`) for `N=1,2,3`. Each "note" is generated by a noise passed through a filter with a specified bandwidth. Loudnesses are equalized and there is no sense of pitch. Each of the three examples uses a different set of bandwidths: all lowpass filters in the first, and two kinds of bandpass filters for the second and third. See p. 97.

[S: 48] *Amplitude Modulations* (`ChangeModAMN.mp3 0:32`) for `N=1,2`. Each "note" is generated by a different rate of amplitude modulation. Loudnesses are equalized and all pitches are the same in the first example. Pitch contours follow those of [S: 46] and [S: 47] in the second example. See p. 97.

[S: 49] *Frequency Modulations* (`ChangeModFMN.mp3 0:32`) for `N= sin, sq, tri`. Each "note" is generated by a different rate of frequency modulation. Loudnesses are equalized and all pitches are the same (but for pitch shifts induced by the FM). The three versions use sine wave, square wave and triangular wave as the carrier. See p. 97.

[S: 50] *Pulsing Silences* (`PulsingSilences.mp3 3:34`) A single harmonic tone enduring throughout the piece is filtered, modulated, made noisy and otherwise manipulated into a "song with one note." Originally from [D: 43]. See p. 97.

[S: 51] *Two Periodic Sequences with Phase Differences* (`PhaseN.mp3 0:10`) for `N=0,2,10,20,30,40,50`. Two regular successions are played each with 0.5 s between clicks. The two sequences are out of phase by `N` percent, so that `N=50` is a equivalent to a double speed sequence with 0.25 s per click. Several rhythmic regimes occur, including the flam (2%), rapid doublets (10%), doublets (20%), and a galloping rhythm (30 and 40%). See Fig. 4.15 on p. 100.

[S: 52] *Two Periodic Sequences* with $T_1 \approx T_2$. (`PhaseLong.mp3 1:40`) Two sequences with periods $T_1 = 0.5$ and $T_2 = 0.503$ are played. Over time, the sound shifts through all the perceptual regimes of [S: 51]: flamming, doublets, galloping, and double speed. See Fig. 4.16 on p. 101.

[S: 53] *Three Periodic Sequences* (`Metro3N.mp3 1:40`) for `N=a,b,c`. In `a`, the three rates are $T_1 = 0.5$ and $T_{2,3} = 0.5 \pm 0.003$ s. In `b`, the three rates are $T_1 = 0.5$, $T_2 = 0.48$, and $T_3 = 0.51$ s. In `c`, the three rates are $T_1 = 0.5$, $T_2 = 0.63$, and $T_3 = 0.29$ s. The latter two contain some patterns that make rhythmic sense and others that are incomprehensibly complex. See p. 102.

[S: 54] *Nothing Broken in Seven* (`Broken7.mp3 3:29`) A single six-note isorhythmic melody is repeated over and over, played simultaneously at five different speeds. See Fig. 4.16 on p. 101.

[S: 55] *Phase Seven* (`PhaseSeven.mp3 3:41`) A single eight-note isorhythmic melody is repeated over and over, played simultaneously at five different speeds. See Fig. 4.16 on p. 101.

[S: 56] *One-hundred Metronomes* (`MetroN.mp3 2:00`) for `N=10,50,100`. These three examples play `N` simultaneous regular successions, each with a randomly chosen period. The appearance is of a random cacophony. Inspired by György Ligeti's *Poeme Symphonique* [D: 31]. See p. 102.

[S: 57] *Listening to Individual Feature Vectors #1* Feature vectors for the *Maple Leaf Rag* are made audible using the technique of Fig. 4.18 on p. 105. See Sect. 4.4.2.
- (i) (`MapleCBFeature9.mp3 0:44`) The feature vector from the ninth critical band (before the derivative) gives one of the clearest rhythmic percepts.
- (ii) (`MapleCBFeature9diff.mp3 0:44`) The feature vector from the ninth critical band (after the derivative) gives one of the clearest rhythmic percepts.
- (iii) (`MapleCBFeature2.mp3 0:44`) The feature vector from the second critical band (before the derivative) gives almost no rhythmic percept.
- (iv) (`MapleCBFeature2diff.mp3 0:44`) The feature vector from the second critical band (after the derivative) gives almost no rhythmic percept.

[S: 58] *Listening to Feature Vectors* Feature vectors for the *Maple Leaf Rag* are made audible using the technique of Fig. 4.18 on p. 105. See Sect. 4.4.2.
- (i) (`MapleFeatureAll.mp3 0:44`) All the feature vectors (before the derivative) from all the critical bands are summed, leaving a clear rhythmic percept.
- (ii) (`MapleFeatureAlldiff.mp3 0:44`) All the feature vectors (after the derivative) from all the critical bands are summed, leaving a clear rhythmic percept.

[S: 59] *Listening to Individual Feature Vectors #2* Feature vectors for the *Maple Leaf Rag* are made audible using the technique of Fig. 4.18 on p. 105. See Sect. 4.4.4.
- (i) (`MapleFeature1.mp3 0:44`) The energy feature vector.
- (ii) (`MapleFeature2.mp3 0:44`) The group delay feature vector.
- (iii) (`MapleFeature3.mp3 0:44`) The spectral center feature vector.
- (iv) (`MapleFeature4.mp3 0:44`) The spectral dispersion feature vector.

[S: 60] *Povel's Sequences* (`PovelN.mp3 0:20`), `N=1,2,...,35`. The thirty-five sequences from Povel and Essens [B: 174] are ordered from simplest to most complex. See [B: 202] for further discussion.

Sound Examples for Chapter 8

[S: 61] *La Marseillaise* (`Marseillaise.mp3 0:09`) The first four bars of the French National anthem. See Fig. 8.1 on p. 196.

[S: 62] *MIDI Beat Tracking with Oscillators I* (`Michelle-MIDIOscN.mp3 0:50`), `N=1,2`. The "jazz" version `jazz1_fast-rep_1.mid` of the Beatles' *Michelle* from the MIDI collection [W: 35] is beat tracked using an adaptive phase-reset oscillator in 1 and using a wavetable oscillator in 2. A burst of white noise is superimposed at each detected beat. See Sect. 8.2.1.

[S: 63] *MIDI Beat Tracking with Oscillators II* (`Maple-DrisceoilOsc.mp3`), (`Maple-RoacheOsc.mp3`), and (`Maple-TrachtmanOsc.mp3`) Three MIDI versions of the *Maple Leaf Rag* by T. O. Drisceoil (at [W: 9]), J. Roache [W: 43], and

W. Trachtman [W: 51] are beat tracked using an adaptive wavetable oscillator. The detected beat locations are indicated by the superimposed stick sound. See Sect. 8.2.1.

[S: 64] *MIDI Beat Tracking with Oscillators III* (`Maple-MIDIOscSlow.mp3`) Trachtman's [W: 51] version of the *Maple Leaf Rag* is beat tracked using an adaptive wavetable oscillator initialized near the eighth-note rate. The detected beat locations are indicated by the superimposed stick sound. See Sect. 8.2.1.

[S: 65] *Statistical MIDI Beat Tracking I* (`Maple-T-Stat.mp3 2:53`) and (`Maple-R-Stat.mp3 2:53`) Two MIDI versions of the *Maple Leaf Rag* by Trachtman [W: 51] (indicated by `T`) and Roache [W: 43] (indicated by `R`) are beat tracked using the statistical approach of Sects. 7.5 and 8.2.2. The detected beat locations are indicated by the superimposed stick sound.

[S: 66] *Michelle-MIDIStat* (`Michelle-MIDIStat.mp3 0:50`) A MIDI version of the Beatles' *Michelle* is beat tracked using the statistical approach. The detected beat locations are indicated by the superimposed stick sound. See Sect. 8.2.2.

[S: 67] *Statistical MIDI Beat Tracking II* (`Maple-T-StatSlow.mp3`) and (`Maple-R-StatSlow.mp3`) Two MIDI versions of the *Maple Leaf Rag* by Trachtman [W: 51] and Roache [W: 43] are beat tracked using the statistical approach. The procedure is initialized near the eighth-note pulse. See Sect. 8.2.2.

[S: 68] *Statistical MIDI Beat Tracking III* (`Maple-T-VerySlow.mp3`) and (`Maple-T-Fast.mp3`) The *Maple Leaf Rag* is beat tracked using the statistical approach. The procedure is initialized near the sixteenth-note tatum (fast) and near the $\frac{2}{4}$ measure (very slow). Observe that the phase of the slow tracking is locked to the syncopated "and" rather than the "one." See Fig. 8.15 on p. 208.

[S: 69] *Three Against Two Polyrhythm* (`Poly32.mp3` 0:15) The polyrhythm of Fig. 8.17 on p. 211 is analyzed from the audio in Fig. 8.18 using the DFT and in Fig. 8.19 using the PT.

[S: 70] *Three Against Two Polyrhythm Accelerating* (`Poly32acc.mp3` 0:15) The polyrhythm of [S: 69] depicted in Fig. 8.17 on p. 211 is slowly accelerated. The audio is analyzed in Fig. 8.20 using various transform methods, which fail due to the unsteady pulse.

[S: 71] *Audio Beat Tracking* is demonstrated on 16 musical works by superimposing a burst of white noise at each detected beat. The excerpts are described in Table A.1 on p. 289 and discussed at length in Sect. 8.3.2.

[S: 72] *Howell's Delight II* (`Tap-Howell2.mp3 0:58`) Audio beat tracking of the song by the Baltimore Consort [D: 1]. In this version, the initial value of the period $\mathcal{T}$ is doubled to $[0.4, 0.8]$ and the method locates the $\frac{6}{8}$ measure at about $\mathcal{T} = 0.72$ s. Observe how much musical activity occurs between each detected timepoint. Compare to [S: 71](12). See Sect. 8.3.2.

[S: 73] *Julie's Waltz II* (`Tap-Julie2.mp3 0:58`) Audio beat tracking of *Julie's Waltz* [D: 40], [S: 8]. In this version, the initial value of the period $\mathcal{T}$ is doubled to $[0.4, 0.8]$ and the method locates the quarter note pulse at $\mathcal{T} \approx 0.61$ s. instead of the eighth note tatum at $\mathcal{T} \approx 0.30$ as in [S: 71](15). See Sect. 8.3.2.

[S: 74] *Ska Tap* (`Tap-SkaOn.mp3 0:44`) and (`Tap-SkaOff.mp3 0:44`) Audio beat tracking of *Lion Says* by Prince Buster and the Ska [D: 10]. When the initial period $\mathcal{T}$ is doubled to $[0.4, 0.8]$, the method can converge to either of two phases which correspond to the "on-" and the "off-" beat. Compare to [S: 71](14). See Sect. 8.3.2.

[S: 75] *Maple Tap* $\frac{3}{2}$ (`MapleTap3-2.mp3 0:44`) Audio beat tracking of the *Maple Leaf Rag* with a tightly constrained initial period causes the algorithm to lock onto a 3 : 2 entrainment at $\mathcal{T} \approx 0.52$ instead of the actual pulse at $\mathcal{T} \approx 0.34$. Compare to [S: 71](5). See Sect. 8.3.2.

[S: 76] *Audio Beat Tracking* of 27 performances of the *Maple Leaf Rag* are demonstrated by superimposing a burst of white noise at each detected beat. The excerpts are described in Table A.2 on p. 290 and discussed at length in Sect. 8.3.2.

Sound Examples for Chapter 9

[S: 77] *Maple Drums* (`MapleDrums.mp3 1:51`) The beat of the *Maple Leaf Rag* is regularized so that each beat interval contains the same number of samples. A preprogrammed drum line is superimposed to emphasize the metronomic regularity. See Fig. 9.2 on p. 225.

[S: 78] *Rhythmic Beat Manipulations* (`MapleBeatMod(N).mp3 1:41`) for `N = a, b, c`. Individual beats of the *Maple Leaf Rag* are stretched and compressed and then played together with the original. See Fig. 9.3 on p. 225.

[S: 79] *Polyrhythmic Rags #1 and #2* (`PolyrhythmicRagsN.mp3 1:36`) for `N = 1, 2`. Individual beats of the *Maple Leaf Rag* are stretched by a factor of $\frac{4}{3}$ and every fourth beat is removed. This is then played simultaneously with the original. See Fig. 9.3(d) on p. 225.

[S: 80] *Changing Tempo* The tempo of the *Maple Leaf Rag* is changed in a variety of ways using the phase vocoder `PV.m`. See Sect. 9.1.

(i) *Maple Leaf Rag* at quarter speed (`MapleQuarter.mp3 1:00`)
(ii) *Maple Leaf Rag* at half speed (`MapleHalf.mp3 1:00`)
(iii) *Maple Leaf Rag* at double speed (`MapleDouble.mp3 1:00`)
(iv) *Maple Leaf Rag* at four times normal tempo (`Maple4x.mp3 0:27`)
(v) *Maple Leaf Rag* at eight times normal tempo (`Maple8x.mp3 0:14`)
(vi) *Maple Leaf Rag* at 16 times normal tempo (`Maple16x.mp3 0:07`)
(vii) *Maple Leaf Rag* at 32 times normal tempo (`Maple32x.mp3 0:03`)
(viii) *Maple Leaf Rag* at 64 times normal tempo (`Maple64x.mp3 0:01`)

[S: 81] *Maple Sleep Rag* (`MapleSleepRag.mp3 3:53`) An elaboration of the half speed version of the *Maple Leaf Rag* [S: 80](ii). See Sect. 9.1.

[S: 82] *Time Stretching I* Extreme time stretching can be an interesting special effect even when not synchronized with the beat. See Sect. 9.1.

(i) *Gong* (`Gong.mp3 0:05`) A single strike of a gong.
(ii) *Long Gong* (`LongGong.mp3 2:26`) The same gong, stretched in a variety of ways to bring out details of the evolution of the sound that are impossible to hear at the normal rate.

[S: 83] *Time Stretching II* Extreme time stretching using beat boundaries. See Sect. 9.1.

(i) *Very Slow #1* (`VerySlow1.mp3 4:00`) A melange of beats chosen from several different songs, all equalized in time and then stretched by a factor of eight.
(ii) *Very Slow #2* (`VerySlow2.mp3 4:00`) The same as (i), but modified with a variety of beat-based filters.
(iii) *Very Slow #3* (`VerySlowInspective.mp3 3:55`) The piece in (iv) is slowed by a factor of eight.
(iv) *Inspective Latency* (`InspectiveLatency.mp3 3:47`) An adaptively tuned piece from [B: 196] is presented here for comparison with its time stretched version in (iii).

[S: 84] *Beat Filtered Rag* (`BeatFilteredRag.mp3 1:51`) A variety of beat-based filters and delay effects are applied to the *Maple Leaf Rag*. See Sect. 9.2.

[S: 85] *Beat Gated Rag* (`BeatGatedRag.mp3 1:51`) A variety of beat-based gates and envelopes are applied to the *Maple Leaf Rag*. See Sect. 9.2.

[S: 86] *Gar Fael Elpam* (`GarFaelElpam.mp3 0:44`) The first 44 seconds of the *Maple Leaf Rag* is played backwards. See Sect. 9.3.

[S: 87] *Beat Reversed Rag* (`BeatReversedRag.mp3 3:44`) The audio in each beat interval is reversed in time so that the sounds are backwards but the piece itself moves forwards. It appears as the *Maple Leaf Rag* is played on an organ or calliope. See Sect. 9.3 and Fig. 9.5 on p. 229.

[S: 88] *Wrongly Reversed Rag* (`WronglyReversedRag.mp3 1:44`) The audio is reversed in approximately beat-sized chunks, but with boundaries that bear no relationship to the beat boundaries. See Sect. 9.3.

[S: 89] *Averaged Sublimes* (`SublimeN.mp3` 0:30) for `N=2,5,30,50`. In each case, `N` successive 8-beat cycles (measures) of the "hip-hop sublime"[S: 31] are averaged together. See Sect. 9.4.

[S: 90] *Noisy Leaf Rag* (`NoisyLeafRag.mp3 1:06`) The noise floor (9.1) is calculated at each beat and all information above the noise floor is removed. When transformed (via the IFFT) back into the time domain, only the noisy parts of the sound remain. See Sect. 9.5 and Fig. 9.6 on p. 231.

[S: 91] *Maple Noise Rag* (`MapleNoiseRag.mp3 1:52`) The *Noisy Leaf Rag* [S: 90] (the only sound source used in this composition) is augmented with a variety of beat-based techniques such as those of Sect. 9.2. See Sect. 9.5.

[S: 92] *Just Noise Rag* (`JustNoiseRag.mp3 1:30`) Another elaboration of the *Noisy Leaf Rag* [S: 90]. See Sect. 9.5.

[S: 93] *Noisy Souls* (`NoisySouls.mp3 2:39`) *Soul* [S: 7] (the only sound source used in this composition) is augmented with a variety of beat-based techniques such as those of Sect. 9.2. See Sect. 9.5.

[S: 94] *Noisy StrangeTree* (`NoisyStrangetree.mp3 1:01`) All information above the noise floor of [S: 95] is removed at each beat. The voice is particularly striking. In the verse, the consonants are (almost) identifiable. In the chorus, where long notes are sustained, the voice effectively disappears. See Sect. 9.5 and Fig. 9.6 on p. 231.

[S: 95] *StrangeTree* (`StrangeTree.mp3 3:36`) An early composition by the author in a straightforward "pop" style used here to demonstrate many of the signal processing techniques such as [S: 94], [S: 106], and [S: 123].

[S: 96] *Signal Leaf Rag* (`SignalLeafRag.mp3 1:06`) All information below the noise floor is removed. When transformed (via the IFFT) back into the time domain, the noisy percussive elements have been removed, leaving the tonal material intact. See Sect. 9.5 and Fig. 9.6 on p. 231.

[S: 97] *Listening to Peaks* (`MapleNPeaks.mp3 1:06`) for `N=1,3,15,50,250`. `N` peaks are identified within each beat interval in the *Maple Leaf Rag*. All other information is removed. See Sect. 9.5.

[S: 98] *Atonal Leaf Rag* (`AtonalLeafRag.mp3 1:38`) The peaks are identified and removed within each beat interval in the *Maple Leaf Rag*, leaving an atonal rhythmic bed. See Sect. 9.5.

[S: 99] *Atonal Leaf Rag #2* (`AtonalLeafRag2.mp3 1:53`) The rhythmic bed with beat-based filters applied. See Sect. 9.5.

[S: 100] *Atonal Soul* (`AtonalSoul.mp3 2:39`) The peaks are identified and removed within each beat interval in the *Maple Leaf Rag*, leaving an atonal rhythmic bed that is processed with beat-based filters. See Sect. 9.5.

[S: 101] *Sixty-Five Hertz Gong* (`Gong65.mp3 0:05`) Spectral mapping of the gong [S: 82](i) into a harmonic template with fundamental at 65 Hz. See Sect. 9.6.1.

[S: 102] *Harmonic Cymbals* (`HarmCymbal.mp3 0:23`) and (`HarmCymbal.avi 0:23`) An inharmonic cymbal is spectrally mapped into a harmonic spectrum. The resulting sound is pitched and capable of supporting melodies and chords. See Sect. 9.6.1.

(i) The original sample contrasted with the spectrally mapped version
(ii) A simple "chord" pattern played with the original sample, and then with the spectrally mapped version

[S: 103] *Maple in 65 Hz* (`Maple65.mp3 1:52`) Spectral mapping of the *Maple Leaf Rag* into a harmonic template with fundamental at 65 Hz. See Sect. 9.6.1.

[S: 104] *Sixty-Five Maples* (`SixtyFiveMaples.mp3 1:57`) Spectral mapping of the *Maple Leaf Rag* into a harmonic template with fundamental at 65 Hz. The only sound source is [S: 103] which is rearranged and post-processed. See Sect. 9.6.1.

[S: 105] *Sixty-Five Souls* (`Soul65N.mp3 2:47`) for `N=PV,FFT`. Spectral mapping of *Soul* [S: 7] into a harmonic template with fundamental at 65 Hz using the phase vocoder and using the beat-synchronous FFT. See Sects. 9.6.1 and 9.8.2.

[S: 106] *Sixty-Five StrangeTrees* (`StrangeTree65.mp3 3:35`) Spectral mapping of *StrangeTree* [S: 95] into a harmonic template with fundamental at 65 Hz. See Sect. 9.6.1.

[S: 107] *Spectral Mappings of Harmonic Sounds to 11-tet Sounds* (`Tim11tet.mp3 1:20`) Several different instrumental sounds alternate with their 11-tet spectrally mapped versions: See Sect. 9.6.2.

(i) Harmonic trumpet compared with 11-tet trumpet
(ii) Harmonic bass compared with 11-tet bass
(iii) Harmonic guitar compared with 11-tet guitar
(iv) Harmonic pan flute compared with 11-tet pan flute
(v) Harmonic oboe compared with 11-tet oboe
(vi) Harmonic "moog" synth compared with 11-tet "moog" synth
(vii) Harmonic "phase" synth compared with 11-tet "phase" synth

[S: 108] *The Turquoise Dabo Girl* (`DaboGirl.mp3 4:16`) The spectrally mapped instrumental sounds of [S: 107] are sequenced into an 11-tet piece. Many of the kinds of effects normally associated with (harmonic) tonal music can occur, even in such strange settings as 11-tet. See Sect. 9.6.2.

[S: 109] *Maple N-tet PV* (`MapleNtetPV.mp3 0:16`) for `N=4,5,10,100`. Spectral mapping of the *Maple Leaf Rag* into a `N`-tet destination template using the phase vocoder. See Sects. 9.6.2 and 9.8.2.

[S: 110] *Maple N-tet FFT* (`MapleNtetFFT.mp3 0:16`) for `N=4,5,10,100`. Spectral mapping of the *Maple Leaf Rag* into a `N`-tet destination template using the beat-based FFT. See Sects. 9.6.2 and 9.8.2.

[S: 111] *Pentatonic Rag* (`PentatonicRag 2:34`) Spectral mapping of the *Maple Leaf Rag* into a 5-tet destination spectrum, augmented with beat-based filters and gates. See Sect. 9.6.2.

[S: 112] *Maple 5-tet Video* (`Maple5tet.avi 0:38`) The first five seconds of the spectral mapping of the *Maple Leaf Rag* into a 5-tet destination template. See Fig. 9.9 on p. 238.

[S: 113] *Pentatonic Souls* (`PentatonicSoulN 2:48`) for `N=PV,FFT`. Spectral mapping of *Soul* [S: 7] into a 5-tet destination spectrum by the phase vocoder and by the beat-synchronous FFT. See Sects. 9.6.2 and 9.8.2.

[S: 114] *Scarlatti 5-tet* (`Scarlatti5tet 1:40`) Spectral mapping of Scarlatti's K517 sonata into a 5-tet destination spectrum. See Sect. 9.6.2.

[S: 115] *Make Believe Rag* (`MakeBelieveRag.mp3 3:37`) Spectral mappings of the *Maple Leaf Rag* into 3, 4, 5, and 7-tet are combined and sequenced in a beat-synchronous manner. Changes in tuning play a role analogous to chord changes in a tonal context. See Sect. 9.6.3.

[S: 116] *Local Anomaly* (`LocalAnomaly.mp3 3:27`) Adaptively tuned from a re-orchestrated standard MIDI file drum track, *Local Anomaly* first appeared in [D: 43] and is discussed at length in Chap. 9 of [B: 196].

[S: 117] *Local Variations* (`LocalVariations.mp3 2:19`) A single fixed spectral band filter is applied to [S: 116]. See Sect. 9.7.1.

[S: 118] *Maple Freeze Rags* (`MapleFreezeRag(N).mp3 1:28`) for `N=a,b`. A spectral freeze is applied to the *Maple Leaf Rag*, with the left and right tracks frozen rhythmically according to the necklace diagrams of Fig. 9.12 on p. 241. See Sect. 9.7.2.

[S: 119] *Soul Freezes* (`SoulFreeze(N).mp3 2:37`) for `N=a,b`. A spectral freeze is applied to *Soul*, with the left and right tracks frozen rhythmically according to the necklace diagrams of Fig. 9.12 on p. 241. See Sect. 9.7.2.

[S: 120] *Frozen Souls* (`FrozenSouls.mp3 2:37`) The *Soul Freezes* of [S: 119] are used as raw material for this elaboration. See Sect. 9.7.2.

[S: 121] *Three Ears* (`ThreeEars.mp3 4:24`) As each new note sounds, its pitch (and that of all currently sounding notes) is adjusted microtonally (based on its spectrum) to maximize consonance. The adaptation causes interesting glides and microtonal pitch adjustments in a perceptually sensible fashion. *Three Ears* first appeared in [D: 42] and is discussed in Chap. 8 of [B: 196].

[S: 122] *Mirror Go Round* (`MirrorGoRound.mp3 3:25`) The harmonic sieve is applied to *Three Ears* [S: 121]. See Sect. 9.7.3.

[S: 123] *Sievetree* (`SieveTree.mp3 3:48`) The harmonic sieve is applied to *Strangetree* [S: 95]. Compare especially to *Sixty-Five StrangeTrees* [S: 106] which spectrally maps the same piece into a harmonic series. See Sect. 9.7.3.

[S: 124] *Phase Space* (`PhaseSpace.mp3 3:10`) Used for comparison with [S: 125].

[S: 125] *Reflective Phase* (`ReflectivePhase.mp3 3:26`) The harmonic sieve is applied to *Phase Space* [S: 124]. See Sect. 9.7.3.

[S: 126] *Instant Leaf Rag* (`InstantLeafRag.mp3 1:51`) A pitch extraction algorithm is applied to each beat interval of the *Maple Leaf Rag*. The sound is spectrally mapped to a destination spectrum that has a fundamental equal to the identified pitch. See Sect. 9.7.4.

[S: 127] *Instant Nightmare* (`InstantNightmare.mp3 3:36`) The periodicity transform identifies the three periodicities with greatest power in each beat interval. These periodicities define the destination spectrum (consisting of all harmonics of the three basic periods). The sound is spectrally mapped to this destination spectrum. Percussion is added and the beats of the piece are rearranged. Ligon comments [B: 129] "Outrageously cool; Beefheart on controlled substances!" See Sect. 9.7.4.

Sound Examples for Chapter 10

[S: 128] *Friend of the Devil of the Friend* (`FriendneirF.mp3 0:40`) The first twenty seconds of the classic song by the Grateful Dead [D: 19] is played forwards and then backwards. Can you hear where the change is made? See Sect. 10.1.

[S: 129] *Devil of a Friend* (`DevilofaFriend.mp3 0:46`) A verse and chorus of the classic song by the Grateful Dead [D: 19] is played backwards. The disjoint and sometimes amusing lyrics indicate a stronger sense of linearity than the instrumental sections alone. See Sect. 10.1.

[S: 130] *Maple Waltzes #1 and #2* (`MapleWaltzN.mp3 1:06`) for `N=1,2`. By removing every fourth beat, the $\frac{4}{8}$ time signature of the *Maple Leaf Rag* is transformed into $\frac{3}{8}$. The `N=1` version removes the final beat of each measure. The `N=2` version removes the third beat of each measure. See Sect. 10.3 and Fig. 10.1 on p. 255.

[S: 131] *Maple Leaf Waltz* (`MapleLeafWaltz.mp3 2:03`) The two waltzes in [S: 130] are merged, combined, and elaborated. See Sect. 10.3 and Fig. 10.1 on p. 255.

[S: 132] *Soul Waltzes* (`SoulWaltzN.mp3 2:18`), `N=1,2,3,4`. By removing every fourth beat, the $\frac{4}{4}$ time signature of *Soul* [S: 7] is transformed into $\frac{3}{4}$. Except for vocal sections where lyrics are truncated, the change is quite smooth. The four versions remove different beats, causing different rhythmic patterns. See Sect. 10.3.

[S: 133] *Bond's Waltz* (`BondsWaltz.mp3 0:33`) By removing every fourth beat, the $\frac{4}{4}$ time signature of the *James Bond Theme* [D: 6] is transformed into $\frac{3}{4}$. See Sect. 10.3.

[S: 134] *Take Four* (`TakeFour.mp3 1:18`) By removing every fifth beat from Grover Washington's classic *Take Five* [D: 45], the $\frac{5}{4}$ time signature is transformed into $\frac{4}{4}$. See Sect. 10.3.

[S: 135] *Julie's March* (`JuliesMarch.mp3 1:06`) By removing every third beat from *Julie's Waltz* ([D: 40], [S: 8]), the $\frac{3}{4}$ time signature is transformed into $\frac{2}{4}$. See Sect. 10.3.

[S: 136] *Howell in* $\frac{2}{4}$ (`Howell24.mp3 0:58`) By removing every third beat from Howell's Delight [D: 1], the $\frac{3}{4}$ time signature is transformed into $\frac{2}{4}$. See Sect. 10.3.

[S: 137] *Half Leaf Rags* (`HalfLeafRag1.mp3 0:49`), (`HalfLeafRag2.mp3 0:49`) Two out of every four beats are removed from the *Maple Leaf Rag*. See Sect. 10.3.

[S: 138] *Half a Soul* (`HalfSoul1.mp3 1:17`), (`HalfSoul2.mp3 1:17`) Two out of every four beats are removed from *Soul* [S: 7]. See Sect. 10.3.

[S: 139] *Quarter Leaf Rag* (`QuarterLeafRag.mp3 0:25`) Three out of every four beats are removed from the *Maple Leaf Rag*. See Sect. 10.3.

[S: 140] *Quarter Soul* (`QuarterSoul.mp3 0:40`) Three out of every four beats are removed from *Soul* [S: 7]. The complete song "fits" into forty seconds. See Sect. 10.3.

[S: 141] *Magic Leaf Rag* (`MagicLeafRag.mp3 2:29`) The *Half Leaf Rags* [S: 137] are processed using beat-based filters and gates in conjunction with the beat-synchronous time delays of Fig. 9.3. See Sects. 9.1, 9.2 and 10.3.

[S: 142] *Make It Brief Rag* (`MakeItBriefRag.mp3 2:01`) The *Quarter Leaf Rag* [S: 139] is processed using the beat-based filters and gates of Sect. 9.2. See Sect. 10.3.

[S: 143] *Backwards Leaf Rag* (`BackwardsLeafRag.mp3 1:39`) The beats of the *Maple Leaf Rag* are played in reverse order: the song begins with the final beat interval and progresses in orderly fashion to the first beat interval. See Sect. 10.4.

[S: 144] *Backwards Soul* (`BackwardsSoul.mp3 2:16`) The beats of *Soul* [S: 7] are played in reverse order: the vocals are especially interesting. See Sect. 10.4.

[S: 145] *Backwards Bond* (`BackwardsBond.mp3 0:52`) The beats of the *Theme from James Bond* [D: 6] are played in reverse order. See Sect. 10.4.

[S: 146] *Random Leaf Rag #1* (`RandomLeafRag.mp3 1:30`) The beats of the *Maple Leaf Rag* are played in random order. All sense of tonal progression is gone. The rhythm and flow of the piece is fundamentally disturbed. Only the timbre of the piano remains. See Sect. 10.5.

[S: 147] *Random Soul #1* (`RandomSoul.mp3 1:27`) The beats of *Soul* [S: 7] are played in random order. The rhythm of the piece is fundamentally disturbed, though some of the feel remains. The timbre of the guitars and voice remain. See Sect. 10.5.

[S: 148] *Permutations of the Rag* The beats of the *Maple Leaf Rag* are permuted on a measure-by-measure basis.
(i) 1234 → 1324 (`Maple1324.mp3 0:43`)
(ii) 1234 → 1423 (`Maple1423.mp3 0:43`)
(iii) 1234 → 1432 (`Maple1432.mp3 0:43`)
(iv) 1234 → 2413 (`Maple2413.mp3 0:43`)
(v) 1234 → 4321 (`Maple4321.mp3 0:43`)
See Sect. 10.5 and Fig. 10.2 on p. 258.

[S: 149] *Permutation Leaf Rag* (`PermutationLeafRag.mp3 0:43`) A randomly chosen permutation is applied to each measure. See Sect. 10.5.

[S: 150] *Permutations of Soul* (`PermutationsofSoul.mp3 2:22`) A randomly chosen permutation is applied to each measure. See Sect. 10.5.

[S: 151] *Random Leaf Rag #2* (`RandomLeafRag2.mp3 0:37`) Choose beats randomly from among those that occupy the same relative location in the measure. See Sect. 10.5.

[S: 152] *Random Souls #2* (`RandomSouls2.mp3 2:37`) Choose beats randomly from among those that occupy the same relative location in the measure. See Sect. 10.5.

[S: 153] *Subdividing the Beat* (`MapleSnippetsN.mp3 0:37`) for `N=2,3,4,8, 12`. For larger `N`, the sound reflects the process of subdivision and destroys the original timbre.

[S: 154] *Maple Leaf Collage* (`MapleLeafCollage.mp3 0:37`) Choose each beat randomly from different versions of the *Maple Leaf Rag.* Ligon [B: 129] says, "Made me laugh out loud." See Sect. 10.6.

[S: 155] *Rag Bag #1* and *Rag Bag #2* (`RagBag1.mp3 1:27`) (`RagBag2.mp3 1:33`) Choose each beat randomly from different versions of the *Maple Leaf Rag.* See Sect. 10.6.

[S: 156] *Beat-Synchronous Cross-Performances* (`nX*Y.mp3 1:23`) `X` and `Y` are performances of the *Maple Leaf Rag* from Table A.2 on p. 290. Beats are chosen sequentially from the two performances and merged using one of the methods `n`, which can be `(i)--(vii)`, as described on p. 261. The 22 soundfiles are listed in Table 10.2 on p. 261 and discussed in Sect. 10.7.

[S: 157] *Grab Bag Rags #1, #2,* and *#3* (`GrabBagRagN.mp3 1:27`) for `N=1,2,3`. Choose each beat randomly from different versions of the beat-synchronous cross-performances of [S: 156]. See Sect. 10.7.

Sound Examples for Chapter 12

[S: 158] *Maple Beats* (`MapleBeatsN.mp3 0:01`), `N=1,2,3,4`. Individual beats from the *Maple Leaf Rag* are readily identifiable once the piece is well known. See Sect. 12.2.

[S: 159] *Beat Game* (`BeatGameN.mp3 0:01`), `N=1,2,...,39`. Individual beats are chosen randomly from 13 of the pieces in Tables A.1 and A.2. Can you tell which beat is from which piece? See Sect. 12.2.

Bibliography

References in the body of the text to the bibliography are coded with [B:] *to distinguish them from references to the discography, sound examples, and websites.*

[B: 1] K. Agawu, *African Rhythm*, Cambridge University Press, 1995.

[B: 2] M. M. Al-Ibrahim, "A multifrequency range digital sinusoidal oscillator with high resolution and uniform frequency spacing," *IEEE Trans. on Circuits and Systems-II,* Vol. 48, No. 9, Sept. 2001.

[B: 3] Safî al-Din al-Urmawî, *Kitâb al-Adwâr* 1252, trans. R. Erlanger in *La Musique arabe*, Paul Geuthner, Paris, 1938.

[B: 4] W. Anku, "Circles and time: a theory of structural organization of rhythm in African music," *Music Theory Online*, Vol. 6, No. 1, Jan. 2000.

[B: 5] Abramowitz and Stegun, *Handbook of Mathematical Functions*, Dover Pubs. Inc. NY 1972.

[B: 6] R. Arora and W. A. Sethares, "Adaptive wavetable oscillators," to appear *IEEE Trans. Signal Processing*, 2007. [A preliminary version is on the CD in `Papers/adaptosc.pdf`.]

[B: 7] American Standards Association, *USA Standard Acoustical Terminology,* New York, 1960.

[B: 8] D. Bañuelos, *Beyond the Spectrum of Music* DMA Thesis, University of Wisconsin, 2005. [Thesis appears (with permission) on the CD in `Papers/BeyondtheSpectrum.pdf`].

[B: 9] I. Barrow, *Lectiones Geometricae* (trans. E. Stone), London, 1735.

[B: 10] B. Beever, *Guide to Juggling Patterns*, `http://www.jugglingdb.com/articles/index.php`, 2000.

[B: 11] W. E. Benjamin, "A theory of musical meter," *Music Perception,* Vol. 1, No. 4, 355–413, Summer 1984.

[B: 12] H. Bergson, *Time and Free Will: An Essay on the Immediate Data of Consciousness*, translated by F.L. Pogson, M.A. London: George Allen and Unwin 1910, reprinted Dover, 2001.

[B: 13] J. M. Bernardo and A. F. M. Smith, *Bayesian Theory*, Wiley, 2001.

[B: 14] W. Berry, "Metric and rhythmic articulations in music," *Music Theory Spectrum*, 7: 7–33, 1985.

[B: 15] J. A. Bilmes "Techniques to foster drum machine expressivity," *Proc. 1993 Int. Comp. Music Conf.*, San Francisco, 276–283, 1993.

[B: 16] J. Blacking, *How Musical is Man?* Univ. Washington Press, Seattle, 1973.

[B: 17] T. L. Bolton, "Rhythm," *American Journal of Psychology,* 6: 145–238, 1894.

[B: 18] A. Bregman, *Auditory Scene Analysis: The Perceptual Organization of Sound*, MIT Press, Cambridge, MA 1990.

[B: 19] G. Brelet, *Le Temps Musical*, Paris, 1949.

[B: 20] J. Brown, "Calculation of a constant Q spectral transform," *J. Acoustical Society of America,* 89, 425–434, 1991.

[B: 21] J. Brown, "Determination of the meter of musical scores by autocorrelation," *J. Acoustical Society of America,* 94 (4), 1953–1957, Oct. 1993.

[B: 22] D. Butler, *The Musician's Guide to Perception and Cognition,* Schirmer Books, Macmillan Inc., 1992.

[B: 23] J. B. Burl, "Estimating the basis functions of the Karhunen-Loeve transform," *IEEE Trans. Acoustics, Speech, and Signal Processing,* Vol. 37, No. 1, 99–105, Jan. 1989.

[B: 24] C. S. Burrus, R. A. Gopinath, and H. Guo, *Wavelets and Wavelet Transforms,* Prentice Hall, NJ 1998.

[B: 25] P. Cariani, "Temporal coding of periodicity pitch in the auditory system: an overview," *Neural Plasticity*, 6, No. 4, 147–172, 1999.

[B: 26] P. Cariani, "Neural timing nets," *Neural Networks*, 14, p. 737–753, 2001.

[B: 27] A. T. Cemgil, B. Kappen, P. Desain and H. Honing "On tempo tracking: tempogram representation and Kalman filtering," *J. New Music Research*, Vol. 28, No. 4, 2001.

[B: 28] A. T. Cemgil and B. Kappen, "Monte Carlo methods for tempo tracking and rhythm quantization," *J. Artificial Intelligence Research,* Vol. 18., 45–81, 2003.

[B: 29] S. Chen and D. L. Donoho, "Basis pursuit," *Proc. 28th Asilomar Conference on Signals, Systems, and Computers,* Pacific Grove, CA, 41–44 Nov. 1994.

[B: 30] J.M. Chernoff, *African Rhythm and African Sensibility,*" Univ. of Chicago Press, Chicago, IL, 1973.

[B: 31] E. F. Clark and C. L. Krumhansl, "Perceiving musical time," *Music Perception*, Vol. 7, No. 3, 213–252, Spring 1990.

[B: 32] T. Clifton, *Music as Heard: A Study in Applied Phenomenology*, Yale University Press, New Haven 1983.

[B: 33] R. R. Coifman and M. V. Wickerhauser, "Entropy-based algorithms for best-basis selection," *IEEE Trans. Information Theory,* Vol. 38, No. 2, March 1992.

[B: 34] E. Condon and T. Sugrue *We Called It Music: A Generation of Jazz* Da Capo Press, 1988.

[B: 35] G. Cooper and L. B. Meyer, *The Rhythmic Structure of Music*, University of Chicago Press, 1960.

[B: 36] D. R. Courtney, *Fundamentals of Tabla*, 3rd Edition, Sur Sangeet Services, Houston, 1998.

[B: 37] H. Cowell, *New Musical Resources* Cambridge University Press 1996. (Original publication Alfred A. Knopf, 1930.)

[B: 38] P. E. Allen and R. B. Dannenberg, "Tracking musical beats in real time," *Int. Computer Music Conf.*, 1990.

[B: 39] I. Daubechies, "Time-frequency localization operators: A geometric phase space approach," *IEEE Trans. Information Theory,* Vol. 34, No. 4, 605–612, July 1988.

[B: 40] S. De Furia, *Secrets of Analog and Digital Synthesis*, Third Earth Productions Inc., NJ, 1986.

[B: 41] G. De Poli, A. Piccialli, and C. Roads Eds, *Representations of Musical Signals*, The MIT Press, Cambridge, MA 1991.

[B: 42] F. Densmore, *Yuman and Yaqui Music*, Washington, 1932.

[B: 43] P. Desain, "A (de)composable theory of rhythm perception," *Music Perception*, Vol. 9, No. 4, 439–454, Summer 1992.

[B: 44] P. Desain and H. Honing, "Quantization of musical time: a connectionist approach," *Computer Music Journal,* 13 (3): 56–66, 1989.

[B: 45] D. Deutsch, ed., *The Psychology of Music*, Academic Press Inc., San Diego, CA 1982.

[B: 46] D. Deutsch, "The processing of structured and unstructured tonal sequences," *Perception and Psychophysics,* 28, 381–389, 1980.

[B: 47] J. Diamond, "Gamelan programs for children from the cross-cultural to the creative," *Ear Magazine*, Vol. VIII, No. 4, Sept. 1983.

[B: 48] S. Dixon, "Automatic extraction of tempo and beat from expressive performances," *J. New Music Research*, Vol. 30, No. 1, 2001.

[B: 49] M. Dolson, "The Phase Vocoder: A Tutorial," *Computer Music Journal*, Spring, Vol. 10, No. 4, 14–27, 1986.

[B: 50] A. Doucet, N de Freitas and N. Gordon, Eds., *Sequential Monte Carlo Methods in Practice*, Springer-Verlag 2001.

[B: 51] A. Doucet, N. de Freitas and N. Gordon. "An introduction to sequential Monte Carlo methods." In *Sequential Monte Carlo Methods in Practice.* Arnaud Doucet, Nando de Freitas and Neil Gordon, Eds, Springer-Verlag. 2001.

[B: 52] A. Doucet, N. de Freitas, K. Murphy and S. Russell. "Rao-Blackwellised particle filtering for dynamic Bayesian networks." *Proceedings of Uncertainty in Artificial Intelligence*, 2000.

[B: 53] W. J. Dowling, "Rhythmic groups and subjective chunks in memory for melodies," *Perception and Psychophysics,* 14, 37–40, 1973.

[B: 54] J. Duesenberry, "The world in a grain of sound," *Electronic Musician*, Nov. 1999.

[B: 55] D. Eck, "A network of relaxation oscillators that finds downbeats in rhythms," in G. Dorffner, ed. *Artificial Neural Networks - ICANN 2001*, 1239–1247, Berlin, Springer, 2001.

[B: 56] D. Eck, "Finding downbeats with a relaxation oscillator," *Psychological Research*, 66 (1), 18–25, 2002.

[B: 57] D. Ehresman and D. Wessel, "Perception of timbral analogies," Rapports IRCAM 13/78, 1978.

[B: 58] W. A. Gardner and L. E. Franks, "Characterization of cyclostationary random signal processes," *IEEE Trans. Inform. Theory*, Vol. IT-21, 4, 1975.

[B: 59] L. J. Eriksson, M. C. Allie, and R. A. Griener, "The selection and application of an IIR adaptive filter for use in active sound attenuation," *IEEE Trans. on Acoustics, Speech, and Signal Processing*, ASSP–35, No. 4, Apr. 1987.

[B: 60] M. C. Escher, *The Graphic Work of M. C. Escher* Harry N Abrams, Pubs., 1984.

[B: 61] R. Fitzhugh, "Impulses and physiological states in theoretical models of nerve induction," *Biophysical Journal*, 1, 455–466, 1961.

[B: 62] J. L. Flanagan and R. M. Golden, "Phase vocoder," *Bell System Technical Journal*, 1493–1509, 1966.

[B: 63] J. E. Flannick, R. W. Hall, and R. Kelly, "Detecting meter in recorded music," Bridges Proceedings, 2005.

[B: 64] N. J. Fliege and J. Wintermantel, "Complex digital oscillators and FSK modulators," *IEEE Trans. Signal Processing,* Vol. 40, No. 2, Feb. 1992.

[B: 65] P. Fraisse, *The Psychology of Time*, Harper & Row, New York, 1963.
[B: 66] P. Fraisse, "Rhythm and tempo," in [B: 45].
[B: 67] J. T. Fraser and N. Lawrence, *The Study of Time*, Springer-Verlag, NY 1975.
[B: 68] D. Gabor, "Acoustical quanta and the theory of hearing" *Nature,* 159, 591–594, 1947.
[B: 69] S. Ganassi, *Fontegara*, Venice 1535. Ed. H. Peter, trans. D. Swainson, Berlin 1956.
[B: 70] T. H. Garland and C. V. Kahn, *Math and Music: Harmonious Connections*, Dale Seymour Publications, 1995.
[B: 71] C. Gerard and M. Sheller, *Salsa: the Rhythm of Latin Music*, White Cliffs Media, Tempe, AZ. 1998.
[B: 72] N. Ghosh, *Fundamentals of Raga and Tala*, Popular Prakashan, Bombay, India 1968.
[B: 73] J. J. Gibson, "Events are perceivable but time is not," in [B: 67], 1975.
[B: 74] R. O. Gjerdingen, " 'Smooth' rhythms as probes of entrainment," *Music Perception*, Vol. 10, No., 4, 503–508, Summer 1993.
[B: 75] N. J. Gordon, D. J. Salmond and A. F. M. Smith, "Novel approach to nonlinear/non-Gaussian Bayesian state estimation," *IEEE Proceedings-F*, 140 (2): 107–113, April 1993.
[B: 76] M. Goto, "An audio-based real-time beat tracking system for music with or without drum-sounds," *J. New Music Research*, Vol. 30, No. 2, 159–171, 2001.
[B: 77] M. Goto and Y. Muraoka, "Real-time beat tracking for drumless audio signals: chord change detection for musical decisions," *Speech Communication*, Vol.27, No. 3–4, 311–335, April 1999.
[B: 78] F. Gouyon and S. Dixon, "A review of automatic rhythm description systems," *Computer Music Journal,* 29:1, 34, Spring 2005.
[B: 79] F. Gouyon and P. Herrera, "A beat induction method for musical audio signals," *Proc. of 4th WIAMIS-Special session on Audio Segmentation and Digital Music* London, UK, 2003.
[B: 80] F. Gouyon and P. Herrera, "Determination of the meter of musical audio signals: seeking recurrences in beat segment descriptors," *Proc. of AES*, 114th Convention, 2003.
[B: 81] F. Gouyon and B. Meudic, "Towards rhythmic content processing of musical signals: fostering complementary approaches," *J. New Music Research*, Vol. 32, No. 1, 159–171, 2003.
[B: 82] N. Gray, *Roots Jam : Collected Rhythms for Hand Drum and Percussion* Cougar WebWorks, 1996.
[B: 83] J. M. Grey, *An Exploration of Musical Timbre* Ph.D. Thesis in Psychology, Stanford, 1975.
[B: 84] J. M. Grey and J. W. Gordon, "Perceptual effects of spectral modifications on musical timbres," *J. Acoustical Society of America,* 65 (5), 1493–1500, 1978.
[B: 85] J. M. Grey and J. A. Moorer, "Perceptual evaluation of synthesized musical instrument tones," *J. Acoustical Society of America,* 62, 454–462, 1977.
[B: 86] G. S. Hall and J. Jastrow, "Studies of rhythm," *Mind*, 11, 55–62, 1886.
[B: 87] R. Hall and P. Klingsberg, "Asymmetric rhythms, tiling canons, and Burnside's lemma," Bridges Proceedings, 189–194, 2004.
[B: 88] S. Handel, "Temporal segmentation of repeating auditory patterns," *J. Exp. Psychol.* 101, 46–54, 1973.
[B: 89] S. Handel and J. S. Oshinsky, "The meter of syncopated auditory polyrhythms," *Perception and Psychophysics,* 30 (1) 1–9, 1981.

[B: 90] S. Handel, "Using polyrhythms to study rhythm," *Music Perception,* Vol. 1, No. 4, 465–484, Summer 1984.

[B: 91] H. E. Harley, S. E. Crowell, W. Fellner, K. Odell, and L. Larsen-Plott, "Rhythm perception and production by the bottlenose dolphin," *J. Acoustical Society of America*, 118 (3), 2005.

[B: 92] S. Haykin, *Adaptive Filter Theory*, Prentice-Hall, Englewood Cliffs, NJ, 1991.

[B: 93] R. J. Heifetz, *On the Wires of Our Nerves*, Bucknell University Press, 1989.

[B: 94] H. Helmholtz, *On the Sensations of Tones*, 1877. (Trans. A. J. Ellis, Dover Pubs., New York 1954.)

[B: 95] P. Hindemith, *The Craft of Musical Composition*, 1937. Trans. A Mendel, Associated Music Publishers, New York, NY, 1945.

[B: 96] D. R. Hofstadter, *Gödel, Escher, Bach: An Eternal Golden Braid* Basic Books, 1979.

[B: 97] E. M. von Hornbostel, "African Negro Music, *Africa*, 1 (1), 1928.

[B: 98] P. J. Huber, "Projection pursuit," *Annals of Statistics,* Vol. 13, No. 2, 435–475, 1985.

[B: 99] A. Hutchinson, "History of dance notation," *The Dance Encyclopedia*, Simon and Schuster, 1967.

[B: 100] A. Hutchinson, *Labanotation: The System of Analyzing and Recording Movement*, Theatre Arts Books, 1970.

[B: 101] C. Huygens, *Horologium*, 1658.

[B: 102] C. R. Johnson, Jr. and W. A. Sethares, *Telecommunication Breakdown: Concepts of Communications Transmitted via Software-defined Radio,* Prentice-Hall 2004.

[B: 103] M. R. Jones, "Time, our lost dimension: toward a new theory of perception, attention, and memory," *Psychological Review*, Vol. 83, No. 5, Sept. 1976.

[B: 104] M. R. Jones and M. Boltz, "Dynamic attending and responses to time," *Psychological Review*, 96 (3) 459–491, 1989.

[B: 105] M. R. Jones, G. Kidd, and R. Wetzel, "Evidence for rhythmic attention," *J. Experimental Psychology,* Vol. 7, No. 5, 1059–1073, 1981.

[B: 106] M. R. Jones, "Musical events and models of musical time," in R. Block, Ed., *Cognitive Models of Psychological Time*, Lawrence Erlbaum Associates, Hillsdale NJ, 1990.

[B: 107] S. Joplin, *Maple Leaf Rag*, John Stark & Sons, 1899. [A rendition of the musical score drawn by Jim Paterson appears on the CD in `Papers/Maple-Leaf-Rag.pdf` [W: 39]. Score used with permission. A standard MIDI file, sequenced by W. Trachtman, is in `Sounds/mapleaf.mid`. [W: 51].]

[B: 108] D. Keane, "Some practical aesthetic problems of computer music," in [B: 93]

[B: 109] R. Kelly, "Mathematics of musical rhythm," Honors Thesis, 2002, see [W: 16].

[B: 110] K. Kharat, "The 'tala' system in Indian music," *Second ASEA Composer Forum on Traditional Music*, Singapore, April 1993.

[B: 111] A. King, "Employments of the "standard pattern" in Yoruba music." *African Music*, 2 (3) 1961.

[B: 112] P. Kivy, *Authenticities*, Cornell University Press, Ithaca NY, 1995.

[B: 113] M. Klingbeil, "Software for spectral analysis, editing, and synthesis," *Proc. of the 2005 International Computer Music Conf.*, Barcelona, 2005.

[B: 114] K. Koffka, *Principles of Gestalt Psychology*, Harcourt, Brace, and Co., New York, NY 1935.
[B: 115] W. Köhler, *Gestalt Psychology*, Liveright, New York, NY 1929.
[B: 116] L. H. Koopmans, *The Spectral Analysis of Time Series,* Academic Press, San Diego 1995.
[B: 117] J. D. Kramer, *The Time of Music*, MacMillan, Inc. 1998.
[B: 118] H. Krim, S. Mallat, D. Donoho, and A. Willsky, "Best basis algorithm for signal enhancement," *Proc. IEEE Conf. on Acoustics, Speech and Signal Processing,* ICASSP–95, 1561–1564, Detroit, May 1995.
[B: 119] A. Krims, *Rap Music and the Poetics of Identity*, Cambridge University Press, Cambridge UK, 2000.
[B: 120] A. B. Kristofferson, "A quantal step function in duration discrimination," *Perception & Psychophysics*, 27, 300–306, 1980.
[B: 121] J. Kunst, *Music in Java*, Martinus Nijhoff, The Hague, Netherlands, 1949.
[B: 122] S. Langer, *Feeling and Form*, Scribners, NY 1953.
[B: 123] E. W. Large and J. F. Kolen, "Resonance and the perception of musical meter," *Connection Science*, 6: 177–208, 1994.
[B: 124] J. Laroche and M. Dolson, "Improved phase vocoder time-scale modification of audio," *IEEE Trans. on Audio and Speech Processing,* Vol. 7, No. 3, May 1999.
[B: 125] M. Leman, *Music and Schema Theory: Cognitive Foundations of Systematic Musicology*, Berlin, Heidelberg: Springer-Verlag 1995.
[B: 126] M. Leman, Ed. *Music, Gestalt, and Computing* Springer, 1997.
[B: 127] M. Leman and A. Schneider, "Origin and nature of cognitive and systematic musicology: an introduction," in [B: 126].
[B: 128] F. Lerdahl and R. Jackendoff, *A Generative Theory of Tonal Music,* MIT Press, Cambridge, 1983.
[B: 129] J. Ligon, private communication, 2006.
[B: 130] D. Locke, *Drum Gahu*, White Cliffs Media, Tempe AZ, 1998.
[B: 131] J.London, "Rhythm," *Grove Music Online*, Ed. L. Macy (Accessed 19–05–2005), `http://www.grovemusic.com`
[B: 132] H. C. Longuet-Higgins and C. S. Lee, "The perception of musical rhythms," *Perception,* Vol. 11, 115–128, 1982.
[B: 133] H. C. Longuet-Higgins and C. S. Lee, "The rhythmic interpretation of monophonic music," *Music Perception,* Vol. 1, No. 4, 424–441, Summer 1984.
[B: 134] D. G. Luenberger, *Introduction to Dynamic Systems: Theory, Models, and Applications,* John Wiley and Sons, Inc. NY, 1979.
[B: 135] D. G. Luenberger, *Optimization by Vector Space Methods*, John Wiley and Sons, Inc. NY, 1968.
[B: 136] E. Mach, *Die Analyse der Empfindungen,* Jena: Fischer, 1886.
[B: 137] D. J. C. MacKay, *Information Theory, Inference, and Learning Algorithms*, Cambridge University Press, 2003.
[B: 138] R. C. Maher and J. W. Beauchamp, "Fundamental frequency estimation of musical signals using a two-way mismatch procedure," *J. Acoustical Society of America*, 95 (4), April 1994.
[B: 139] S. G. Mallat and Z. Zhang, "Matching pursuit with time-frequency dictionaries," *IEEE Trans. Signal Processing,* Vol. 41, No. 12, Dec. 1993.
[B: 140] W. P. Malm, *Music Cultures of the Pacific, the Near East, and Asia,* Prentice-Hall, NJ, 1996.

[B: 141] M. V. Mathews and J. R. Pierce, "Harmony and inharmonic partials," *J. Acoustical Society of America*, 68, 1252–1257, 1980.

[B: 142] M. V. Mathews, J. R. Pierce, A. Reeves, and L. A. Roberts, "Theoretical and experimental explorations of the Bohlen-Pierce scale," *J. Acoustical Society of America*, 84, 1214–1222, 1988.

[B: 143] J. D. McAuley, *Perception of time as phase: toward an adaptive-oscillator model of rhythmic pattern processing*, Ph.D Thesis, Indiana University, 1995.

[B: 144] J. D. McAuley, Time as phase: a dynamic model of time perception," *Proc. Sixteenth Conf. of Cognitive Science Society,* Lawrence Erlbaum, 607–612, 1994.

[B: 145] M. K. McClintock, "Menstrual synchrony and suppression," *Nature*, 229:244–245, 1971.

[B: 146] R. Meddis and L. O'Mard, "A unitary model of pitch perception," *J. Acoustical Society of America*, 102 (3): 1811–1820, Sept. 1997.

[B: 147] E. Meumann, "Untersuchungen zur psychologie und ästhetik des rhythmus," *Phiosophische Studien*, 10, 1894.

[B: 148] G. A. Miller, "The magical number seven, plus or minus two: some limits on our capacity for processing information," *The Psychological Review*, Vol. 63, 81–97, 1956.

[B: 149] J. D. Miller, L. P. Morin, W. J. Schwartz, and R. Y. Moore, "New insights into the mammalian circadian clock," *Sleep*, Vol. 19., No. 8., 641–667, 1996.

[B: 150] B. C. J. Moore, *An Introduction to the Psychology of Hearing*, Academic Press, Inc., 1982.

[B: 151] R. D. Morris and W. A. Sethares, "Beat Tracking," *Int. Society for Bayesian Analysis*, Valencia Spain, June 2002. [On CD in `Papers/valenciaposter.pdf`.]

[B: 152] M. Nafie, M. Ali, and A. Tewfik, "Optimal subset selection for adaptive signal representation," *Proc. IEEE Conf. on Acoustics, Speech and Signal Processing,* ICASSP–96, 2511–2514, Atlanta, May 1996.

[B: 153] U. Neisser, *Cognitive Psychology,* Appleton Century-Crofts, New York, 1967.

[B: 154] A. M. S. Nelson, *This is How We Flow*, Univ. S. Carolina Press, 1999.

[B: 155] J. H. K. Nketia, *Drumming in Akan Communities of Ghana*, Edinburgh: Thomas Nelson and Sons, Ltd., 1963.

[B: 156] L. van Noorden and D. Moelants, "Resonance in the perception of musical pulse," *J. New Music Research*, Vol. 28, No. 1, March 1999.

[B: 157] C. Palmer and C. Krumhansl, "Mental representations for musical meter," *J. Exp. Psychology: Human Perceptual Performance*, 16, 728–741, 1990.

[B: 158] R. Parncutt, "A perceptual model of pulse salience and metrical accent in musical rhythms," *Music Perception,* Vol. 11, No. 4, Summer 1994.

[B: 159] R. D. Patterson, M. H. Allerhand, and C. Giguere, "Time domain modeling of peripheral auditory processing: a modular architecture and a software platform," *J. Acoustical Society of America*, 98: 1890–1894, 1995.

[B: 160] R. D. Patterson and B. C. J. Moore, "Auditory filters and excitation patterns as representations of frequency resolution," *Frequency Selectivity in Hearing*, Ed. B. C. J. Moore, Academic Press, London 1986.

[B: 161] L. S. Penrose and R. Penrose, "Impossible objects: a special type of visual illusion," *British J. of Psychology*, 1958.

[B: 162] D. N. Perkins, "Coding position in a sequence by rhythmic grouping," *Memory and Cognition,* 2, 219–223, 1974.

[B: 163] A. Pikovsky, M. Rosenblum, and J. Kurths, *Synchronization: a Universal Concept in Nonlinear Science*, Cambridge University Press, Cambridge, UK, 2001.

[B: 164] J. R. Pierce, "Periodicity and pitch perception," *J. Acoustical Society of America*, 90,No. 4, 1889–1893, 1991.

[B: 165] H. Platel, C. Price, J. C. Baron, R. Wise, J. Lambert, R. S. J. Frackowiak, B. Lechevalier, and F. Eustache, "The structural components of music perception: a functional anatomical study," *Brain*, 120, 229–243, 1997.

[B: 166] R. Plomp and W. J. M. Levelt, "Tonal consonance and critical bandwidth," *J. Acoustical Society of America*, 38, 548–560, 1965.

[B: 167] R. Plomp, "Timbre as a multidimensional attribute of complex tones," in *Frequency Analysis and Periodicity Detection in Hearing,* ed. R. Plomp and G. F. Smoorenburg, A. W. Sijthoff, Lieden, 1970.

[B: 168] R. Plomp, *Aspects of Tone Sensation*, Academic Press, London, 1976.

[B: 169] I. Pollack, "The information of elementary auditory displays," *J. Acoustical Society of America,* 24, 745–749, 1952.

[B: 170] B. Porat, *Digital Signal Processing*, Wiley 1997.

[B: 171] M. R. Portnoff, "Implementation of the digital phase vocoder using the fast fourier transform," *IEEE Trans. Acoustics, Speech, and Signal Processing,* Vol. ASSP–24, No. 3, June 1976.

[B: 172] D. J. Povel, "Internal representation of simple temporal patterns," *J. of Experimental Psychology*, Vol. 7, No. 1, 3–18, 1981.

[B: 173] D. J. Povel, "A theoretical framework for rhythmic perception," *Psychological Research*, 45: 315–337, 1984.

[B: 174] D. J. Povel and P. Essens, "Perception of temporal patterns," *Music Perception,* Vol. 2, No. 4, 411–440, Summer 1985.

[B: 175] D. J. Povel and H. Okkerman, "Accents in equitone sequences," *Perception and Psychophysics,* Vol. 30, 565–572, 1981.

[B: 176] D. Preusser, "The effect of structure and rate on the recognition and description of auditory temporal patterns," *Perception and Psychophysics,* Vol. 11 (3) 1972.

[B: 177] M. S. Puckette and J. C. Brown, "Accuracy of frequency estimates using the phase vocoder," *IEEE Trans. Speech and Audio Processing,* Vol. 6, No. 2, March 1998.

[B: 178] J. P. Rameau, *Treatise on Harmony*, Dover Pubs., New York 1971, original edition, 1722.

[B: 179] C. Raphael. "Automated rhythm transcription" *Proc. Int. Symposium on Music Inform. Retriev.*, IN, Oct. 2001.

[B: 180] L. G. Ratner, *The Musical Experience*, W, H, Freeman and Co. New York, 1983.

[B: 181] C. Roads, *Microsound*, MIT Press, 2002.

[B: 182] D. A. Robin, P. J. Abbas, and L. N. Hug, "Neural responses to auditory temporal patterns," *J. Acoustical Society of America,* Vol. 87, 1673–1682, 1990.

[B: 183] J. G. Roederer, *The Physics and Psychophysics of Music*, Springer-Verlag, New York, 1994.

[B: 184] D. Rosenthal, "Emulation of rhythm perception," *Computer Music Journal,* Vol. 16, No. 1, Spring 1992.

[B: 185] T. D. Rossing, *The Science of Sound*, Addison Wesley Pub., Reading, MA, 1990.

[B: 186] F. L. Royer and W. R. Garner, "Perceptual organization of nine-element temporal patterns," *Perception and Psychophysics,* Vol. 7 (2) 1970.

[B: 187] C. Sachs, *Rhythm and Tempo*, W. W. Norton and Co., Inc. NY 1953.

[B: 188] E. D. Scheirer, "Tempo and beat analysis of acoustic musical signals," *J. Acoustical Society of America,* 103 (1), 588–601, Jan. 1998.

[B: 189] J. Schillinger, *The Schillinger System of Musical Composition,* Carl Fischer, Inc. NY, 1946.

[B: 190] J. Schillinger, *Encyclopedia of Rhythms,* Da Capo Press, NY, 1976.

[B: 191] A. Schoenberg, *Fundamentals of Music Composition,* Faber and Faber, London, 1967.

[B: 192] C. H. Sears, "A contribution to the psychology of rhythm," *American Journal of Psychology,* Vol. 13 (1) 28–61, 1913.

[B: 193] A. K. Sen, *Indian Concept of Tala*, Kanishka Pubs. 9/2325 Kailash Nagar, Delhi, India 1984.

[B: 194] J. Seppänen, "Computational models of musical meter recognition," MS Thesis, Tempere Univ. Tech. 2001.

[B: 195] X. Serra, "Sound hybridization based on a deterministic plus stochastic decomposition model," in *Proc. of the 1994 International Computer Music Conf.,* Aarhus, Denmark, 348, 1994.

[B: 196] W. A. Sethares, *Tuning, Timbre, Spectrum, Scale*, Springer-Verlag, 1997. Second edition 2004.

[B: 197] W. A. Sethares, "Repetition and Pseudo-Periodicity," *Tatra Mt. Mathematics Publications*, Dec. 2001. [This paper appears on the CD in `Papers/pnorm.pdf`].

[B: 198] W. A. Sethares, "Specifying spectra for musical scales," *J. Acoustical Society of America*, 102, No. 4, Oct. 1997.

[B: 199] W. A. Sethares, "Consonance-based spectral mappings," *Computer Music Journal,* 22, No. 1, 56–72, Spring 1998.

[B: 200] W. A. Sethares, "Local consonance and the relationship between timbre and scale," *J. Acoustical Society of America*, 94, No. 3, 1218–1228, Sept. 1993. [On CD in `Papers/consonance.pdf`.]

[B: 201] W. A. Sethares, "Automatic detection of meter and periodicity in musical performance," *Proc. of the Research Society for the Foundations of Music*, Ghent, Belgium, Oct. 1999.

[B: 202] W. A. Sethares, "Clock Induction," See `Papers/ clockinduction.pdf`.

[B: 203] W. A. Sethares, "Some Statistical Models of Periodic Phenomenon," see `Papers/statmodels.pdf`.

[B: 204] W. A. Sethares and R. D. Morris, "Performance measures for beat tracking," *Int. Workshop on Bayesian Data Analysis, Santa Cruz, Aug. 2003.*

[B: 205] W. A. Sethares and R. D. Morris, "Performance measures for beat tracking," (Technical Report, Univ. of Wisconsin, 2005, available on the CD in `Papers/beatqual.pdf`).

[B: 206] W. A. Sethares, R. D. Morris and J. C. Sethares, "Beat tracking of audio signals," *IEEE Trans. on Speech and Audio Processing*, Vol. 13, No. 2, March, 2005. [This paper appears on the CD in `Papers/beatrack.pdf`].

[B: 207] W. A. Sethares and T. Staley, "The periodicity transform," *IEEE Trans. Signal Processing*, Vol. 47, No. 11, Nov. 1999. [This paper appears on the CD in `Papers/pertrans.pdf`].

[B: 208] W. A. Sethares and T. Staley, "Meter and periodicity in musical performance," *J. New Music Research*, Vol. 30, No. 2, June 2001. [This paper appears on the CD in `Papers/jnmr2001.pdf`].

[B: 209] W. A. Sethares, `TapTimes`: A Max [W: 30] patch designed to allow listeners to tap along with a piece of music and to record the times of the tapping. Available by request from the author.

[B: 210] P. Schaeffer, *A la Recherche d'une Musique Concrète*, Paris, Seuil, 1952.

[B: 211] R. N. Shepard, "Circularity in the judgments of relative pitch," *J. Acoustical Society of America*, 36, 2346–2353, 1964.

[B: 212] D. S. Sivia and J. Skilling, *Data Analysis: A Bayesian Tutorial*, Oxford University Press, Second Edn., 2006.

[B: 213] A. Slutsky and C. Silverman, *James Brown Rhythm Sections*, Manhattan Music, Inc., 1997.

[B: 214] P. Singh, "The role of timbre, pitch, and loudness changes in determining perceived metrical structure," *J. Acoustical Society of America*, 101 (5) Pt 2, 3167, May 1997.

[B: 215] A. M. Small, "An objective analysis of artistic violin performances," *University of Iowa Studies in Music Psychology IV*, p. 172–231, 1936.

[B: 216] A. M. Small and M. E. McClellan, "Pitch associated with time delay between two pulse trains," *J. Acoustical Society of America*, 35 (8), 1246–1255, Aug. 1963.

[B: 217] L. M. Smith, "Listening to musical rhythms with progressive wavelets," *IEEE TENCON Digital Signal Processing Applications,* 1996.

[B: 218] S. W. Smith, *The Scientist and Engineer's Guide to Digital Signal Processing*, California Technical Publishing, Second Ed., 1999. [See website at www.DSPguide.com.]

[B: 219] B. Snyder, *Music and Memory*, MIT Press, Cambridge, MA 2000.

[B: 220] J. Snyder and C. L. Krumhansl, "Tapping to ragtime: cues to pulse finding," *Music Perception*, Summer 2001.

[B: 221] N. Sorrell, *A Guide to the Gamelan*, Faber and Faber Ltd., London, 1990.

[B: 222] M. J. Steedman, "The perception of musical rhythm and metre," *Perception* Vol. 6, 555–569, 1977.

[B: 223] T. Stilson and J. Smith, "Alias-free digital synthesis of classic analog waveforms," `http://www-ccrma.stanford.edu/~stilti/papers`

[B: 224] K. Stockhausen, "...How Time Passes ...," *die Reihe* 3:10–43, English edition trans. by C. Cardew, 1959.

[B: 225] E. Terhardt, "Pitch, consonance, and harmony," *J. Acoustical Society of America*, 55, No. 5, 1061–1069, May 1974.

[B: 226] E. Terhardt, G. Stoll, and M. Seewann "Algorithm for extraction of pitch and pitch salience from complex tone signals," *J. Acoustical Society of America*, 71, No. 3, 679–688, March 1982.

[B: 227] B. Tiemann and B. Magnusson, "A notation for juggling tricks," *Jugglers World*, 1991.

[B: 228] N. P. M. Todd, D. J. O'Boyle and C. S. Lee, "A sensory-motor theory of rhythm, time perception and beat induction," *J. of New Music Research,* 28:1:5–28, 1999.

[B: 229] P. Toiviainen "Modeling the perception of metre with competing subharmonic oscillators," *Proc. 3rd Triennial ESCOM Conf.* Uppsala, Sweden, 1997.

[B: 230] P. Toiviainen, "An interactive MIDI accompanist," *Computer Music J.*, 22 (4):63, 1998.

[B: 231] B. Truax, "Real-time granular synthesis with a digital signal processor," *Computer Music Journal*, 12 (2), 14–26 1988.

[B: 232] G. Tzanetakis and P. Cook, "Musical genre classification of audio signals", *IEEE Trans. Speech and Audio Processing*, 10 (5), July 2002.

[B: 233] E. Uribe, *The Essence of Afro-Cuban Percussion and Drum Set*, Warner Bros. Pubs., Miami FL, 1996.

[B: 234] B. van der Pol and J. van der Mark, "The heartbeat considered as a relaxation oscillation and an electrical model of the heart," *Philos. Mag. Suppl.* No. 6, 763–775, 1928.

[B: 235] J. D. Vantomme, "Score following by temporal pattern," *Computer Music Journal*, 19:3, 50–59, Fall 1995.

[B: 236] J. Vermaak, C. Andrieu, A. Doucet, and S. J. Godsill, "Particle methods for Bayesian modeling and enhancement of speech signals," *IEEE Trans. Speech Audio Processing*, 10 (3):173–185, 2002.

[B: 237] R. A. Waterman, " 'Hot' rhythm in Negro music," *J. of the American Musicological Society,* 1, Spring 1948.

[B: 238] M. Wertheimer, "Principles of perceptual organization," in D. Beardslee and M. Wertheimer, Eds, *Readings in Perception,* Van Nostrand, Princeton, NJ 1958.

[B: 239] N. Whiteley, A. T. Cemgil, S. Godsill, "Sequential inference of rhythmic structure in musical audio," ICASSP, Hawaii 2007.

[B: 240] C. G. Wier, W. Jesteadt, and D. M. Green, "Frequency discrimination as a function of frequency and sensation level," *J. Acoustical Society of America*, 61, 178–184, 1977.

[B: 241] L. Wilcken, *The Drums of Vodou*, White Cliffs Media, Tempe, AZ, 1992.

[B: 242] T. Wishart, *Audible Design*, Orpheus the Pantomine, York, 1994.

[B: 243] H. Wong and W. A. Sethares, "Estimation of pseudo-periodic signals," *IEEE International Conference on Acoustics, Speech and Signal Processing,* Montreal 2004. [This paper appears on the CD in `Papers/wong2004.pdf`].

[B: 244] H. Woodrow, "A quantitative study of rhythm," *Archives of Psychology,* Vol. 14, 1–66, 1909.

[B: 245] H. Woodrow, "The role of pitch in rhythm," *Psychological Review*, Vol. 11, 54–76, 1911.

[B: 246] E Wold, T. Blum, D. Keislar, and J. Wheaton, "Content-based classification, search and retrieval of audio," *IEEE Trans. Multimedia,* 3 (3):27, 1996.

[B: 247] I. Xenakis, *Formalized Music*, Indiana University Press, Bloomington, IN, 1971.

[B: 248] R. K. R. Yarlagadda and J. E. Hershey, *Hadamard Matrix Analysis and Synthesis*, Kluwer Academic, 1997.

[B: 249] L. D. Zimba and D. A. Robin, "Perceptual organization of auditory patterns," *J. Acoustical Society of America,* Vol. 104, No. 4, Oct. 1998.

[B: 250] V. Zuckerkandl, *Sound and Symbol,* Princeton University Press, Princeton, NJ, 1956.

[B: 251] E. Zwicker and H. Fastl, *Psychoacoustics*, Springer-Verlag, Berlin, 1990.

[B: 252] E. Zwicker, G. Flottorp, and S. S. Stevens, "Critical bandwidth in loudness summation," *J. Acoustical Society of America*, 29, 548, 1957.

Discography

References in the body of Rhythm and Transforms *to the discography are coded with* [D:] *to distinguish them from references to the bibliography, websites, and sound examples.*

[D: 1] Baltimore Consort, *A Baltimore Consort Collection*, Dorian Recordings, 98101, 1995.
[D: 2] The Beatles, *Rubber Soul*, Capitol, CD #46440, 1990.
[D: 3] S. Bechet, *Legendary Sidney Bechet*, RCA, CD #6590, 1990.
[D: 4] Bhundu Boys, *True Jit*, Mango Records CCD 9812, 1988.
[D: 5] C. Bolling, *Original Ragtime*, Polygram Int., CD #558024.
[D: 6] *Best of Bond James Bond*, Capitol, ASIN: B00006I0BO, 2002.
[D: 7] The Byrds, *Mr. Tambourine Man*, Sony CD #64845 1996.
[D: 8] James Brown, *Foundations Of Funk: A Brand New Bag: 1964-1969*, Polydor, 1996.
[D: 9] Dave Brubeck Quartet, *Time Out*, Sony/Columbia, CK65122, 1997.
[D: 10] Prince Buster and the Ska
[D: 11] Canadian Brass, *Red Hot Jazz: The Dixieland Album*, Philips, ASIN: B000004140, 1993.
[D: 12] F. Cramer, *Piano Masterpieces (1900-1975)*, RCA, CD #53745, 1995.
[D: 13] T. Dorsey, *1935-1939*, Spv U.S., CD #31102, 2000.
[D: 14] S. Earl, *The Very Best Of Steve Earle*, Telstar TV, ASIN: B00000INIX, 1999.
[D: 15] Gamelan Gong Kebyar, *Music from the Morning of the World*, Elektra/Nonesuch 9 79196-2, 1988.
[D: 16] E. Glennie, *Rhythm Song*, RCA, CD #RD60242, 1990.
[D: 17] Bali: Golden Rain, Nonesuch, 1969.
[D: 18] *The Sting: Original Motion Picture Soundtrack* MCA, ASIN: B00000DD02, 1998.
[D: 19] Grateful Dead, *American Beauty*, Rhino Records, 1970.
[D: 20] C. Halaris, *Byzantine secular classical music*, Orata, Ltd., Athens, Greece.
[D: 21] Handel, *Water Music; Concerto Grosso*, Delta, CD #15658, 1990.
[D: 22] I. Hayes, *Shaft: Music From The Soundtrack*, Stax, 88002, 1971.
[D: 23] J. Hendrix, *Electric Ladyland*, 1968.
[D: 24] A. J. M. Houtsma, T. D. Rossing, and W. M. Wagenaars, *Auditory Demonstrations* (Philips compact disc No. 1126-061 and text) Acoustical Society of America, Woodbury NY 1987.
[D: 25] T. Hosokawa, *deep silence*, WER 6801 2, Mainz, Germany, 2004.
[D: 26] D. Hyman, *Joplin: Piano Works 1899-1904*, RCA #87993, 1988.

[D: 27] Ice Cube, *AmeriKKKa's Most Wanted*, Priority Records, 1990.
[D: 28] Kronos Quartet, *Pieces of Africa*, Nonesuch, CD #79275, 1992.
[D: 29] KRS-One, Jive Records, 1995.
[D: 30] Kukuruza, *Where The Sunshine Is*, Solid Records, 1993. http://www.kukuruza.info/
[D: 31] P. Charial and J. Hocker, *György Ligeti Edition 5: Mechanical Music*, Sony, CD #62310, 1997.
[D: 32] Jelly Roll Morton, *The Complete Library of Congress Recordings*, Disc 3, Rounder Records, ASIN: B000AOF9W0, 2003.
[D: 33] S. Reich, *Steve Reich 1965-1995*, Nonesuch, 1997.
[D: 34] M. Reichle, *Scott Joplin Complete Piano Works Volume One 1896-1902*, ASIN: B00004U2HE, 2000.
[D: 35] J. Rifkin, *Ragtime: Music of Scott Joplin*, Angel Records, ASIN: B0009YA43A, 2005.
[D: 36] M. Roberts, *Joy of Joplin*, Sony, CD #60554, 1998.
[D: 37] D. Van Ronk, *Sunday Street*, Philo Records, ASIN: B00000GWY9, 1999.
[D: 38] Todd Rundgren, *No World Order*, Rhino Records, 1993.
[D: 39] D. Scarlatti, *Complete Keyboard Sonatas, Vol. 2*, Michael Lewin, Naxos, CD #8553067, 1999.
[D: 40] M. Schatz, *Brand New Old Tyme Way*, Rounder, ASIN: B0000002N8, 1995. See also [W: 47].
[D: 41] *Spirit Of Ragtime*, ASV/Living Era, ASIN: B000006167, 1998.
[D: 42] W. A. Sethares, *Xentonality*, Odyssey Records XEN2001 1997.
[D: 43] W. A. Sethares, *Exomusicology*, Odyssey Records EXO2002 2002.
[D: 44] Butch Thompson, *The Butch Thompson Trio Plays Favorites*, Solo Art, ASIN: B00005Y894, 1994.
[D: 45] Grover Washington, Jr. *Prime Cuts: The Greatest Hits 1987-1999*, Sony 1999.

World Wide Web and Internet References

This section contains all web links referred to throughout Rhythm and Transforms. *References in the body of the text to websites are coded with* [W:] *to distinguish them from references to the bibliography, discography, and sound examples. The web examples may also be accessed using a web browser. Open the file* `html/weblinks.html` *on the CD-ROM and navigate using the html interface.*

[W: 1] *9 Beet Stretch*, Ludwig van Beethoven's 9th Symphony stretched to 24 hours http://www.notam02.no/9/

[W: 2] *Alternate tuning mailing list*, http://groups.yahoo.com/group/tuning/

[W: 3] D. Bañuelos, *Beyond the Spectrum of Music* http://eceserv0.ece.wisc.edu/~sethares/banuelos/

[W: 4] Big Mama Sue, *Big Mama Sue and Mr. Excitement* http://www.bigmamasue.com/

[W: 5] D. Blumberg, *tunesmith 2002* http://freeaudioplayer.freeservers.com/radio2002.html

[W: 6] *David Bowie's Mashup Contest*, http://www.davidbowie.com/neverFollow/

[W: 7] S. Brandorff and P. Møller-Nielsen, *Sound Manipulation in the Frequency Domain*, http://www.daimi.au.dk/~pmn/sound/pVoc/index.html

[W: 8] `CamelSpace` http://www.camelaudio.com/

[W: 9] *Classical MIDI Archives*, http://www.classicalarchives.com/

[W: 10] P. Copeland, *Classic Cat*, http://www.classiccat.net/performers/copeland_paul.htm

[W: 11] *U. S. Copyright Office*, http://www.copyright.gov/

[W: 12] *Dance Notation Bureau*, http://dancenotation.org/DNB/

[W: 13] *Database of drums tablature*, http://drumbum.com/drumtabs/

[W: 14] D. Ellis, *A phase vocoder in MATLAB®*, http://www.ee.columbia.edu/~dpwe/resources/matlab/pvoc/

[W: 15] *Gnutella File Sharing Software*, http://www.gnutella.com

[W: 16] R. Hall, *Mathematics of Musical Rhythm* http://www.sju.edu/ rhall/research.html

[W: 17] *Heftone Banjo Orchestra*, http://heftone.com/orchestra

[W: 18] *Institute for Psychoacoustics and Music*, http://www.ipem.rug.ac.be/

[W: 19] *Internet Juggling Database*, http://www.jugglingdb.com/

[W: 20] S. Jay, *The Theory of Harmonic Rhythm*, http://www.stephenjay.com/hr.html

[W: 21] *Keyfax Software*, http://www.keyfax.com

[W: 22] M. Klingbeil, *SPEAR: Sinusoidal Partial Editing Analysis and Resynthesis*, http://klingbeil.com/spear/

[W: 23] *Kyma from Symbolic Sound Corp.*, http://www.symbolicsound.com/

[W: 24] *Introduction to Labanotation*, http://user.uni-frankfurt.de/~griesbec/LABANE.HTML

[W: 25] A. Maag, *Tango*, http://www.maagical.ch/maagicalGmbH/andreasmaagprivat_laban.html

[W: 26] *Make Micro Music mailing list*, http:// groups.yahoo.com/ group/ MakeMicroMusic/

[W: 27] *Making Microtonal Music Website*, http://www.microtonal.org/

[W: 28] *Mark of the Unicorn*, http://www.motu.com/

[W: 29] Matlab, http://www.mathworks.com/

[W: 30] *Max 4.0 Reference Manual*, http://www.cycling74.com/products/dldoc.html

[W: 31] Metasynth, http://www.uisoftware.com/

[W: 32] *MIREX 2005*, Music Information Retrieval Contest in Audio Onset Detection http://www.music-ir.org/mirex2005/index.php/Audio_Onset_Detection

[W: 33] P. Møller-Nielsen and S. Brandorff, *Sound Manipulation in the Frequency Domain*, http://www.daimi.au.dk/~pmn/sound/

[W: 34] Motta Junior, http://music.download.com/mottajunior/

[W: 35] *Music, Mind, Machine,* http://www.nici.kun.nl/mmm

[W: 36] *Musical Dice Game* with a computer implementation, http://www.nationwide.net/ amaranth/MozartDiceGame.htm

[W: 37] *Name That Tune* http://en.wikipedia.org/wiki/Name_That_Tune

[W: 38] *Native Instrument's* Traktor implements extensive looping features. http://www.native-instruments.com/

[W: 39] J. Paterson, *Music Files* http://www.mfiles.co.uk/

[W: 40] *A. Pertout's samba lessons*, http://pertout.customer.netspace.net.au/

[W: 41] *Propellerhead software's* ReCycle is a tool for working with sampled loops. http://www.propellerheads.se/

[W: 42] M. S. Puckette, *Theory and Techniques of Electronic Music,* March 2006, http://www.crca.ucsd.edu/~msp/techniques/v0.08/book-html/

[W: 43] *John Roache's Ragtime MIDI Library*, http://www.johnroachemusic.com/mapleaf.html

[W: 44] M. Op de Coul, *Scala Homepage*, http://www.xs4all.nl/~huygensf/scala/

[W: 45] *Rhythm and Transforms*, http://eceserv0.ece.wisc.edu/~sethares/

[W: 46] SFX Machine http://www.sfxmachine.com/

[W: 47] Mark Schatz, http://www.yellowcarmusic.com/markschatz/

[W: 48] *Siteswap Juggling Notation*, http://jugglinglab.sourceforge.net/html/ssnotation.html

[W: 49] *Standard MIDI File Specification*, http://www.sonicspot.com/guide/midifiles.html

[W: 50] *Sony's* Acid looping software http://www.sonymediasoftware.com/products/acidfamily.asp

[W: 51] *Ragtime Piano MIDI files by Warren Trachtman*, http://www.trachtman.org/ragtime/

[W: 52] *Dynamic Spectrograms of Music*, http://nastechservices.com/Spectrograms

[W: 53] MATLAB® routines for the calculation of the Periodicity Transforms are available at http://eceserv0.ece.wisc.edu/~sethares/

[W: 54] Sunspot data is available at http://www.spaceweather.com/java/archive.html

[W: 55] R. Van Niel, http://www.soundclick.com/

[W: 56] T. Wishart, *Composer's Desktop Project*, http://www.bath.ac.uk/~masjpf/ CDP/CDP.htm

Index